Dmitry Yu. Murzin
Formulation Product Technology

Also of Interest

Product-Driven Process Design.
From Molecule to Enterprise
Edwin Zondervan, Cristhian Almeida-Rivera, Kyle Vincent Camarda, 2023
ISBN 978-3-11-101490-6, e-ISBN 978-3-11-101495-1

Engineering Catalysis
Dmitry Yu. Murzin, 2020
ISBN 978-3-11-061442-8, e-ISBN 978-3-11-061443-5

Product and Process Design.
Driving Innovation
Jan Harmsen, André B. de Haan, Pieter L. J. Swinkels, 2018
ISBN 978-3-11-046772-7, e-ISBN 978-3-11-046774-1

Formulations.
In Cosmetic and Personal Care
Tharwat F. Tadros, 2016
ISBN 978-3-11-045236-5, e-ISBN 978-3-11-045238-9

Dmitry Yu. Murzin

Formulation Product Technology

2nd Edition

DE GRUYTER

Author
Prof. Dmitry Murzin
Laboratory of Industrial Chemistry and Reaction Engineering
Åbo Akademi University
Henriksgatan 2
FIN-20500 Turku/Åbo
Finland

ISBN 978-3-11-078844-0
e-ISBN (PDF) 978-3-11-079796-1
e-ISBN (EPUB) 978-3-11-079824-1

Library of Congress Control Number: 2023949894

Bibliographic information published by the Deutsche Nationalbibliothek
The Deutsche Nationalbibliothek lists this publication in the Deutsche Nationalbibliografie; detailed bibliographic data are available on the internet at http://dnb.dnb.de.

Cover image: jevelin/iStock/Getty Images Plus
Typesetting: Integra Software Services Pvt. Ltd.
Printing and binding: CPI books GmbH, Leck

www.degruyter.com

Preface

The previous textbooks of the author (*Engineering Catalysis*, de Gruyter, 2013, and *Chemical Reaction Technology*, de Gruyter, 2015) were focused mostly on production of commodities and corresponding enabling tools (catalysis). While synthesis of fuels and basic chemicals is still a huge business in terms of tonnage and turnover, one of the current trends in process industries is that many chemical companies are exiting commodities markets and position themselves as manufacturers of products different from (chemical) commodities. These changes often result in rebranding even dropping the word “chemical” from the companies’ names. More consumer-oriented approach implies that products are manufactured with apparently lower volume but significantly higher values added.

It has been recognized some time ago that chemical engineering education should also follow this trend. In several chemical engineering departments across the world either courses on chemical product design have been introduced or even full MSc programs have been developed. Few books on chemical product design and development have already appeared highlighting their strong market orientation. In fact, understanding the needs of customers is extremely important in this field. The author of this text at some point of his professional life has spent several years dealing with technical marketing and sales of a range of chemical products; the author has taken part in almost hundreds of negotiations and sales meetings related to implementation of, for example, catalysts to dozens of chemical processes. Numerous discussions with customers were concentrated not only on performance (activity, selectivity and pressure drop) but also on comparison with products from alternative supplies, economics and logistics of supplying sometimes hundreds of tons of catalysts. The customers rarely based their decision barely on the price and often knew more about the quality and performance of a particular catalyst than some catalyst manufacturers, who did not have own experience in running a process with that catalyst. Discussions involved people with different skills and background (plant managers, technical personnel, plant engineers, financial directors, economists, etc.). From personal experience, it can be said that building trust is essential as customers should feel a genuine interest in solving their problems preferably in an optimal way.

Many chemical engineering students might not feel comfortable doing or studying product design for the reason that it requires close interactions with potential customers and understanding their needs, which the students apparently might not know. The current chemical engineering curriculum is still centered mainly on chemical process engineering, with efficient production, process economics and safety being the key aspects. Classical process design for specification products is done using optimization criteria production costs as well as some other factors, such as safety, environmental footprint and thus waste minimization and liability. Product purity specification is typically given and is not debated by potential customers.

https://doi.org/10.1515/9783110797961-202

While all chemical engineers still need these traditional skills, it is becoming apparently clear that more knowledge of product engineering and product technology from chemical engineering students is required, in particular, in areas such as physical and colloidal chemistry, interfacial engineering, rheology, pharmaceutical, and medical sciences. Unit operations of relevance for product technology are also different from classical chemical process technology, involving complex media and particulate solids (e.g., granulation, extrusion, precipitation, crystallization, spray drying, coating and emulsification).

This textbook was conceived with the aim of introducing future chemical engineers to the field of chemical product technology. Chapter 1 deals with the basics of chemical product design and its relation to the market, while Chapter 2 covers the theoretical basis and relevant unit operations. The two subsequent chapters address several types of performance products (plastics and polymer composites; paints and coatings; laundry detergents; adhesives; lubricants; ceramics; catalysts) and personal chemicals (absorbent hygiene products; drug delivery systems; cosmetic skin and hair care products). The author had to limit the description to the essentials omitting many specific details, which are available in a more specialized literature. The author used a variety of sources, textbooks, encyclopedia, original scientific literature and internet sites of products manufacturers to get a reliable and concise compilation. The textbook is thus by no means a replacement of books on polymer composites, surface coatings, glass, cement, shampoo or hair styling products. The author hopes that this book will rather spark interest and increase curiosity of students in particular topics of chemical product technology and relationship of these topics to other fields of science as well as to the needs of customers.

Turku, December 2017

Preface to the second edition

Several years have passed after the first edition of this book, entitled originally “Chemical Product Technology,” has appeared. It became even more apparent that chemical engineering graduates in their daily professional life deal more and more with product technology and design. A very recent example is an exponential growth in application of hand sanitizers, which became very popular due to the global outbreak of COVID-19. The alcohol-based hand sanitizers are containing primarily one or more alcohols in addition to different excipients and humectants. Such consumer products should be carefully formulated, taking into account potential interference of different substituents in the overall composition.

Considering, therefore, a request from the publisher to prepare a second edition of this book, the decision was taken to modify the title, emphasizing the formulation aspects of the product technology. The corresponding sections have been subsequently expanded. Chapters on other products have also been revised, benefiting from the experience of the author in teaching product technology using the first edition of the book. Moreover some additional products, not covered in the first edition, have been introduced, such as dyes and agrochemicals.

The book is dedicated to the late Elena Murzina, who sadly passed away after the first edition was completed.

Turku/Åbo, June 2023

https://doi.org/10.1515/9783110797961-203

Contents

1 Product design

1.1 Basics

Product design and subsequent product technology requires not only optimization against cost as in classical process design for specification products but also understanding of the needs and demands of customers. In fact, the latter does not necessarily desire the cheapest product, as formulation products should provide a certain function being otherwise worthless. A myriad of diverse chemical/formulation products are available, which makes it impossible to cover all of them under a single roof in one textbook. To illustrate this point, one can just list formulation products currently manufactured by various companies: cosmetics, detergents, surfactants, bitumen, paints, paper, rubber, plastic composites, drugs, adhesives, lubricants, textiles, inks and so on. Classification of these products can still be done. Some of them belong to functional chemicals (coatings, catalysts, plastics, detergents, energy storage materials, agrochemicals and fertilizers), whereas others are related to human health and well-being, including pharmaceuticals, cosmetics and personal (skin, hair and body) care products.

Some common features of these products are discussed later. Many of them are complex media and particulate solids, such as non-Newtonian liquids, gels, foams, polymers, glass, colloids, dispersions, suspensions, emulsions and microemulsions. Rheology and interfacial phenomena are therefore decisive in manufacturing and applications. Not only chemical properties but also appearance, feel, odor, taste, smell, handling properties, along with biocompatibility are of importance.

Many of these so-called formulated products require batch operations contrary to continuous technology encountered with commodities. For the latter, classical unit operations are applied, namely, distillation, extraction and absorption, in addition to chemical processing per se in (predominantly catalytic) reactors.

There are substantial differences in handling solid products compared to gases and liquids. For solid products, the particle size distribution and morphology are of importance, which could be difficult to maintain. For chemical products involving complex media and particulate solids, operations other than classical ones are decisive: granulation, extrusion, precipitation, crystallization, prilling, gelation, spray drying, coating, emulsification, shaping, calcination and so on. Some of these operations might lead to unwanted or premature solidification of particles in the places from where solid removal after or during manufacturing can be difficult.

Some examples of complex products are provided in Table 1.1. Introducing these products to the market is different from the case of classical chemical specification products such as nitric acid or ammonia. In fact, it should be first realized that there is a need for these new products with presumably improved properties. Thereafter, a particular technology can be developed for their production. An alternative to this

https://doi.org/10.1515/9783110797961-001

market pull approach when there is a need for a particular product with specific features is the so-called technology push. The latter implies development of a novel technology, which might bring new features to a product even if a market need is not clearly identified or the market potential has yet to be explored.

Invention of electricity in the nineteenth century or more recently lasers, television or smart phones/smart watches can be seen as an example of the technology push. While these examples are not directly related to formulation products, they are, however, worth mentioning. Introduction of a new product into the market is risky as it is unclear if or not customers would be ready to accept the product or the manufacturing process will be profitable.

An interesting recent example is a Pop-It fidget toy consisting of a colored silicone tray with pokable bubbles that can be flipped and reused. The toy game became very popular among children potentially due to boredom and stress originating from the COVID-19 pandemic. The toy per se was invented in 1975 and was too expensive to manufacture it from rubber at that time.

If the product launch is, however, successful, there is a chance to earn a huge financial reward by becoming the first company to enter the market, thereby acquiring a substantial market share. The followers should come up with a better product that usually has a lower market potential and a lower risk.

Table 1.1: Examples of chemical products.

Type of products	Particulate solids	Dispersed liquid	Soft solids
Form	Granules, tablets, powders	Emulsions, dispersion, latex	Pastes, gels
Continuous phase	Gaseous	Liquid	Solid network
Dispersed phase	Solids	Solid/liquid	Liquid/gaseous
Examples	Pharmaceutical tablet	Pesticides	Coatings
Composition	Active ingredient	Active component	Pigment
	Filler	Antifoam agent	Viscosity enhancer
	Lubricant	Antifreeze	Gelling agent
		Dispersing agent	Antifoam agent
		Wetting agent	Dispersing agent
		Solvent	Solvent

Sometimes, even the name of a company that was first on the market can be used instead of a particular product name. Xerox was so successful in introducing copying machines that in many languages the verb "to xerox" is used instead of "to make a photocopy." Another example related to information technology or more precisely search engines includes the term "googling," which became widespread because of Google's success in acquiring a dominant market share.

An example related to chemistry is various practices of removing skin wrinkles. Introduction (injection) of botulinum (Botox), one of the most powerful poisons, paralyzes the facial muscles. Flexibility of these muscles causes skin wrinkles, and despite huge controversy in the recent years with regard to such side effects as facial paralysis, blurred vision and severe allergy, application of Botox in cosmetics has attracted many customers who wish to look young. Many consider Botox injection as dangerous; consequently, other antiaging products have been introduced in the market, including various creams, gels and powders, claiming to be the fountain of youth.

The above example highlights several issues related to launching of a successful product. There should be a certain demand (i.e., from people who would like to avoid wrinkles and have a younger appearance); the product should have desired properties and composition; there should be adequate production and administration technology; the product should adhere to safety and environmental regulations and it should be financially rewarding.

Products mentioned above are directly marketed to an ordinary consumer, whose preferences significantly depend on country, geographical location, age, gender and so on. Many other formulation products to be covered in this textbook are aimed at industrial and government buyers. It is apparently clear that different strategies are thus needed for marketing of a washing aid for dishwashers, pesticides, diapers containing superadsorbents, or a catalyst for the removal of NO_x emissions. Products that are oriented toward so-called ordinary consumers should be readily available in large and reliable quantities at affordable prices in local stores. The general public should be informed about a product and its key feature without having to disclose about its composition or mode of action.

There are obviously exceptions in this practice, when customers are becoming more and more interested in the chemical composition of the formulation products as can be illustrated in a recent trend to introduce so-called sulfate-free hair care products. In particular, sodium lauryl sulfate and sodium laureth sulfate are used as surfactants in many skin and hair products, from shampoos to body washes. There have been claims that sodium lauryl sulfate, which in fact is a better surfactant than currently available sulfate-free alternatives, is a carcinogen and can lead to serious health issues like cancer or infertility. Other sources denounce such claims as a myth still acknowledging that people with sensitive skin or having hair on the drier end might prefer sulfate-free products, even if the latter do not lather as well as sulfate-containing shampoo.

Apart from such examples of products directly aimed at regular consumers, technical discussions related to industry-oriented performance products should be handled at a completely different level. In this context exhaust emissions aftertreatment systems for waste incineration or nitric acid plants can be mentioned.

Typically, chemical engineers deal with technologies of fuels and production of basic chemicals, which involve well-defined molecules or their mixtures. While fuels such as gasoline and diesel are sold directly to customers, marketing of basic chemical products and a special type of fuels (e.g., jet fuel) is different. Commodity or bulk chemicals are manufactured in large-scale continuous processes in very large quantities (thousands or even millions of tons) being virtually independent on a particular manufacturer. There might be some differences in the properties depending on the geographical location: for instance, sulfur content in diesel in Nordic countries was below 10 ppm already in the mid-2000s, whereas in many other places throughout the world it was substantially higher. Other products with high sales volume (methanol, ammonia and sulfuric acid) are sold globally based on their specifications and have the same properties independent on who is manufacturing the product and where. Cost-effective manufacturing of specification products is thus decisive. Manufacturers and engineering companies in a quest for earning their desired profit at low selling prices are focusing heavily on process efficiency and cost optimization; they pay significant attention to the feedstock cost and are willing to design novel processes starting from cheap raw materials if they are available. Commodities are capital-intensive with low profit margins, and it is difficult to expect product innovation and patent protection for products per se. Thus, it is not surprising that many chemical companies, which established themselves in the twentieth century as global players in manufacturing bulk commodity chemicals or primary chemicals, have been gradually exiting this market, turning more to performance products (secondary chemicals).

An apparent switch in strategy is related to higher profit margins of fine and specialty chemicals, which are mainly but not necessarily manufactured in small quantities (below hundred tons per annum) in mostly batch operations. Although fine chemicals are mainly defined by their chemical structure and content, specialty or performance chemicals should have desired properties to satisfy customer needs. Performance chemicals are often mixtures with proprietary chemical composition, meaning that these high-profit margin chemicals require dedicated research and development (R&D) efforts. Product innovation comes hand-in-hand with not only patent protection but also customer satisfaction and timely introduction to the marketplace.

A specific feature of performance products is that while some properties (density, viscosity, microstructure, particle size distribution, etc.) are genuinely required for the performance, others might not be necessary from the functional viewpoint; however, they still should be present in order to satisfy, for example, sensorial or aesthetic customer needs.

A few examples of current development in high-value specialty chemicals either available on the market or rather close to their implementation will be discussed in this chapter. Fundamental issues related to manufacturing of these products and a more systematic discussion on functional chemicals and products related to human well-being will be considered in the subsequent chapters.

1.2 Examples of advanced high-value-added specialty products

Advanced coating compositions incorporating pigments that absorb, reflect and transmit solar radiation to a different degree have been developed to mimic colors appearing in nature (Figure 1.1).

Figure 1.1: Coating mimicking nature [1].

The type of inorganic pigments used for coatings, including their combinations, and the coating thickness determines color (Figure 1.2).

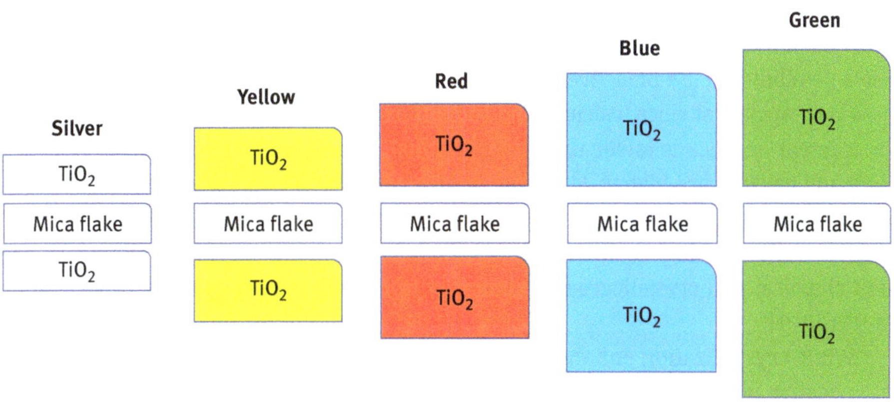

Figure 1.2: Color dependence on coating thickness [1].

Another example of an advanced inorganic product is bioactive glass [2]. An interest in their application is related to some drawbacks of competitive materials (Figure 1.3). For instance, because of low mechanical strength of polymers, they cannot withstand the stresses required in many applications. An apparent shortcoming of metals is high corrosion rate. Moreover, low biocompatibility undesirable for living tissues and high diffusion of metal ions potentially leading to allergic reactions limit application of metals despite high wear resistance, strength and ductility. On the other hand, good biocompatibility of ceramics along with resistance to corrosion and compression is undermined by small resilience, high density and low fracture strength of these brittle materials.

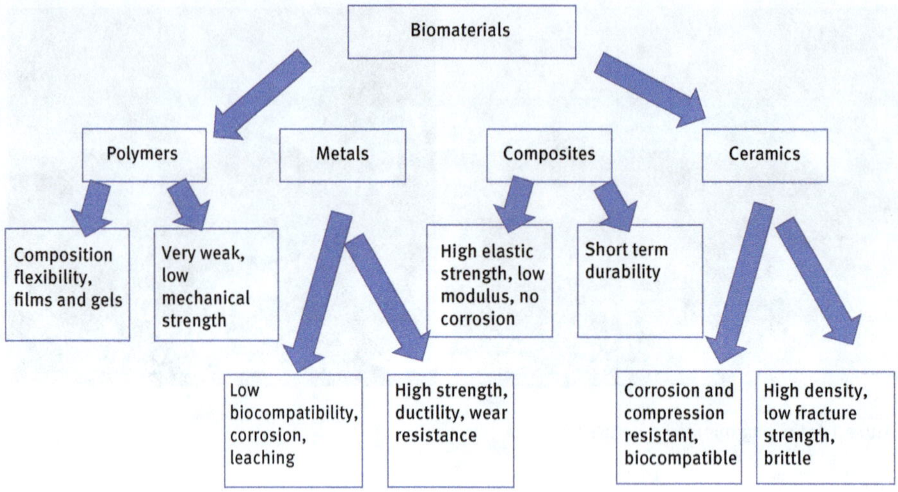

Figure 1.3: Schematic of classification of biomaterials (modified from [2]).

The scaffolds fabricated from calcium phosphate-based inorganic materials or bioceramics such as bioactive glass usually provide a higher mechanical strength. There is also a possibility to use bioactive glass and glass–ceramics in bone reparation. Bioactive glass is composed of silica, calcium and sodium oxides and P_2O_5 in specific proportions. The content of silica is lower than in traditional glasses, whereas the amounts of sodium and calcium are higher. Moreover, a high calcium/phosphorus ratio promotes the formation of apatite crystals with calcium and silica ions acting as crystallization nuclei. Such materials as bioglass when used for bone replacement should preferentially exhibit chemical and crystallographic similarity to natural bone mineral hydroxyapatite $Ca_5(PO_4)_3(OH)$.

While crystallization enhances the glass mechanical strength, it affects bioactivity in a negative way. This could be challenging as thermal heat treatment is required for making scaffolds leading to nucleation and subsequent growth of crystalline phases embedded in a matrix of glass. Such crystallized phases should not induce any cytotoxicity.

The role of tissue scaffolds is to provide a temporary structure for cells to synthesize new tissue. When the glasses are in contact with the body fluids, they should lead to the formation of a hydroxyapatite layer. They must degrade into nontoxic products, which can be easily removed or excreted from the human body. The mechanical properties of bioglass are related to a need of withstanding pressure or strain preventing any kind of structural failures. For bone engineering, bioglass should possess controllable interconnected porosity allowing cells to grow into the required physical structure and regulating diffusion of nutrients. Finally, bioglass should be economically attractive.

Another example of innovation is a functional food product Benecol [3] developed in Finland in the 1990s. An active ingredient is sitostanol ester, which is obtained by chemical transformations of sitosterol (Figure 1.4), a byproduct of pulping. The mode of action is related to lowering the cholesterol level of human serum, which is considered as one of the main risk factors for heart and coronary diseases. As shown in Figure 1.4, the structure of sitostanol resembles one of cholesterol. This is conceptually similar to the utilization of many pharmaceutical drugs acting as enzyme inhibitors being structurally close to the substrate.

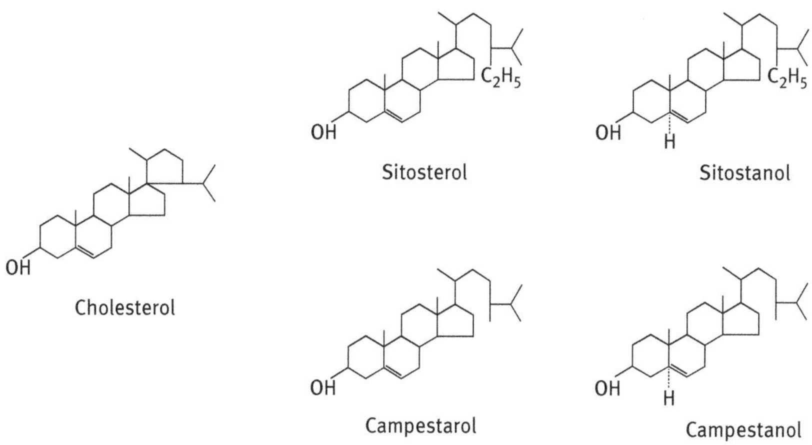

Figure 1.4: Structure of cholesterol and plant sterols.

It was known that sitostanol obtained by hydrogenation of sitosterol has a larger effect than the latter, which in addition is adsorbed to some extent to the blood vascular system. One of the problems with sitostanol application in vegetable fat products such as margarine or yogurt was connected to its poor solubility. This hurdle was mitigated by producing sitostanol ester, which is fat-soluble and can be mixed easily with vegetable fats.

Production of a line of products under the trade name Benecol therefore involves hydrogenation of a complex mixture of sterols available as a by-product of pulping. Besides sitosterol, this mixture also contains campesterol as the second largest component in addition to some other minor sterols. The subsequent step is esterification of a mixture of stanols with fatty acids (60% oleic acid) making the product, which can easily be incorporated into a variety of foods. Obviously, before introducing Benecol to the consumer market, extensive medical testing of the products had been conducted.

Recently, a significant attention has been devoted to nanocellulose, which is abundant and relatively cheap [4]. The global nanocellulose market was close to 300 million USD in 2019 with a projected annual growth rate of 20%. Ultrathin nanofibers with diameters of 2–5 nm allow nanocellulose to display extraordinary optical, mechanical and thermal properties (Figure 1.5).

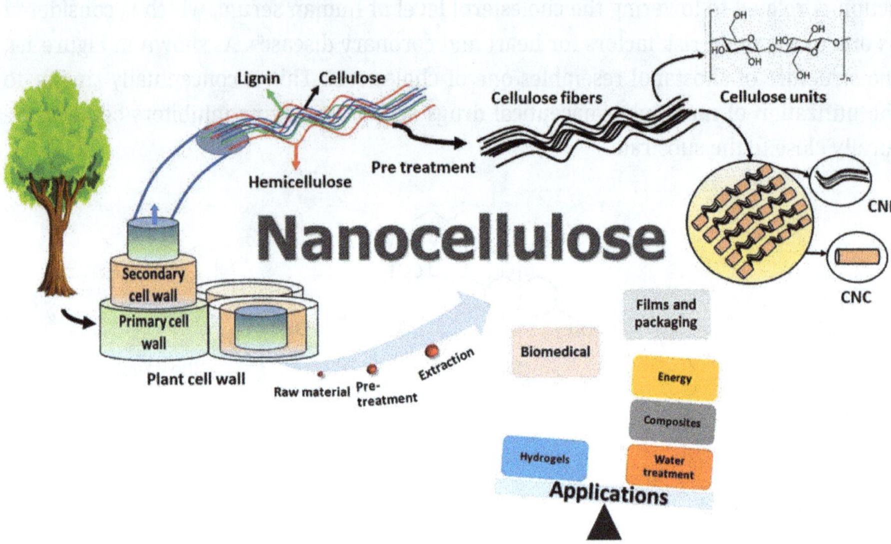

Figure 1.5: Nanocellulose [5]. CNC and CNF correspond to cellulose nanocrystals and nanofibrils, respectively. Reporduced with permission from Elsevier.

Nanocellulose-based nanocomposites have a number of advantages. For example, when interwoven, nanocellulose can form highly porous and mechanically strong bulk materials such as nanocellulose papers, films and aerogels. Metals (Ag, Au and Ni), minerals (calcium carbonates or phosphate) and carbon nanomaterials (carbon nanotubes and graphene) have been incorporated into nanocellulose substrates showing very promising electrical and optical properties. Nanocellulose-based nanocomposites can also be applied in solar cells, which convert solar energy to electricity. As nanocellulose paper is transparent, smooth and mechanically strong, it is a promising candidate to be used as a solar cell substrate. Transparent nanopaper increases the path length of light through the absorbing layer, thus resulting in greater solar light absorption. The downside of nanocellulose paper utilization for photovoltaics is connected to lower overall power conversion efficiency compared with the glass-based solar cell. Nanocellulose can also be used in Li-ion batteries providing an opportunity to design thin, flexible and high-performance batteries.

While metals with a sea of delocalized electrons have high electrical and heat conductivities, organic compounds do not possess such delocalized electrons, being poor electrical and heat conductors. In fact, coating of electrical wires with insulators is required. Electroconducting polymers in this context would be useful as they can be processed at low temperatures and pressures. Serendipity discovery of conducting polymers in the early 1970s was because of a human mistake when a graduate student preparing a new form of polyacetylene inadvertently added 1,000 times more catalyst than required by the recipe. As a result, *trans*-polyacetylene film produced in the reaction reflected light like aluminum foils. Further research was focused on doping polyacetylene with iodine giving a polymer with conductivity improved by more than a billion times. In 2000, the Nobel Prize in Chemistry was jointly awarded to Alan J. Heeger, Alan G. MacDiarmid and Hideki Shirakawa. The best electrically conducting polymers cannot, however, compete with copper in conductivity. Moreover, they have poor processability limiting large-scale applications. Although conductive polymers have been incorporated into commercial displays and batteries, manufacturing costs, material inconsistencies, toxicity, poor solubility in solvents and inability to directly melt are obvious obstacles. Nevertheless, currently there are attempts to use them in organic solar cells, printing electronics, antistatic substances for photographic films, shields for computer screens against electromagnetic radiation, sensors and actuators, to name a few potential application areas. However, it remains to be seen in which applications they will be eventually used.

The final example in this section is related to an interesting recent application of sensors and actuators in electronic textiles [6, 7]. Such textiles are fabrics that contain woven into them material electronics with physical flexibility and size which cannot be achieved with other available electronic manufacturing techniques. Components and interconnections in e-textiles should be intrinsic to the fabric. Fabric-based sensing can be used for biomedical and safety purposes, such as electrocardiogram or temperature sensing (Figure 1.6).

Integration of carbon electrodes into fabrics allows detection of specific features such as oxygen or moisture. A challenging aspect of e-textiles is power generation, which can be achieved through piezoelectric elements harvesting energy from motion or photovoltaics. Another challenge apart from technical issues of integrating microchip and computer systems into clothing and a need of overcoming washability issues could be reluctance of the eventual (elderly) consumers to buy e-textiles because of costs, safety concerns or general lack of acceptance.

Technological approaches to electronics and textile industry are very different. Clothing fabrication is still labor-intensive, located in countries with low labor costs and does not have very accurate requirements in terms of minimum dimension and yarn placement. On the contrary, electronic processes require precision, accuracy and specific skills that are more expensive. Overall, the industry of e-textiles is still emerging facing many challenges from technical, business, regulatory and marketing perspectives.

Involvement of the end user in all phases of product development, from design to validation, seems to be important to ensure in this case not only technical but also commercial success.

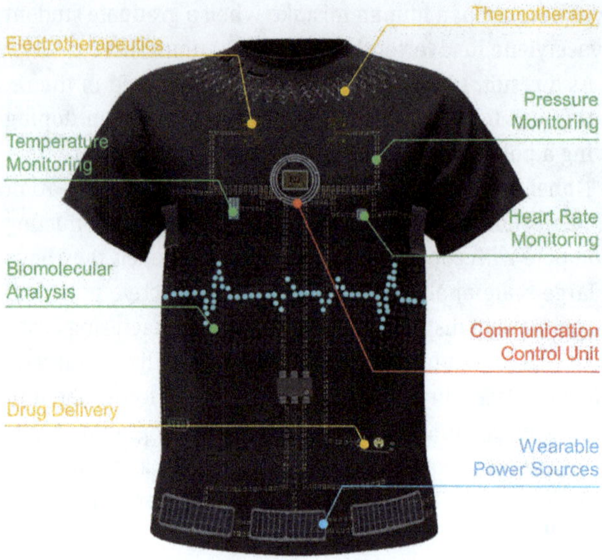

Figure 1.6: Representative functional units of intelligent wearable point-of-care textile systems, involving diagnostic devices (e.g., biomechanical sensors, biopotential sensors, temperature sensors and biochemical sensors), therapeutic devices (e.g., electrotherapeutics, drug delivery and phototherapy), protective devices (e.g., thermoregulation, electromagnetic shielding and toxicant degradation), wearable power sources and communication control units [8]. Reproduced with permission.

1.3 General aspects of product design

As already clear from the discussion above, product design is related to product performance. Process design more familiar to chemical engineers is linked to process performance (Figure 1.7).

For high-value and low-volume chemical products, there is a need to monitor product quality mainly in batch operations. When the process conditions are established, there is practically no room for changing them or optimizing the process parameters because such changes can result in off-spec products. For other chemical products, which belong to specification rather than performance products, processing conditions can be altered to control product quality.

A general structure of product design is presented in Figure 1.8, which implies that in product design besides physicochemical characteristics of a particular product, performance in a particular application and attractiveness to consumers should be considered. A more extended list for disperse products is shown in Table 1.2.

Among dispersed products, cosmetics, detergents, surfactant foams, inks, paints and coatings, as well as drugs can be mentioned combining several functions and properties in a single product. Such unit operations as crystallization, precipitation,

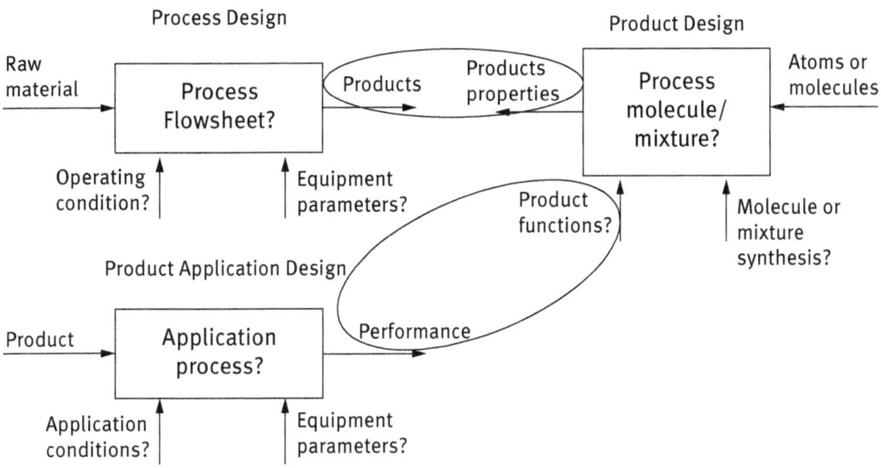

Figure 1.7: Links between product-process design and product-application design (reproduced with permission from [9]).

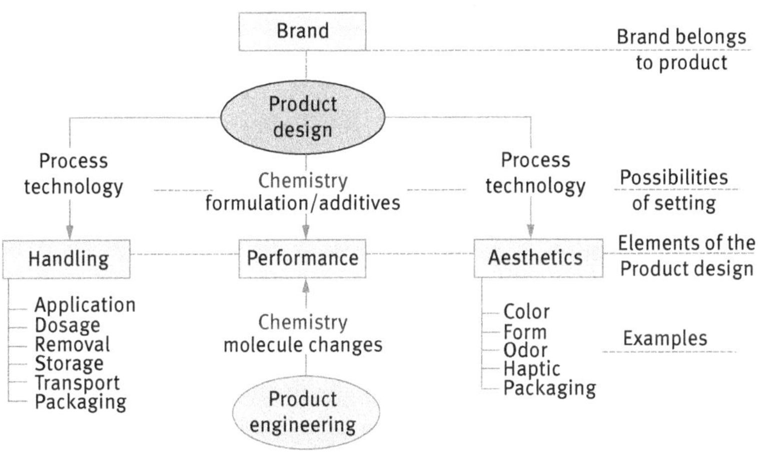

Figure 1.8: General structure of product design (reproduced with permission from [10]).

agglomeration, grinding and compaction are needed to control the particle size and its distribution, particle morphology and the final shape. For such complex media as polymers, colloids, microemulsions, suspensions and particulate solids, rheology and interfacial phenomena play an important role in achieving the desired product quality and performance.

Design of structured formulation products starts with identification of the desired end-use properties. Then, research is initiated on how to control the microstructure formation in order to obtain these desired features. Fundamental issues of interfacial phenomena, phase equilibria, kinetics etc should be considered along with specific

Table 1.2: General framework for specification lists of disperse products (adopted from [10]).

Particle characterization	Aesthetics	Physicochemical properties
Size and distribution	Size	Content of water
Flowability	Color	Bulk density
Abrasion resistance	Shape	Surface area
Explosive dust	Odor	Rate of dissolution
Pneumatic conveying	Taste	Pore structure
Strength	Whiteness	Melting point
Elasticity		Flocculation point
Composition		Glass temperature
Degree of crystallinity		Stability to storage
Toxicity		Wettability
		Content of dust

product design issues (nucleation growth, stabilization, additive, etc.). Along with this work, process design issues related to equipment sizing, mass and energy balance calculations and process control aspects (process parameters' regulation, sensors and quality assurance) should be addressed.

Often, new formulation products are required if the functionality of existing products is not sufficient. Thus, product development following the market pull starts with a need to improve properties of a current product if some undesirable properties are calling for modifications and performance improvements. New inventions in alternative technology push (lasers, television, smartphones or nylon fibers more related to chemical products) are creating markets and finding their applications in society with no apparent prior awareness of consumers about the benefits of a particular invention. In the case of technology push with unclear market prospects, it is of immense importance to receive feedback from the market by testing the product with potential customers.

Similar testing is done in seemingly different areas. For instance, new movie releases are shown to specific target audience sometimes even influencing the content (e.g., a potential happy ending) or the way a certain movie is marketed and distributed. Negative response from the target audience can lead to a direct release to DVD without even showing a movie in theaters.

In the times of so-called quarter economy, research outside of the main product lines might be difficult to conduct especially if there is absence of apparent synergy of new products with existing manufacturing and marketing capabilities. At the same time, some companies allow their scientists to allocate approximately 10–15% of time on projects not directly related to immediate business needs.

More freedom of independent research organizations and academia funded by government and foundations in selecting research topics can promote research discoveries. This eventually leads to spin-off companies that either develop a certain product using venture capital or offer their technology to enterprises able to arrange manufacturing and marketing.

Development of materials with superior properties can be hindered by unwillingness of customers to pay for novel features, even if they are superior compared with existing ones.

Other important issues are related to intellectual property rights as well as time, brand image, legal and regulatory constraints limiting the design space for product innovation. For instance, time to market is extremely important as companies lucky enough to be first in a new market can sometimes receive a lion's share approaching 50–70%.

In addition to skills related to technology, product development also requires marketing skills because a product's success is not measured when it is manufactured, but rather when customers who buy it are satisfied with the product. The phases in product development are illustrated in Figure 1.9 and also include the product introduction phase.

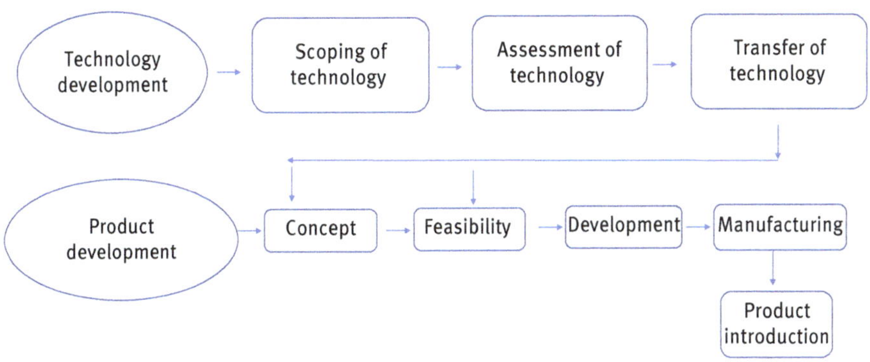

Figure 1.9: Phases in product development.

It should be noted that while engineering skills rely on information from textbooks and reference books based on fundamental laws of nature or/and well-established correlations, information obtained from marketing surveys of a limited number of (sometimes poorly informed) consumers, might be inaccurate and even contradictory.

For structured and complex products, the product life is much shorter than for specification chemicals, ranging sometimes from months to few years in comparison to decades for simple molecules. An important concept of the product life cycle (Figure 1.10) is used in marketing describing various stages of a product from cradle to grave. The introduction phase includes developing and then launching the product. During the growth phase, sales are increasing at their fastest rate coming at some point to the highest sales. In the maturity phase while sales are near their maximum, the growth rate is slowing down because of market saturation or competitors entering the market. The final stage of the cycle is decline, when sales start to fall.

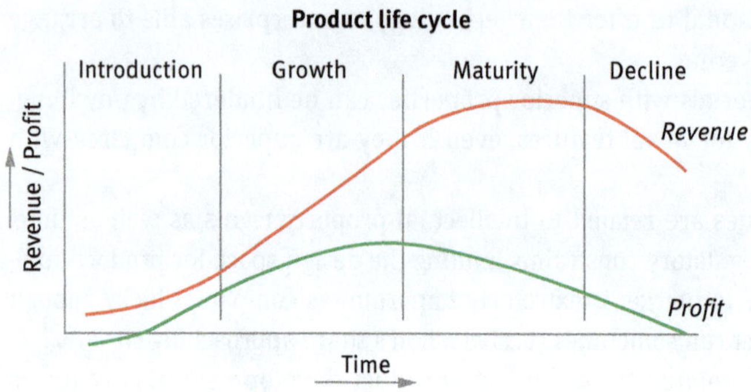

Figure 1.10: Product life cycle (from [11]).

The product life cycle can be extended by several marketing techniques, such as advertising to attract new customers, price reduction, exploring new (geographical) areas and market segments and new packaging. Another approach is a more technology-oriented one and relies on adding value, for example, introducing new features to a current product.

The following typical steps in product development can be mentioned: (a) identification of needs, (b) determination of the key parameters necessary for success and (c) developing ideas that can be potential solutions to the problems.

The needs are typically defined by customers or can come from societal trends. For example, catalysts for treating NO_x emissions from locomotives started to be developed more than a decade before the regulations came to power.

One of the ways to get the required information is to conduct interviews with customers, which might result in listing positive and negative experiences with a particular product, some other relevant and irrelevant products. As an outcome, promising as well as irrelevant ideas emerge containing wishes about the desired product properties and statements about what the product should not do.

Requirements of customers should not contain solutions or problems and should be sufficiently specific to allow finding a proper solution. Too specific requirements can, on the other hand, substantially limit the range of potential solutions.

It also recommended when developing a new product to analyze and understand behavior and performance of products from competitors.

While determining the key parameters necessary for success, the product developer should clearly establish properties that are imperative, properties that can be compromised and possibility of performance optimization by compromising the customer wishes to a certain degree. Ranking of needs from undesirable to essential through several levels (not important; extra bonus; desirable, strongly desirable) is thus useful. Such ranking can be done for each product attribute (composition, size, shape, flowability, particle size distribution, viscosity, yield stress, color, texture, elasticity, etc.).

Sometimes a product attribute might not be important for the performance. For instance, customers might not be eager to get a slightly yellowish product even with a better performance if they are used to a product that is transparent.

In addition to technical properties, including performance per se and product manufacturing, intellectual property issues, safety and health aspects should be considered.

It is important to realize that customer needs are related to quality of life and improvement of standards of living rather than appreciation of the efforts and innovations made during the product development.

When developing ideas for potential solutions obviously all relevant literature should be properly studied, including patent literature. The latter is often neglected in academic world, being, however, one of the most important sources for information in the industry defining the limitations for solutions.

Brainstorming is a useful method for generating ideas and should be properly organized. Often it is good to have not only one brainstorming group, but at least two comprising people with different background and education holding similar positions. In the case of large companies, the groups include people not only working in different departments but also in different locations or even countries.

Some people might be shy to express their opinions in the presence of their superiors. The task of the first group might be related to generation of ideas and thus people who are able to generate new ideas and of abstract thinking should be involved. The second group can contain experts with a more analytical and critical approach who can evaluate advanced ideas from the viewpoint of their originality, economic and technological feasibility. The number of people involved in brainstorming can be different ranging from individual and pairwise versions to small groups of 5–7 people and even larger ones (14–15).

The brainstorming might result not only in ideas, which sustain a certain product, just making it better (e.g., treating cancer by chemotherapy with another better performing drug), but also in a disruptive development (e.g., application of monoclinical antibodies).

Usually, the chemotherapy drugs work by damaging the RNA or DNA that tells the cell how to copy itself in division. The cells die if they are unable to divide. Chemotherapy does not know the difference between the cancerous cells and the normal cells. The treatment with monoclonal antibodies, that is, laboratory-produced molecules, Y-shaped proteins, is conceptually different as they are designed to recognize and bind to antigens (pathogens) that are generally more numerous on the surface of cancer cells than healthy cells.

Among other examples of disruptive inventions tea-bags and pod coffee-machines using capsules for home customers can be mentioned.

Obviously in the case of health care products and pharmaceuticals the threshold for a potential user to accept a new type of treatment especially for a life-threatening decease is lower; in other cases, the customers might not be willing to change easily their habits.

When Illinois introduced laws requiring motorcyclists to wear helmets in 1969, the Illinois Supreme Court ruled it an unconstitutional restriction of personal liberty, subsequently in this state there is no law that oblige the motorcyclists to wear a helmet, as the ruling still stands today. In fact this example illustrates that even within one country the laws can be very different (Figure 1.11).

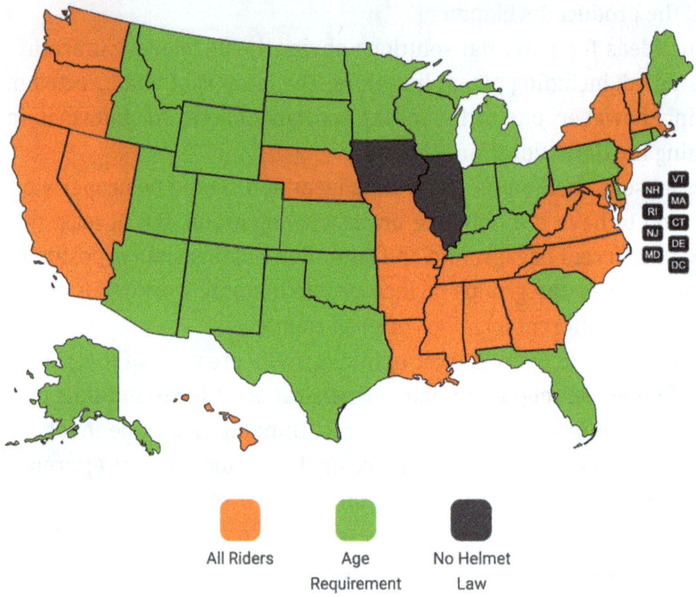

Figure 1.11: Requirements for wearing helmets in different states of the USA [12].

Another reflection of the persistent habits is the seat-belt law in New Hampshire, which does not require drivers and adult front-seat passengers to wear seat belts contrary to the all other states in the USA.

Securing intellectual property is very important in product development and manufacturing. Patenting is typically a preferred option of securing intellectual rights. The inventor has to show that the invention is not described in an existing patent and is not known from the publicly available literature. Thus, a defensive way of preventing others from patenting when patenting is considered not appropriate because of financial or other reasons is to write papers in scientific journals, popular magazines and even show results at international conferences in the form of oral or poster presentations.

The inventor in a patent should show that the invention is not obvious to a person skillful in the field disclosing to the public certain novel things (novel product); how they should be prepared (a new process) and/or a new application areas of known products, while the government gives a right to the inventor to stop others from using a particular invention, typically for a period of 20 years. Keeping a patent might be expensive especially if patents should be maintained in many different countries requiring annual payments.

Alternatively, the inventor can sell the patent or license it against a fee. This is often done by academic researchers or small companies, who cannot manufacture a particular product due to obvious reasons. Rights enforcement is an expensive and complicated procedure because it is the task of the inventor to find patent infringement, approach the infringer, negotiate with lawyers and bring the infringer to court.

Information provided in the patents should not be known to the public prior to patenting. Therefore, a product cannot be sold before patenting. There is an apparent danger of releasing too sensitive information in a patent application or infringing patents of competitors without realizing it.

An alternative to patenting is keeping know-how as a trade secret without disclosing the details of how to make or sell a product.

Formulation product design is often done by using temporary project teams recruited from different departments of a particular company. Teams typically have people with different expertise and can include people from different geographic locations.

In the early stages of product development, utilization of complex models should be avoided as in a later stage another idea might be pursued. Large efforts to develop an apparently wrong idea with subsequent time and financial investments could prevent from abandoning an idea that is not feasible or economic.

At a certain point when realistic solutions are proposed for a new product, they should be evaluated from the technical, scientific and economic feasibility viewpoints. It is not enough that a certain idea is consistent with thermodynamics and does not violate the fundamentals laws of nature. It is important that it can be implemented in a fast way and there are required capabilities for manufacturing a product in a profitable way.

When product concepts and property requirements are established, a certain formulation should be made and tested first at the lab scale and then preferable by customers for establishing if the marketplace is satisfied with properties of the product. A product is recommended (or not) for commercialization only after evaluating its potential.

Small- and full-scale testing answers questions related to manufacture of required quantities such as production location, technology equipment for manufacturing and packaging, supply chain, production costs, logistics of manufacturing and shipment to customers and marketing.

Equipment used in small-scale manufacturing is different from those used in large-scale production; therefore, scaling up can be a serious issue. In this phase of product development, the product engineers should work together not only with process engineers but also with specialists in marketing, making the product more suitable for the marketplace.

In the phase of commercialization, engineers are involved in establishing the plant location, raw materials, manufacturing technology and required equipment and designing disposal strategy.

The marketing of a product starts from understanding reasons of buying, characteristics, habits, budget of customers and an assessment of all competing products. A value proposition highlighting compelling attributes of a particular product should be done from the customers' perspective and is thus not related to product manufacturing technologies but to the product per se.

A particular product can be aimed at millions of individual buyers and/or at a very limited number of professional buyers. The former buy a product based on a limited knowledge on the technical details, being rather much influenced by the price, design, brand, advertising in media, experience of friends and so on. Aggressive marketing by advertising a particular product through newspapers, TV and mass media is essential for increasing the identifiability of a product by a consumer, who should get a desire for the product being also able to afford it. Individual buyers have different needs and habits depending on country, region, place of living (country, suburban and urban), climate and landscape, age, marital status, children, gender, income, religion and race, to name a few. There is also a priority of buying with the resources first allocated for food and then for housing and clothing. In the framework of the Maslow's hierarchy of needs (Figure 1.12) such needs are physiological, which should be satisfied before progressing to higher level needs.

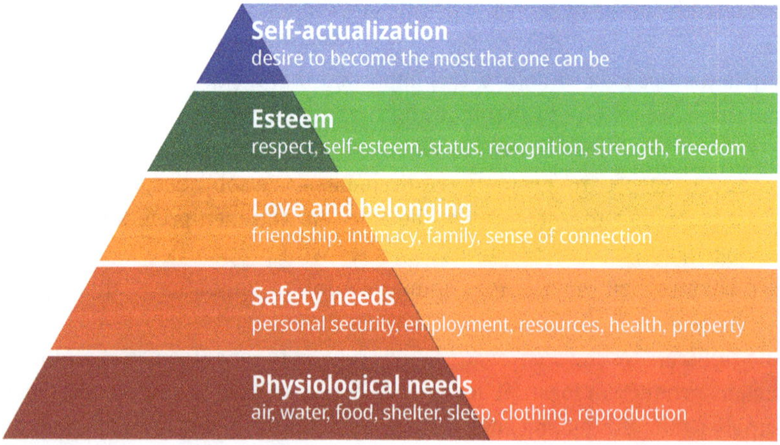

Figure 1.12: Maslow's hierarchy of needs [13].

Professional buyers purchase in large quantities at regular intervals or by long-term contracts and can evaluate the quality and performance quantitatively. Contrary to individual buyers, professionals have a possibility to compare alternative products in terms of performance, price, payment conditions, delivery terms and so on. Selling and marketing to business (or Business to Business, B2B) is clearly different from selling to individual customers (i.e., Business to Customers, B2C). In some market segments, such as the pharmaceutical industry, while the buyers are individual customers – the patients, decision-makers are not only the medical doctors but also health insurance companies/organizations, who have substantial influence in this business by deciding which type of drugs and to which extent will be covered. A certain percentage of sales for several less serious diseases is related to individual consumers. Thus, advertising announcements for few drugs can be also seen on television; however, marketing of drugs for life-threatening diseases is aimed at the physicians.

Government buyers (at all levels) purchase products related to national defense, police, fire, healthcare, transportation (roads and bridges), education and environmental protection (sewage and solid waste). They typically operate with lower budgets being constantly under public scrutiny. In some countries, an open and transparent bidding process with at least three potential suppliers is required before awarding a purchase contract.

In the domains where purchasing is done by professional buyers, if a consumer is not satisfied with a product, it might significantly influence the product image; therefore, often sellers provide technical support that help resolve consumer problems and modify their product accordingly.

Chemical engineers are more familiar with technology analysis (what can be done with a particular technology, how products are currently manufactured, what are the raw materials, what is the environmental impact) than with marketing analysis (who are the customers, why are these products needed, how are the products used, which properties are required, how a certain market is served and with which products, what are safety and environmental regulations).

At the same time, evaluation of the product against the competing ones available on the market requires a financial analysis to assess the profit and justify the investment along with the risks involved. The final decision on commercialization is usually made based on a solid and reliable business plan, which includes marketing (ability to solve customer problems, quality, pricing, projected sales volume), manufacturing (i.e., secure raw material supply, reliable manufacturing process, qualified staff and reliable manufacturing), environmental, organizational (internal synergism and risk diversification) and economic factors (requirement for investment capital, rate of return on investment, risks).

An important issue related to marketing and sales is product pricing, which can be set to get a certain profit margin, cover the fixed and variable costs and ensure a desired return on investment time. In some markets, pricing is more related to the willingness of the customers to pay what they think is a fair price and to the price of alternative products offered by the competition. A particular example is so-called sustainable or environmentally friendly products. It is a common belief that such products should be more expensive as they might use more expensive raw materials of higher quality or have more complex manufacturing adhering to green practices. A driving force for a customer to buy “eco” products in this case can be a desire to support sustainable resources or green practices.

Prices might change depending on demand, the stage of the product life cycle, appearance of products with new features and so on.

The business success of commercialization depends not only on an adequate selection of a new product but also on the reduction of the development time, including outsourcing technology development to other companies and university laboratories, while keeping the core competence inside a company.

At the end of this chapter, it is worth discussing some of the current world market mega-trends that can influence the chemical and formulation product technology in the foreseeable future. Such trends include increase of life expectancy and growing population of senior citizens because of better health, longer life and lower birth rates not only in economically well-developed countries but also in so-called emerging economics. An increase in life expectancy by 2050 with 10+ years will come along with a strong economic growth in East and South Asia. Another trend is creation of new mega-cities. In general, it is projected that 70% of the population are supposed to live in cities by 2050. An increase in population in general will lead to an increase of the middle class (estimation projection is more than 3 billion people by 2050), which in turn will require 30% more food and 50% more primary energy. Moreover, a 100% increase in global road transportation is anticipated by 2050. Greater globalization of trade, manufacturing, marketing and labor migration will bring more geographically uniform environmental requirements. Environmental protection will be of strong concern not only in more developed countries but will be seriously considered by emerging economies, whose citizens will require the same quality of air and soil as achieved in other countries. Environmental protection involves, as mentioned above, not only the concept of tightening emission regulations and improving the air and soil quality by emissions cleaning but also avoiding such emissions through incorporating the principles of green chemical engineering. This will allow improving process efficiency by increasing the yields of the desired products, improving feedstock utilization efficiency as well as separation efficiency.

References

[1] http://projects.nfstc.org/trace/2009/presentations/5-perry-new.pdf, accessed on 23.12.2017.

[2] G. Kaur, O. P. Pandey, K. Singh, D. Homa, B. Scott, G. Pickrell, A review of bioactive glasses: Their structure, properties, fabrication and apatite formation, J. Biomed. Mater. Res. Part A, 2014, 102A, 254.

[3] J. Lehenkari, Studying innovation trajectories and networks: The case of benecol margarine, Sci. Stud., 2000, 13, 50.

[4] H. Wei, K. Rodriguez, S. Renneckar, P. J. Vikesland, Environmental science and engineering applications of nanocellulose-based nanocomposites, Environ. Sci. Nano., 2014, 1, 302.
[5] C. Zinge, B. Kandasubramanian, Nanocellulose based biodegradable polymers, Eur. Polym. J., 2020, 133, 109758.
[6] K. Cherenack, L. Van Pieterson, Smart textiles: Challenges and opportunities, J. Appl. Phys., 2012, 112, 091301.
[7] M. Stoppa, A. Chiolerio, Wearable electronics and smart textiles: A critical review, Sensors, 2014, 14, 11957.
[8] G. Chen, X. Xiao, X. Zhao, T. Tat, M. Bick, J. Chen, Electronic textiles for wearable point-of-care systems, Chem. Rev., 2022, 122, 3259.
[9] R. Gani, K. Dam-Johansen, K. M. Ng, Chemical Product Design – A Brief Overview, in *The Chemical Product Design: Toward a Perspective through Case Studies*, Ed. K. M. Ng, R. Gani, K. Dam-Johansen. Elsevier, Amsterdam, 2007, 1–20.
[10] W. Rähse, Chemical Product Design – A New Approach in Product and Process Development, in *Industrial Product Design of Solids and Liquids: A Practical Guide*, Wiley-VCH, 2014, doi: 10.1002/9783527667598.ch1.
[11] http://2012e.igem.org/wiki/index.php/Team:UIUC_Illinois/business-plan, accessed on 23.12.2017.
[12] http://www.knightsonbikesdallas.org/motorcycle-helmet-laws-by-state/
[13] https://www.simplypsychology.org/maslow.html

2 Fundamentals and unit operation

2.1 Crystallization and precipitation

2.1.1 Growth and nucleation

Crystallization and precipitation are widely used in chemical and processing industries for production or separation purposes. The vast majority of pharmaceutical, fine and specialty chemical products, for instance, pharmaceutical pills, fats, pigments and catalysts are in crystalline form.

Crystalline forms, that is, polymorphs, differ in their lattice arrangement and thus exhibit different physical and chemical properties, such as density, refractive index, thermal and electrical conductivity, hygroscopicity, melting point, heat capacity, vapor pressure, solubility and dissolution rate, thermal stability, reactivity, crystal habit, morphology, surface area, particle size distribution, as well as mechanical properties.

As an example, calcium carbonate has three different forms. The most common arrangement for both precipitated and ground calcium carbonates is calcite of the hexagonal form. In fact, several different calcite crystal forms are possible: scalenohedral, rhombohedral and prismatic. A less common form of calcium carbonate is aragonite, which has a discrete or clustered needle-like orthorhombic crystal structure. The calcium carbonate mineral vaterite is rare and generally unstable. Precipitated calcium carbonate is a material finding multiple industrial applications. For example, it is used as a pigment, filler or extender in the production of paper, plastics, paints, adhesives, textiles, pharmaceuticals, cosmetics or food. These applications are strictly determined by the characteristic properties of the material, such as the average particle size, particle size distribution, specific surface area, morphology or chemical purity. Physicochemical properties of precipitated calcium carbonate are highly influenced by different synthesis methods or synthesis conditions.

Crystal growth of calcium carbonate can result in the formation of either metastable spherical valerite or the most thermodynamically stable calcite (Figure 2.1).

Moreover, an intermediate aragonite phase with a needle-like shape can also be stabilized depending on the process conditions. The maximum rate of calcite formation is observed at ca. pH 8.6, while lower pH results in the formation of valerite. Besides pH, a high concentration of calcium ions also favors formation of calcite. An increase of calcium hydroxide concentration brings changes of $CaCO_3$ morphology from rectangular plates to spherical particles. An increase in the CO_2 flow rate is favorable from the viewpoint of forming plate-like calcium carbonate particles. Such an increase is responsible for the acceleration of the nucleation rate which, consequently, leads to a shorter reaction (growth) time, thereby leading to nanocrystals rather than microparticles. Rectangular particles can be achieved at low temperatures and diluted suspensions.

https://doi.org/10.1515/9783110797961-002

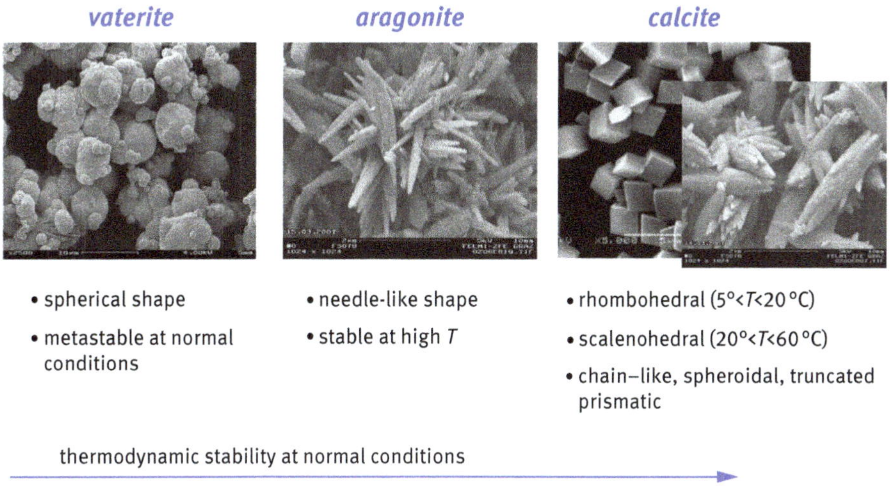

Figure 2.1: Polymorphs of calcium carbonate.

As apparently clear from the example above only one polymorph is thermodynamically stable at a specified temperature and pressure, while other polymorphs are capable of being transformed into the stable one. The rates of such polymorphic transformations are case specific.

In general, kinetics of solid–solid phase transformations tend to be much slower than liquid–solid transformations because of higher activation energy related, among other reasons, to larger solid–solid interfacial energies, smaller changes of free energy per unit volume and slower solid-state diffusion. As a consequence, solid–solid phase transformations rarely reach equilibrium, resulting in formation of metastable phases.

Polymorphs of a pharmaceutical compound may have very different solubility, bioavailability and therefore pharmacological action. Alterations in the preparation procedure might thus lead to the formation of such polymorphs of a pharmaceutical compound, which are no longer capable of delivering the required pharmaceutical performance.

At a fixed temperature and pressure, the more stable phase always has a lower solubility and higher melting point in any given solvent. Identification of possible polymorphs thus includes crystallization from a broad range of solvents with different polarity and hydrophilicity/hydrophobicity at different temperatures: crystallization from the melt or by sublimation, precipitation by rapid quenching with a liquid in which precipitate is not dissolved; and so on.

With regard to thermodynamic properties, there exist monotropic and enantiotropic systems. In the former case, plots of the free energy curves of the various polymorphs against temperature do not cross before all polymorphs melt, meaning that transitions from one polymorph to another below the melting point are irreversible. For an enantiotropic system, a plot of the free energy versus temperature exhibits a crossing point threshold before various melting points. Moreover, solid-phase transitions are reversible upon heating or cooling, or through physical contact with a lower energy polymorph. For such enantiotropic systems, the temperature range of the crystallization process is determined by the particular polymorph required. Crystallization begins just above the transition temperature in the domain of slow kinetics if the desired form is metastable below the transition temperature. The crystallization at this point is assisted by seeding with the desired polymorph. For the polymorph, which is stable below the transition temperature, seeding starts just below the transition temperature.

For monotropic systems, seeding is also recommended. The operation procedure to get the stable polymorph includes keeping temperature constant after seeding for the time needed for desupersaturation, followed by appropriate cooling while maintaining a constant controlled supersaturation.

Crystallization is a highly complex process in a multiphase and multicomponent system involving heat and mass transfer and is strongly dependent on fluid and particle properties, such as time-dependent size and size distribution.

Design of crystallization methods requires knowledge on solubility, phases and their stability, nucleation and growth characteristics as well as hydrodynamics of crystal suspensions.

Nucleation and growth depend not only on temperature and supersaturation but also on impurities. Supersaturation must first be achieved independent of the phase in which crystals can be grown. In fact, such growth can be done either from the liquid (solution or melt) or vapor phases (desublimation). For the industrial crystallization of inorganic substances from solution, water is used as a solvent.

The degree of supersaturation is expressed by the concentration difference Δc:

$$\Delta c = c - c^{*} \tag{2.1}$$

where c is the actual solution concentration and c^{*} is the equilibrium saturation value. The supersaturation ratio S defined as

$$S = c/c^{*} \tag{2.2}$$

is commonly used in explaining crystallization.

The fundamental dimensionless driving force for crystallization is expressed through the chemical potential difference of a given substance in the transferring (solution) and the transferred (solid) states

$$\Delta\mu/RT = \ln(a/a^{*}) = \ln S \tag{2.3}$$

where a^{*} is the activity of a saturated solution, implying that

$$S = \exp(\Delta\mu/RT) \tag{2.4}$$

Solubility–supersolubility diagram (Figure 2.2) is often used to illustrate how concentration depends on temperature.

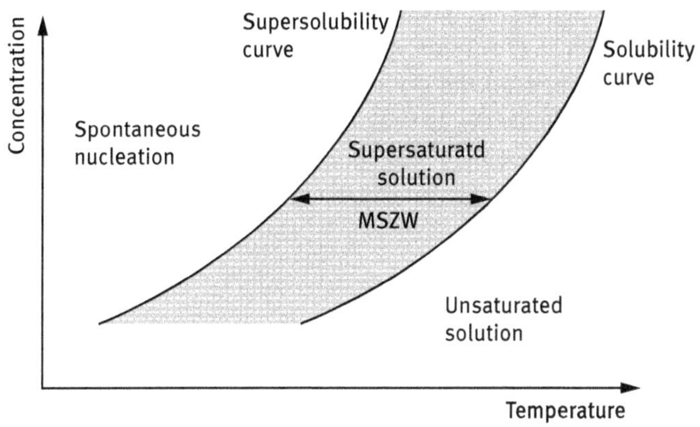

Figure 2.2: Solubility–supersolubility diagram with MSZW denoting the metastable zone width.

While position of the lower equilibrium solubility curve can be accurately determined, location of the upper supersolubility curve is less clear as it is influenced by many factors such as the rate of supersaturation, agitation intensity and presence of crystals or impurities. The metastable zone width visible in Figure 2.2 is thus characteristic of a particular crystallization system.

An important concept in crystallization is the critical crystal size r_{crit} (0.5–5 nm) starting from which there is a fast growth of nuclei, resulting in a large number of crystals of different sizes. For lower r_{crit}, the degree of supersaturation should be more extensive.

Typically, two basic modes of nucleation are considered: a primary one in the absence and a secondary one in the presence of crystals. Homogeneous primary nucleation does not require the presence of foreign particles, while heterogeneous nucleation typical for industrial crystallization is induced by the foreign solid particles present in the working solutions. The mechanism of heterogeneous nucleation starts with adsorption of the crystallizing species on the surface of solid particles. When crystalline bodies larger than the critical nucleus size are formed, they continue to grow into macrocrystals.

For homogeneous nucleation, spherical nucleus represents the most likely shape minimizing the amount of interfacial area per unit volume. The total Gibbs energy change, ΔG_{tot}, involved in forming a spherical nucleus of radius r is

$$\Delta G_{tot} = \Delta G_{volume} + \Delta G_{int} = \frac{4}{3}\pi r^3 \Delta G_V + 4\pi r^2 \gamma \tag{2.5}$$

where ΔG_V is the free energy per unit volume released upon creating the new second-phase particle. This term is negative to allow a phase transition. The interfacial energy per unit area γ is positive as energy is expanded when an interface is created. This term is associated with the creation of new interfacial area.

Figure 2.3 displays the dependence of volume, surface and free energy changes as a function of the nucleus radius r. Beyond the critical nucleus size r^*, the total free energy of the system is decreasing. Below the critical size, it is energetically unfavorable for nuclei to continue growing, and they will spontaneously decompose back. When nuclei are formed above the critical size ($r > r^*$), they will continue to grow.

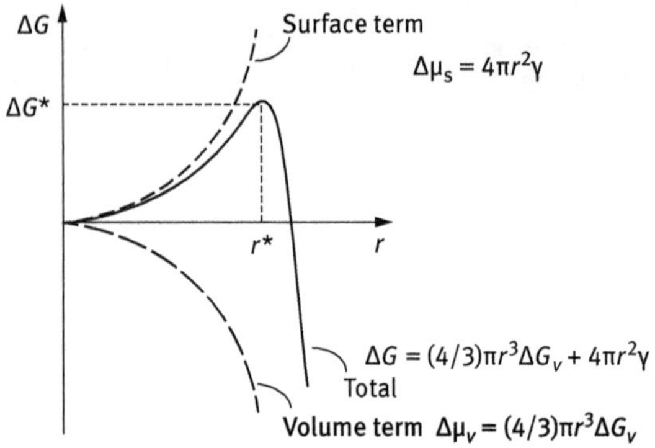

Figure 2.3: Volume, surface and total free energy changes associated with the homogeneous nucleation of a spherical particle of radius r [1].

Taking the derivative of eq. (2.5) and setting it equal to zero allow to calculate the critical nucleus size r^* as $r = r^*$ when

$$\frac{\partial \Delta G_{\text{tot}}}{\partial r} = 0 = \frac{\partial \frac{4}{3}\pi r^3}{\partial r}\Delta G_V + \frac{\partial 4\pi r^2}{\partial r}\gamma = 4\pi (r^*)^2 \Delta G_V + 8\pi r^* \gamma \tag{2.6}$$

Therefore,

$$r^* = -\frac{2\gamma}{\Delta G_V} \tag{2.7}$$

It is more common for nucleation to occur heterogeneously at specific sites (surfaces, interfaces or grain boundaries). Secondary nucleation takes place when crystals of the final species are present, and is operative in industrial crystallizers.

The total free energy associated with the nucleating phase is

$$\Delta G_{\text{tot}} = \Delta G_{\text{volume}} + \Delta G_{\text{int}} = \frac{4}{3}\pi r^3 \Delta G_V f(\theta) + 4\pi r^2 \gamma f(\theta) \tag{2.8}$$

which is different from that with homogeneous nucleation by the $f(\theta)$ factor, accounting for the geometry of the cap versus a sphere.

The critical nucleus size is $r = r^*_{\text{het}}$ for heterogeneous nucleation which can be calculated in a similar fashion as for homogeneous nucleation, that is, $r = r^*_{\text{het}}$ when

$$\frac{\partial \Delta G_{\text{tot}}}{\partial r} = 0 = \frac{\partial \frac{4}{3}\pi r^3}{\partial r}\Delta G_V f(\theta) + \frac{\partial 4\pi r^2}{\partial r}\gamma f(\theta) \tag{2.9}$$

Therefore,

$$0 = 4\pi (r^*_{\text{het}})^2 \Delta G_V f(\theta) + 8\pi r^*_{\text{het}} \gamma f(\theta) \tag{2.10}$$

thus giving the same expression as for homogeneous nucleation (i.e., eq. (2.7))

$$r^*_{\text{het}} = -\frac{2\gamma}{\Delta G_V} \tag{2.11}$$

even if the heterogeneous nucleation activation barrier, ΔG^*_{het}, is different from homogeneous nucleation defined as

$$\begin{aligned} \Delta G_{\text{tot}} &= \Delta G_{\text{volume}} + \Delta G_{\text{int}} = \frac{4}{3}\pi (r^*)^3 \Delta G_V f(\theta) + 4\pi (r^*)^2 \gamma f(\theta) = \\ &= \frac{4}{3}\pi \left(-\frac{2\gamma}{\Delta G_V}\right)^3 \Delta G_V f(\theta) + 4\pi \left(-\frac{2\gamma}{\Delta G_V}\right)^2 \gamma f(\theta) = \frac{16\pi\gamma^3}{3(\Delta G_V)^2} f(\theta) \end{aligned} \tag{2.12}$$

Concentration of nuclei (n^*) that can be formed at any given temperature T is calculated through the change in ΔG^*:

$$n^*_{\text{het}} = n_0 e^{-N_A \Delta G^*_{\text{het}}/RT} \tag{2.13}$$

Since ΔG^*_{het} strongly decreases with decreasing θ (being $1/2\Delta G^*$ for $\theta = 90°$ and $\approx 1/75\Delta G^*$ for $\theta = 30°$), n^*_{het} increases exponentially with decreasing θ, explaining why the concentration of viable heterogeneous nuclei is orders of magnitude higher than the corresponding value for homogeneous nuclei.

Stable nucleus once created in a solution is capable of growing into a crystal. Crystal growth involves mass transport, either by diffusion or convection from the bulk solution to the crystal face, followed by integration of growth units into the crystal lattice by a surface reaction.

In most crystal growth processes, not only diffusion but also surface reactions are important. The two most common surface processes are referred to as spiral (formation of screw dislocations on a crystal face) and polynuclear growth processes. The latter develops from monolayer nucleation on various parts of a crystal such as faces, corners or edges.

As a result, the crystal growth kinetics often depends not only on the crystal size but also on the surface structure or perfection. Small crystals (<50 μm) typically have a slower growth rate than larger ones. It is important to predict the behavior of small crystals since it has a practical consequence in continuously operated industrial crystallizers. When new crystals with a size of 1–10 μm are constantly generated by secondary nucleation, their share in the overall crystal size distribution (CSD) decreases significantly. Therefore, the ability to predict the growth rates of small crystals is useful in assessing the performance of crystallizers.

Another parameter influencing the growth rate is the presence of impurities, which because of their adsorption can block the crystal surface, thereby decreasing the growth rate. Moreover, impurities may change the crystal shape (or habit) by altering growth at different crystal faces. Habit modifications can also be influenced in some cases by adjusting operating conditions such as the rate of cooling or evaporation, and the degree of supersaturation or crystallization temperature. This might be important from the process technology viewpoint and help to improve rheological properties of the slurry, downstream unit operations (filtration or washing) and handling properties of the dried product and its storage stability. Other parameters, which can be used to influence the crystal shape, are solvent and pH. Moreover, some impurities can be added to the system to influence the crystal growth habit or alternatively can be removed from the crystallization milieu.

These precautions are done because of undesired caking (formation of hard materials) upon storage of crystalline materials. Such caking depends not only on storage per se, including its duration, humidity, pressure and temperature as well as fluctuations in some parameters, but also on the crystal size, shape and moisture content. Some measures to diminish caking include prevention of air contact with the crystals by careful drying, efficient packaging and coating the crystals with a protective layer of an inert dust.

Despite these measures, the leading factor in prevention of cake formation is the size of crystals and more importantly the size distribution and shape. Larger crystals are less prone to caking, due to a smaller number of contact points per unit mass. Caking can be also minimized by having a narrower size distribution and more granular shape. Operating conditions during crystallization and application of habit modifiers can influence the CSD and shape, respectively.

Nucleation and growth are not separated in time as once few viable nuclei are formed they start to grow. The Johnson–Mehl equation describes the overall transformation rate

$$F(t) = 1 - e^{-(\pi/3)\dot{N}\dot{G}^3 t^4} \tag{2.14}$$

It relates the overall fraction of the transformed material as a function of time [$F(t)$] with the nucleation ($\dot{N}$) and the growth ($\dot{G}$) rate, which are considered to be time independent. The third power in the linear growth rate follows because of three-dimensional (3D, spherical) growth, resulting in the transformed volume which increased in proportion to $\dot{G}^3$.

Equation (2.14) corresponds to sigmoidal transformations, when [$F(t)$] first increases exponentially in time, slowing down thereafter and finally asymptotically approaching complete transformation. A simplified version of the Johnson–Mehl equation, known as the Avrami or Avrami–Erofeev equation, is very often used in practice

$$F(t) = 1 - e^{-kt^n} \tag{2.15}$$

where n is known as the Avrami exponent and k is a parameter comprising both the nucleation and growth rates associated with the phase transformation. Obviously, larger values of k result in faster transformations. For 3D phase transformations, n typically varies between 4 and 3, while for alternative 2D (thin films) or 1D (wire or rod) transformations, n typically varies between 3 and 2 and between 2 and 1, respectively. Moreover, the value of the Avrami exponent could reflect strong deviations from spherical growth or time-dependent nucleation and growth rates. Figure 2.4 illustrates the dependence of this function [$F(t)$] as a function of time for different values of the Avrami exponent.

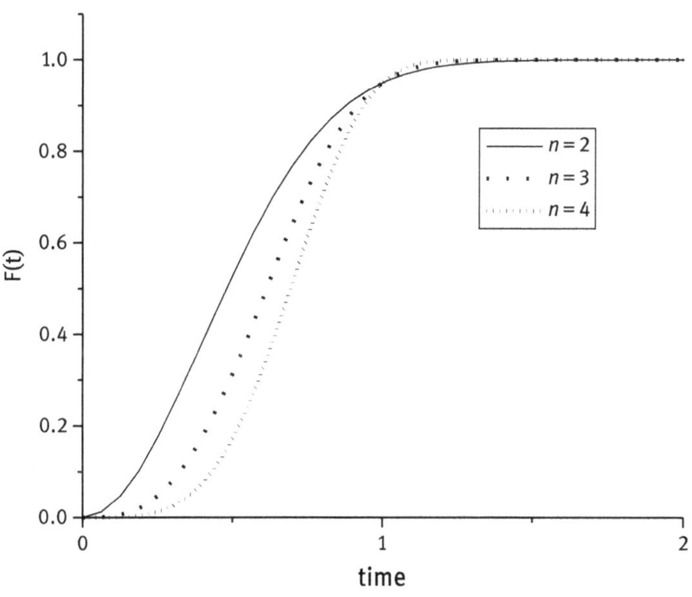

Figure 2.4: Fraction-transformed *F(t)* during a phase transformation as a function of the time elapsed at different values of the Avrami parameter, $k = 3$.

CSD reflects how the size is distributed over the size range (Figure 2.5). The dominant crystal size is most often used as a representation of the product size.

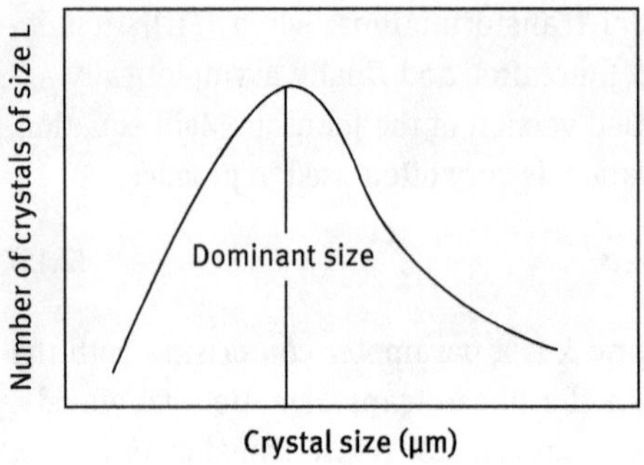

Figure 2.5: Crystal size distribution.

The form of the CSD curve in a continuous perfectly mixed crystallizer is determined by the residence time distribution and can be modified by selective removal of crystals altering such residence time distribution.

2.1.2 Crystallization from solutions

Several methods exist to initiate crystallization. One of the preferred approaches is to lower the temperature of the feed solution by direct or indirect cooling. Fast cooling and low molecular mass of the dissolved substrate give smaller crystals, which are more difficult to separate. However, there will be fewer impurities. Another option, when solubility is less sensitive to temperature, is to remove the solvent by alternatively adding heat. For the case of strong temperature dependence of solubility, vacuum cooling of the feed solution without external heating can be applied. Vacuum cooling can be combined with external heating when there is an intermediate dependence of solubility on temperature. Finally, a nonsolvent can be added for precipitating solute from a solution. Smaller crystals are obtained at higher temperature and supersaturation, as well as with more efficient stirring.

In the industry, crystallization for high-volume production is often done in batch crystallizers taking advantage of precipitation from a solvent in a controlled way to ensure not only the phase separation but also the particle size and morphology as discussed above. This can be done by a careful selection of operation parameters related to cooling, evaporation and dilution. One of the downsides of batch crystallization is product quality variations.

If crystallization is done without any stirring, the technology is very simple, since after introduction of a hot feedstock solution into an open vessel, cooling is done by natural convection, often taking several days. Introduction of foreign metal particles promotes growth of crystals on their surfaces, thereby diminishing the amount of product which will end up at the bottom of the crystallizer. Slow cooling results in broad CSD with large interlocked crystals retaining some of the mother liquor and leading to impure crystals after drying.

Capital, operating and maintenance costs are low in such type of crystallization; at the same time, low productivity, high space requirements and labor costs make this method economical mainly for small batches.

Agitation in open-tank crystallizers allows a decrease of batch time and formation of more uniform and pure crystals of smaller size. Better purity is possible since less mother liquor is retained, and washing is more efficient. Apparently, introduction of an impeller increases costs, which is overcompensated by much higher productivity. In a large agitated cooling crystallizer, an upper conical section is needed to decrease the upward velocity of liquor, thereby preventing carryover of the crystals with the spent liquor. Cooling is done by using preferentially water jackets to avoid potential accumulation of crystals on cooling coils.

Evaporation of a part of solvent can be used to obtain the required degree of solution supersaturation if lowering the temperature is not sufficient. In a vacuum crystallizer (an insulated vessel, operating at reduced pressure), supersaturation is achieved by simultaneous evaporation and adiabatic cooling of the feedstock.

Large vessels and poor control over crystal form in batch crystallizers have sparked an interest in fully continuous crystallization, resulting in the development of different crystallization options, with the most popular ones being forced-circulation (Figure 2.6), fluidized beds and draft-tube-agitated crystallizers (Figure 2.7).

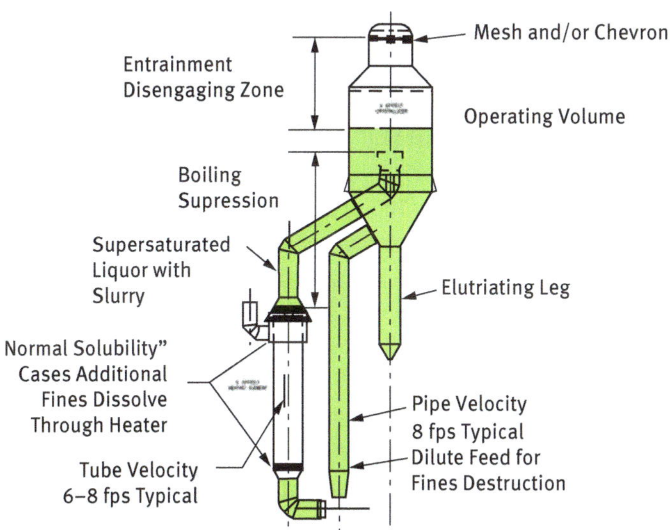

Figure 2.6: Forced-circulation Swenson crystallizer [2].

Swenson forced-circulation crystallizer (Figure 2.6) with an external heat exchanger operates under reduced pressure. High recirculation rate provides a good heat transfer, minimizing encrustation. The crystal magma passing through the vertical tubular

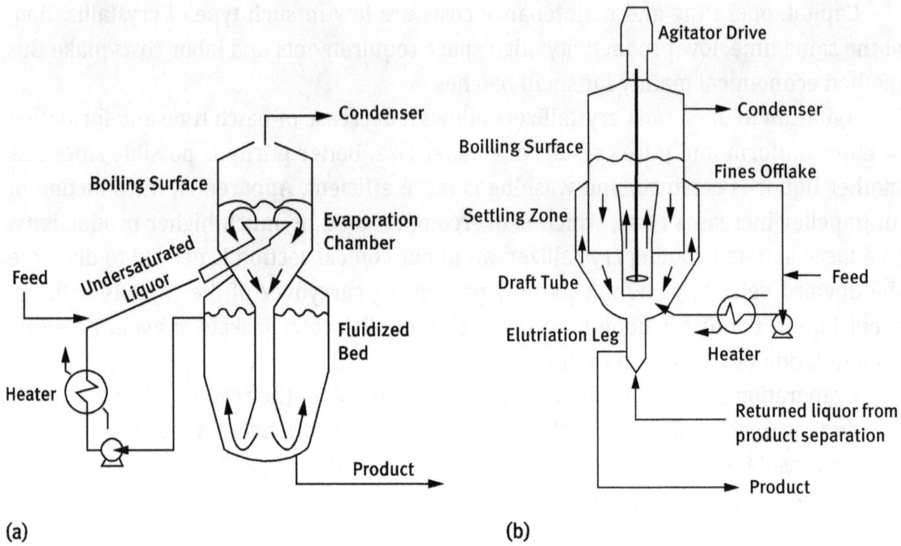

Figure 2.7: Continuous crystallizers: (**a**) Oslo cooling crystallizer [3] and (**b**) Swenson draft-tube-baffled (DTB) crystallizer [4].

heat exchanger is reintroduced in the evaporator below the liquor level, preventing sudden evaporation.

Intensive mixing in fixed-bed crystallization enhances mass transfer, leading to better crystal growth, whose rate can be faster than that of nucleation. This allows to use solutions with a low degree of supersaturation. In an Oslo fluidized bed crystallizer, there is an upflow of the supersaturated liquor, and a bed of crystals is suspended in space around a central downcomer (Figure 2.7a).

Another option for continuous crystallization is to use draft-tube-agitated vacuum crystallizers. One example (the so-called Swenson draft-tube-baffled vacuum unit) is illustrated in Figure 2.7b. Location of a relatively low-speed agitator is in a draft tube somewhat below the liquor level. At the base of the draft tube, there is an entrance of concentrated hot feedstock.

In chemical process engineering, continuous, steady-state operation is regarded as the most efficient. For crystallization processes, this conventional thinking might not hold true as besides operational advantages related to equipment simplicity and minimization of encrustation on heat-exchanger surfaces the required crystal properties (purity, form and distribution) can be produced only by batch crystallization. As mentioned above, batch crystallization suffers, however, from batch-to-batch variability and higher operating costs.

Continuous crystallization requires a holdup tank to allow equilibrium to be reached as the product removal is not done at equilibrium. Without installation of a holdup tank further unwanted crystallization leading to downstream deposits or undesired phase transformation can be difficult to avoid. In continuous crystallization, frequent shutdowns and washing are required to prevent self-seeding.

Design of industrial crystallizers is far from being straightforward as it is difficult to characterize particle-fluid hydrodynamics and extrapolate data in lab-scale reactors on nucleation and crystal growth to large-scale units. Crystal suspension velocity is an important parameter for scaling up of fluidized bed crystallizers, while for agitated vessels, the minimum rotational speed necessary to keep all crystals in suspension should be elucidated. Scaling up of agitated vessel crystallizers is often done based on constant power input per unit volume.

2.1.3 Crystallization from melts

Melt crystallization is commercially attractive, compared with operating at much higher temperature distillation for separation of close-boiling organic substances, especially if they are thermally unstable. For melt crystallization, the composition of the crystallized solid should be different from that of the liquid mixture from which it is deposited.

An example of crystallization from melt is the Czochralski process widely used to obtain single crystals of semiconductors and metals (Figure 2.8). To high-purity silicon melted in a crucible at 1,425 °C, some dopants can be added to achieve different electronic

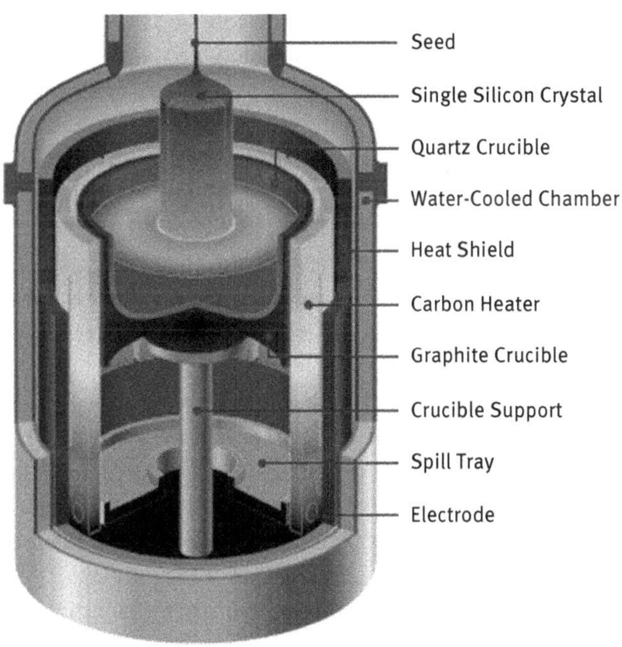

Figure 2.8: Crystallization from melt. Schematic setup of a Czochralski crystal puller [5].

properties. The seed crystal's rod is first dipped into the molten silicon and thereafter slowly pulled upward and rotated simultaneously. Controlling temperature gradients, rate of pulling and rotation speed is needed to obtain a large, single-crystal, cylindrical ingot from the melt. The overall process is done under an inert atmosphere (e.g., Ar).

2.1.4 Precipitation

Precipitation is closely related to crystallization from solution, and differs from crystallization in that a precipitate is usually amorphous, not pure and has a poorly defined size and shape, contrary to crystallization. In manufacturing various chemical products, including catalysts, paints, pigments and pharmaceuticals, precipitation is often applied.

Precipitation can be carried out in one or in several steps, gradually removing the desired components by such fractional precipitation, which however, cannot be very sharp.

Precipitation can be considered as a first step in the overall process comprising precipitation and crystallization, where the latter is used for final purification.

Similar to crystallization, precipitation consists of supersaturation, nucleation and growth of nuclei. Supersaturation can be done by, for example, cooling, solvent evaporation, salting out, introduction of antisolvents and pH alterations.

In saturated solutions, particle coarsening or the so-called Ostwald ripening occurs by dissolution of smaller particles and deposition from the solute on the larger particles. The driving force for ripening is the difference in solubility between small and large particles. In formulation of pastes, Ostwald ripening can be undesirable, for example, in flowable concentrates for seed treatment formulations when large particles exhibit lower effectiveness or in the case of pigment-containing paints. In the latter example, larger crystals have weaker color. Alternatively, recrystallization of pigments giving a powder of high hiding power might be beneficial for that purpose.

Precipitation is a complex phenomenon, which involves besides nucleation, ripening, agglomeration, phase transformations and habit modifications also. Supersaturation, concentration of active impurities and to a certain extent pH are the parameters influencing precipitation.

Several rules apply for precipitation. A reactant concentration increase gives first an increase of the median particle size of the precipitate at a certain time after mixing. This is followed by a subsequent decrease of the median crystal size. With an increase of the time interval, the maximum is shifted toward lower initial supersaturation and higher median particle sizes (Figure 2.9). For completed precipitation, the median size of the precipitate crystals decreases with an increase in the initially created supersaturation.

Highly supersaturated solutions can be made by mixing two reacting solutions in a very fast way, which might be challenging if reasonably uniform conditions should be maintained. The sequence of reactants mixing can be thus important. The induction period that starts after mixing and continues until appearance of nuclei depends

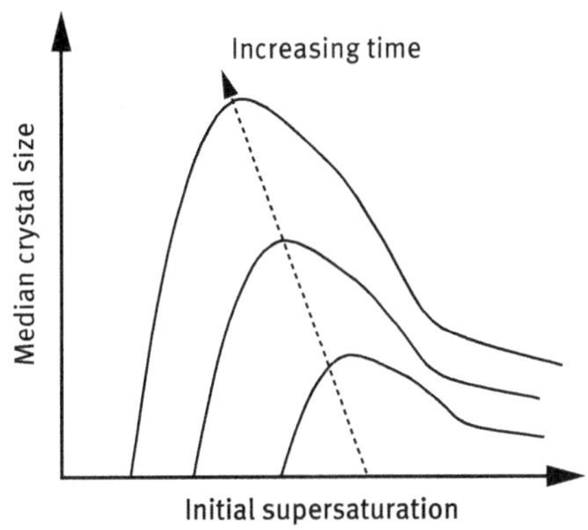

Figure 2.9: Illustration of precipitation rules.

on a number of parameters, such as supersaturation, mixing characteristics, temperature and impurities. This induction period is followed by rapid desupersaturation comprising primary as well as secondary nucleation and nuclei growth. Thereafter, particle coarsening might start because of ripening or agglomeration.

Similar to crystallization, a simple batch operation can be more advantageous than a continuous, steady-state operation mode. In industrial precipitation, good mixing should be ensured, which is essential to smooth out supersaturation peaks in local regions. Both micromixing at the molecular level, which is influenced by physical properties, and macromixing, which is influenced by the impeller geometry and speed, as well as vessel geometry and other parameters, are important.

For instance, the feedstock entry point can influence the precipitate quality. In fact, two reactants can be introduced simultaneously near the surface or the impeller blades. Alternatively, the second reactant can be introduced when the first is already in the precipitation vessel. This can be done near the surface of the first reactant or near the impeller blades. In the latter case due to good local mixing, nucleation is less prominent and larger primary crystals are formed.

2.2 Size changing and particles shaping

2.2.1 Size reduction

Size reduction or comminution of solids is a unit operation to produce a desired particle size distribution needed for a final application. The ground product after milling is separated, and a coarser stream is returned to the mill feed. Such recycling helps to avoid overgrinding.

Several options are used for size reduction. The necessary stress applied between two solid surfaces (crushing) either for single particles or a bed of particles is determined by the force applied to the solid surfaces. Size reduction can also be achieved by the impact of a particle against a solid surface or other particles. Two variants can be used, either a solid surface is accelerated to impact the particle or alternatively the particle is accelerated against a surface.

For plastic materials, stressing can be done by cutting in a scissor-type way. Finally, agglomerates can be broken using shear forces or pressure waves (Figure 2.10).

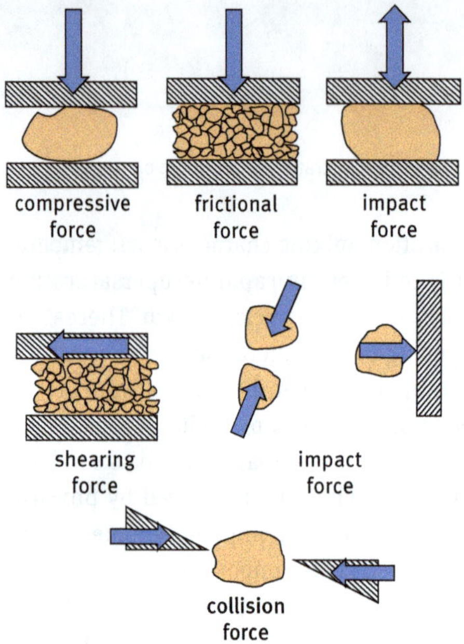

Figure 2.10: Various stressing options [6].

One example of a crusher that can process particles up to 1.5 m in edge length and applied in mining and production of road construction materials is illustrated in Figure 2.11a.

Such impact-type double roll crusher consists of two rollers rotating toward each other, being separated by a gap. The latter can be adjusted depending on the needed size. In this way, rather abrasive materials can be processed. Nevertheless, resurfacing of the rolls should be done at certain intervals because of abrasive wear imposed by abrasive materials. In grinding media mills (Figure 2.11b), there is a grinding medium (ball, rods, etc.), which moves because of rotation of a cylindrical or conical tube filled with this grinding medium. As a result, the balls roll against each other, imposing a very high local stress and leading to significant size reduction.

The feed material is comminuted between the grinding media because of impact and shear forces. Ball milling can be organized in a continuous mode (Figure 2.12), when the product is pumped to the mill, flows through the agitated bed of grinding beads and finally exits the mill through a suitable discharge. The volume of the grinding beads can be ca. 70–80% of the mill volume. The size of the mills used in, for example, catalyst manufacturing is ca. 1–3 m^3.

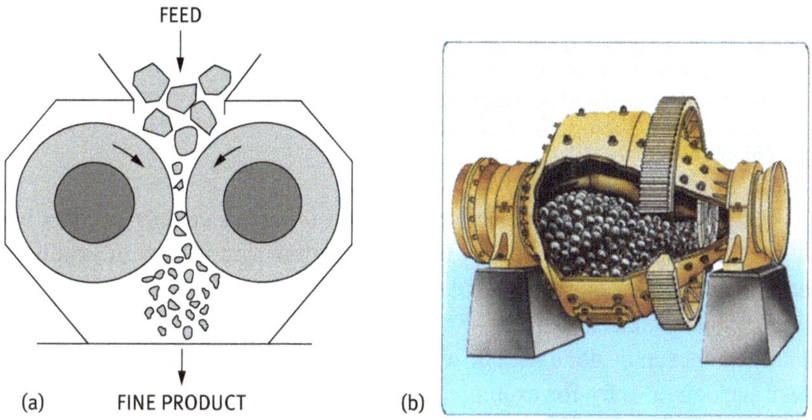

Figure 2.11: Examples of size reduction equipment: (**a**) double roll crusher and (**b**) ball mill [7].

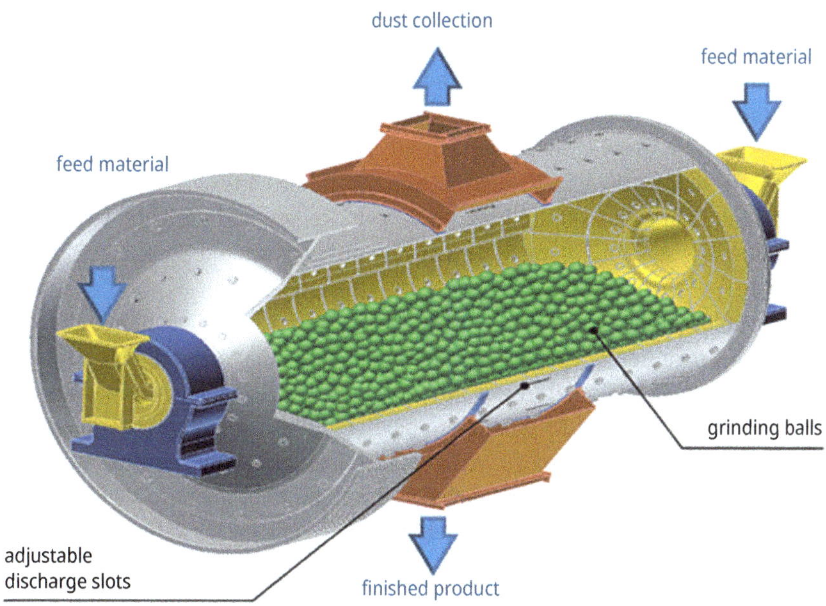

Figure 2.12: Continuous ball milling [8].

Other ways of size reduction include, for example, impact and cutting mills. In the former case, the stress is applied by machine–particle contact when the particles fly against an impact plate. Cutting mills imply cutting between rotating and static sharp edges with a narrow clearance and are used for such nonabrasive materials as polymers, rubber or paper, which are too tough to be processed with other types of mills.

In such cutting mills equipped with stationary knife bars and a rotor, the latter has several knife blades for cutting the product.

2.2.2 Size enlargement

Contrary to size reduction, size enlargement is needed to form a coarser product by agglomeration. Such size enlargement is employed for making a variety of products in manufacturing of fertilizers, pesticides, catalysts or pharmaceutical products in different shapes (spheres, tablets, etc.).

Size enlargement can be done with or without external forces allowing the formation of larger agglomerates by, for example, pressure compaction and extrusion, tumbling and other agitation methods, using chemical reactions or such physical methods as drying. In some methods, to achieve desired mechanical properties, binders are added to the feed. Catalysts made by extrusion can be one of the cases when binders are very often necessary as otherwise shaped materials will fall apart.

There are different agglomeration methods reflecting a variety of feedstock, which can be a dry solid or a wet paste or a fluid. In some cases, pumps can be used to transport the starting material, while in some other cases this can be challenging. Some materials can be sprayed, while others not. Thus, the initial state as well as the final agglomerate shape, size and its distribution should be considered. For some applications, the final materials should have sufficient strength and porosity, such as in catalysis. Finally, the size enlargement or shaping methods should be selected based on the production capacity.

When size enlargement is done by growth agglomeration, capillary binding forces hold the particles together in the presence of aqueous media, and binders allow to make typically spherical agglomerates of the size between 0.5 and 20 mm. Binders are often added in order to enhance particle-to-particle adhesion, helping to keep the strength after drying. In the case of catalytic materials, better mechanical strength provided by the inorganic binders is achieved at the expense of the pore size. Alternatively, application of organic binders necessitates their removal by calcination, leading to lower mechanical strength.

Inclined devices (drums, cones and pans) can be used for size enlargement (Figure 2.13).

In an inclined agglomerator (Figure 2.13), the feed enters from the top, while the product agglomerates of a rather uniform size are discharged over the rim. Large capacity, longer residence type, easy operation with dusty materials and robustness are clear advantages of drum agglomerators.

Obviously, after a certain size, no further agglomeration is possible in such agitated system as the destructive forces become more prominent. Temporarily bonded conglomerates should undergo a curing step in tumble agglomeration methods to generate more permanent bonding. An example of such curing is drying, which in gen-

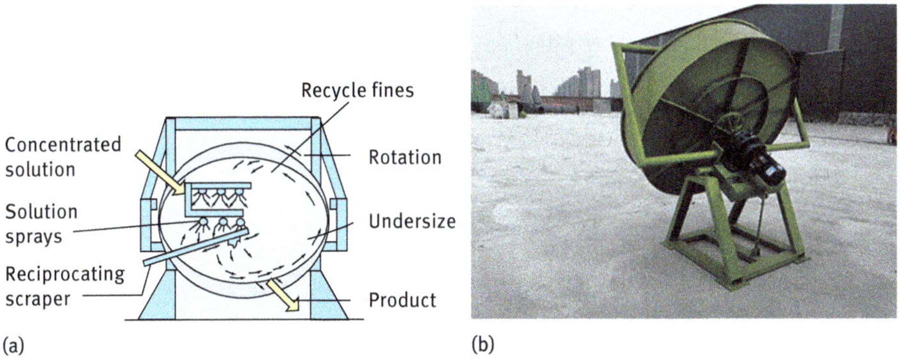

Figure 2.13: Continuous agglomerator: (a) scheme [9] and (b) photo [10].

eral is expensive. Nevertheless, tumble agglomeration can have a high throughput, making fine particles of high surface area and porosity.

Spray methods are most commonly used in chemical and pharmaceutical processing. The technology is based on atomizing pumpable suspensions and evaporating of liquids from the droplets by applying hot air. Fine agglomerates of 20–500 μm in size are formed, held first by capillary forces and then crystal bridges at the contact points.

Spraying will be discussed in more detail in a separate section on drying.

2.2.3 Pressure agglomeration, including pilling and tableting

Particles with only low amounts of moisture can be agglomerated in tablets of few millimeters and larger briquettes of several centimeters in different types of presses. Advantages of this method include, besides operation with dry solids, the uniform shape of tablets, independence on the feed particle size, rather high throughput (up to 30 t/h) and a possibility to avoid cutting.

Pill production, pressing, molding and extrusion (which will be discussed later in the section related to rheology) are commonly used methods to produce agglomerates of well-defined shape, dimension and uniformity.

Pilling by either single-punch or rotary presses requires high mechanical force. In a roller press, the particulate material is introduced between two oppositely turning rollers, being compressed when it is squeezed through the gap between two rotating rolls. Binders can also be added to assist size enlargement. Screw feeders can be applied when feeding fine materials (Figure 2.14).

Large void volumes of the feedstock implies that a large volume of air is expelled during agglomeration; thus, a path of air release opposite to the downward flow of the feed should be provided. A need for air removal limits the rotational speed of roller presses and their throughput.

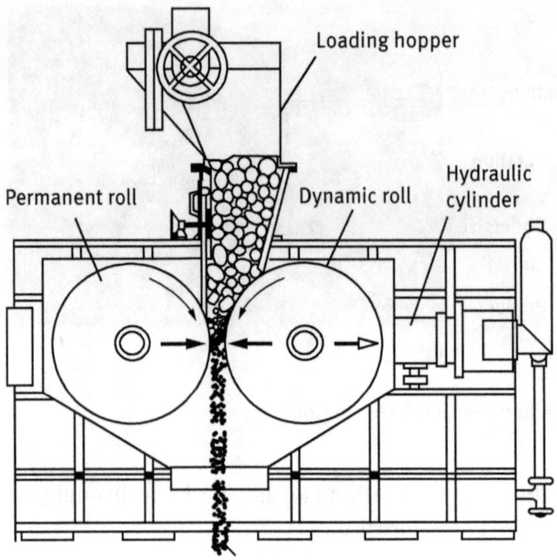

Figure 2.14: Operation of a roller press with a feeder [11].

The screw feeder is installed when the flow property of a material or the loading capacity is not adequate and as a result gravity feeding through a simple hopper is not sufficient because of undesired caking or formation of bridges.

Often, tableting is considered primary in connection with production of pharmaceuticals; however, it is also widely used in other areas, such as production of catalysts.

Compression of the material loaded into a mold is done using two punches (upper and lower ones). In eccentric press, lower punch is usually stationary even if it can be adjusted. After completion of pressing, the tablet is removed by an expulsion stroke of the lower punch (Figure 2.15). Such tableting of relatively low capacity (3,000 tablets/h) can be used for production of catalysts.

Tableting problems are related to potentially poor flow properties of the substrate, too high moisture content or too low lubrication. As a result of sticking during the tablet release, surface defects can appear. The interaction of the particle size distribution, the crystal shape and other physical properties of the substrate are the main factors affecting the compressive strength, the disintegration time and the release of the active ingredient.

Improvement of tableting can be done by modification of the fillers' properties, changes in a binder or a lubricant, pressure and speed, and more importantly, humidity and compression rate.

The steps are: (1) the feed cup fills the die; (2) in order to fill the same volume, the feed cup removes the excess material from the surface of the die and moves away to ensure the free movement of the upper punch; (3) the upper punch compresses the particle aggregate inside the die; (4) the upper punch returns to its initial upper dead-

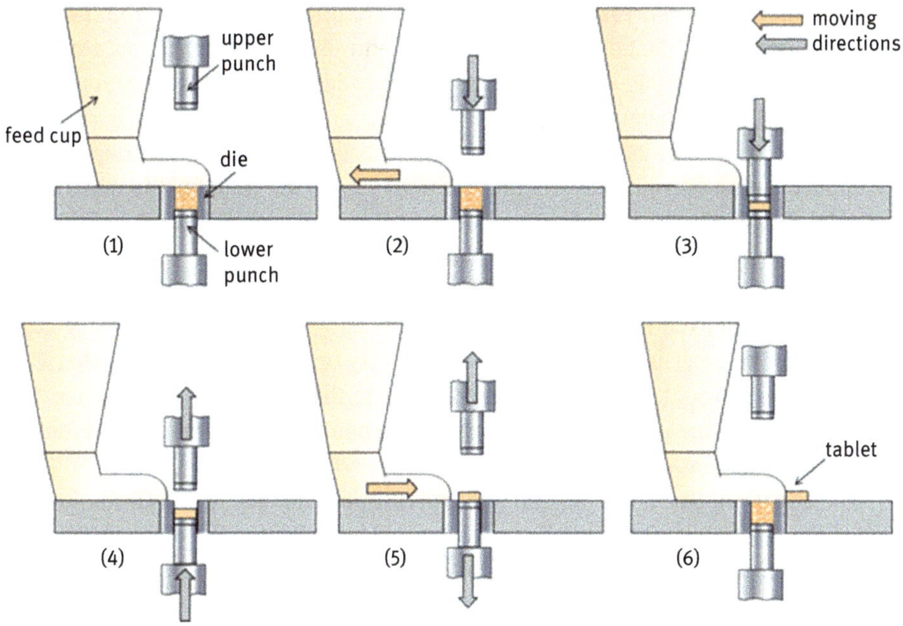

Figure 2.15: Schematic illustration of the compression process in an eccentric tablet press [12].

point position, while the lower punch ejects the tablet from the die – reaching its upper dead-point position; (5) the feed cup rolls off the tablet; and (6) the lower punch returns to its lower dead-point position and the feed cup fills the die.

As an alternative to eccentric tablet presses, rotating ones are used where compressive forces are generated between an upper and a lower pressure roll (Figure 2.16).

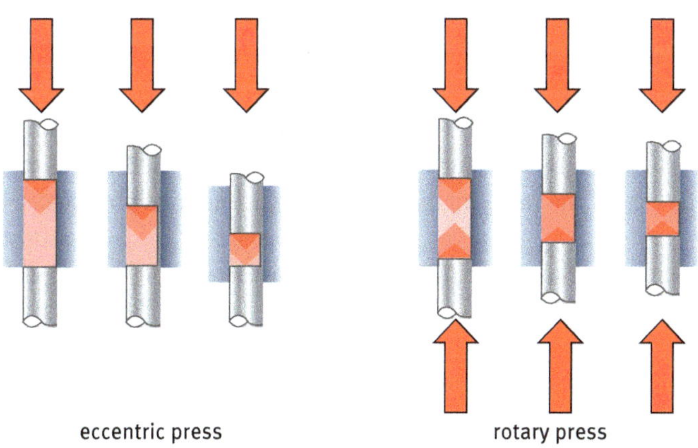

Figure 2.16: Schematic illustration of the compression process in an eccentric and rotary press [12].

In a rotating press, compression is done not only from the top but also from beneath the tablet, producing one tablet per revolution. Centrally rotating machines are capable of producing ca. 10,000–700,000 tablets per hour depending on the rotation speed and a number of punch pairs.

The important parameter in tableting is plastic deformation, which depends on the applied stress, overall time, the rate of compression and time for the material to be under maximum stress. The pellet hardness and integrity are influenced by the maximum pressure.

High mechanical stress during tableting has the downside of potential crushing of crystals, dislocations of crystal planes and low porosity, typically not exceeding 30%. In the case of heterogeneous catalysts where porosity can be a serious issue otherwise preventing an efficient transport of the reactants to the catalytically active sites, the formulation contains porosity improvers, like carbon, which can be burned during calcination treatment.

The important parameter in tableting is plastic deformation. As the elastic component of the response to the applied pressure stored in the material is relaxed after the pressure release, if such component is too high, unstable tablets are formed.

In pharmaceutical industry, tableting is performed at pressures of 25–250 MPa with granulated compounds as illustrated in Figure 2.17.

Low pressures can lead to insufficient mechanical strength, which can be also observed at insufficient concentrations of a binder, an adhesive and a lubricant, as well as low plasticity of substances. Subsequently if the tablets fall apart when ejected, the compression pressure should be increased or more binders added. On the contrary, too high pressure can lead to delamination of the tablets. The latter can also happen when the granulate is too dry. If the granulate is too humid, the particles of the tablet mass might have a higher adhesion to the punch than to each other. As a consequence, tablets are ejected with a damaged or uneven surface.

Some of the defects of the tablets are more related to the equipment rather than the composition of the tableting mass. If, for example, the tablet has edges, the diameter of the used punches is small. When the lower punch does not rise high enough, the tablet surface is knocked off requiring an adjustment of the low punch position. Inaccuracy in dosing of various auxiliary substances diminishes the shelf life of the drug, while inaccurate dosage of the main ingredient can compromise its pharmaceutical efficiency.

Direct compression without granulation allowing to bypass some unit operations, on the other hand, can lead to delamination of the tablet mass and changes in dosage when pressing with a small amount of active ingredients requiring also a need to use high pressure. Efficient utilization of direct compaction imposes certain restrictions on the material, namely, an isodiametric crystal shape, good flowability (at least 5–6 g/s) and low adhesion to the pressing tool of the tablet machine. The vast majority of drugs are not able to spontaneously fill the matrix of a tablet machine due to a significant (>70%) content of small particles and their irregular shape, causing strong inter-

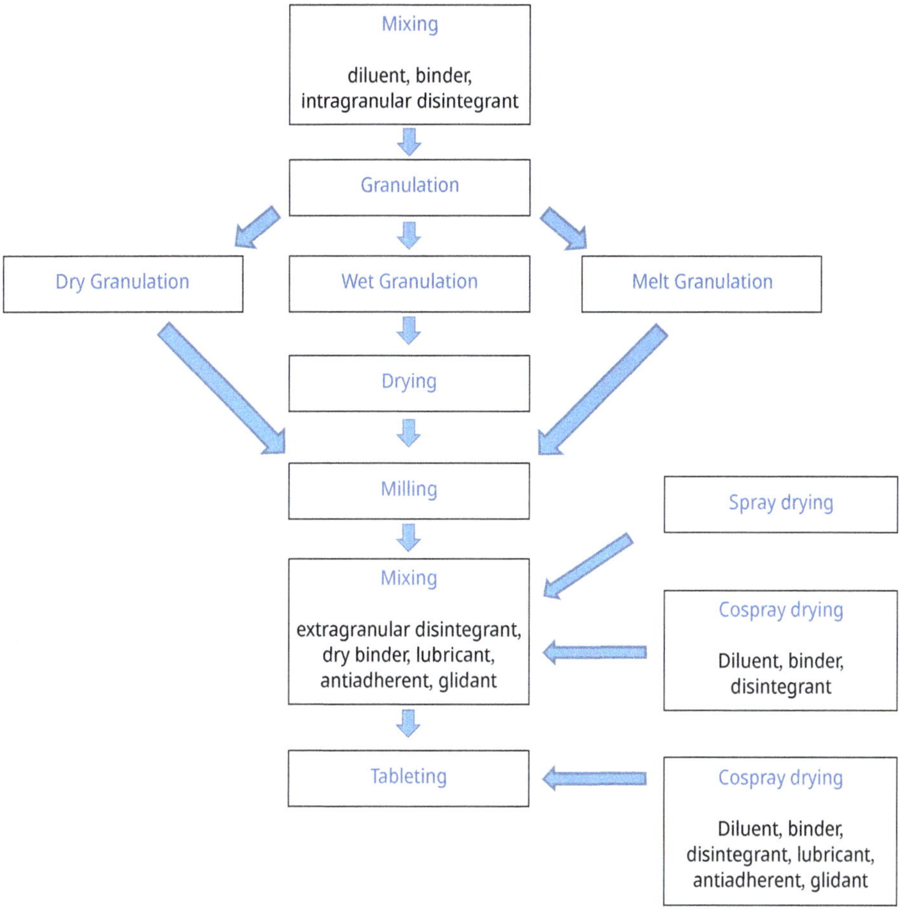

Figure 2.17: Unit operations employed in the production of tables by granulation and spray/co-spray drying [13].

particle friction. Subsequently rheology improvers as well as organic binders are added including, for example, microcrystalline cellulose or polyethylene oxide.

The lower the concentration in the tablet mass of a particular component, the smaller should be its particles. To obtain the optimal composition of a mixture comprising several compounds, their particle sizes must correspond to their concentration, and the density should preferably be close to each other and the shape close to spherical.

2.3 Drying

2.3.1 Drying for agglomeration

Drying can be done for several purposes, such as removal of humidity or for agglomeration. In the latter case, drying and agglomeration proceed simultaneously.

Drying is used thus not only to eliminate a volatile liquid but also to make a dry solid of a desirable shape, size, porosity, density and texture. From the process technology viewpoint, drying can be either batch or continuous under steady-state conditions.

For batch operations, typically, energy consumption and labor costs are higher, which is somewhat compensated by lower capital costs. Such operation mode is suitable for lower capacity and multiple products. Drying of a well-specified product at a high production rate is preferably done in a continuous mode, allowing optimization of energy requirements and a decrease of labor costs.

After reaching the equilibrium temperature, drying proceeds at an almost constant rate for a certain time, when a continuous water film exists on the initially wet surface. Evaporation proceeds in the same way as from a free liquid surface independent of the presence of a solid and is governed by heat transfer to the wet surface. When the surface of the solid materials becomes dry, evaporation starts to proceed also from the pores and is controlled by transport of the moisture from the solid interior. The rate of drying in this so-called falling period decreases; thus, despite a relatively small amount of moisture removed, the removal time is rather long. Finally, only the strongly bound moisture is left, leading to a rapid decrease of the drying rate.

Selection of the type of dryer depends on the wet material properties and production capacity. Lumpy and granular materials can be dried using a stationary approach in, for example, simple drying chambers or are moved by some supporting elements as, for instance, in belt, air-lift, drum dryers or cyclones. Materials that are liquefied or pasty can be processed to fine drops with spray drying. Roller drying in this case results in thin layers, while partial drying in a grooved roller dryer gives a crumbly material. Vacuum drying is applied for temperature-sensitive materials. The residence time in the dryer can be adjusted by selecting the drying type, namely forced convection dryers, spiral tube dryers and spray dryers allow for shorter residence times.

Hot drying medium, such as air, can be used for drying of chemically insensitive materials, allowing a direct contact with the drying agent, while otherwise an inert gas or other methods should be used.

For the purpose of organic solvent recovery or when toxic/flammable liquids are removed from the solid, special care should be taken, and drying requires completely enclosed dryers to prevent escape of the liquid to atmosphere.

For moving solids, there are different options for arranging the air flow (Figure 2.18).

When the flow is co-current, obviously dry air is put in contact with the solid bearing the largest moisture content. This implies high initial drying rate due to large differences in temperature and humidity. Along the dryer the driving forces decrease

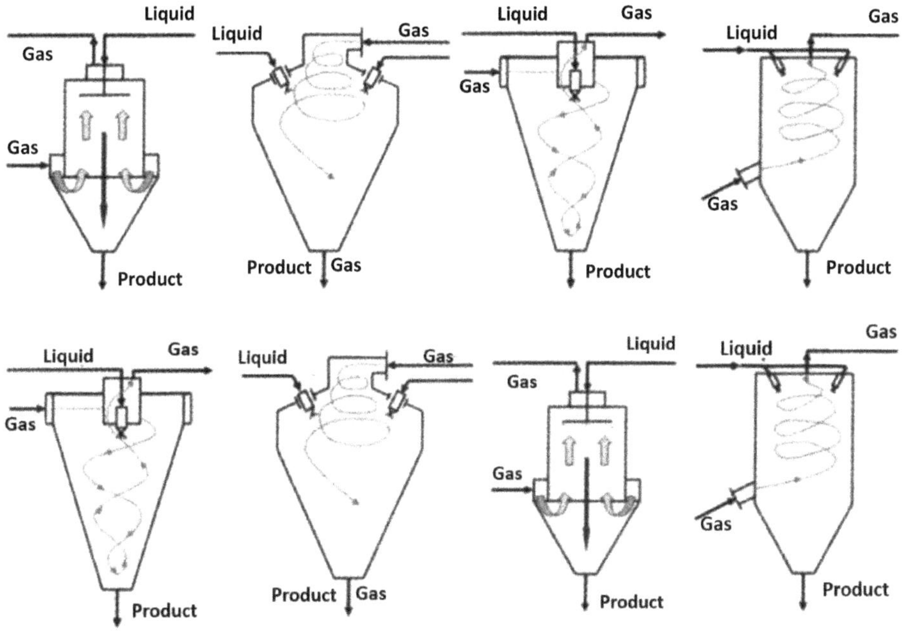

Figure 2.18: Different arrangement of the flow (co-current and countercurrent) in spray drying.

in strength, and as a result of falling drying rates, there is a certain amount of moisture left. Such flow arrangements are done when the materials to be dried are not sensitive to high temperature, and the requirements for the quality of drying are not very high, allowing some hygroscopic moisture in the final product.

In counter-current dryers not suitable for temperature-sensitive products, there is a contact of the hot air with moisture-free solid at a high temperature. Air is gradually cooled and incorporates moisture eventually coming in contact with the moist solid that enters the dryer. This operation mode can be used for materials that require slow initial drying and are capable of withstanding high temperatures without deterioration when the moisture content is very low.

It should be noted that drying can be subdivided into several zones, and in fact heating the drying agent at several places along the dryer is possible, thereby regulating the temperature of the solid in each zone can be easily accomplished. Humidity can be adjusted when the drying agent is circulated in each zone by adding small amounts of air from the next zone.

Drying methods related to size enlargement include atomization (simple and fluidized spray drying), contact (roller and belt) drying and vacuum drying.

In spray drying (Figure 2.19) used for drying of pastes, suspensions or solutions, the starting material is sprayed into the drying agent (e.g., hot air) by a suitable atomizer. This allows formation of spray-dried particles of the size 5 μm to 1 mm.

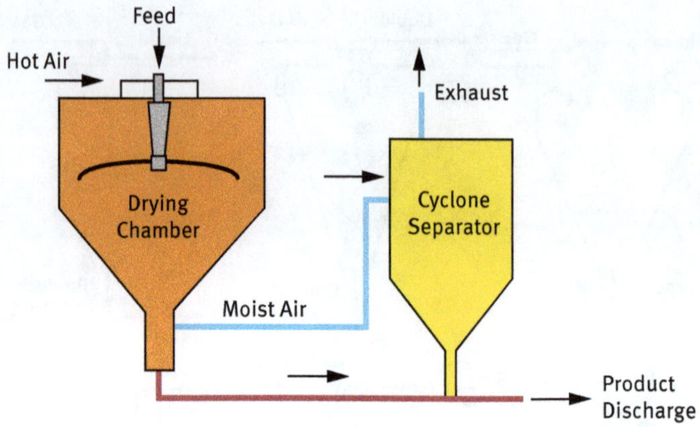

Figure 2.19: Illustration of spray drying [14].

The following parameters influence the properties of the spray-dried particles: composition and nature of the ingredients, solid content, solvent type and viscosity, atomization pressure, feed rate, inlet air temperature, gas feed rate and humidity of air used for drying. Rapid solvent evaporation arranged by high inlet air temperature and selection of a suitable solvent can give amorphous powders that are more moisture sensitive. Complex molecules exhibit lower chances to crystallize during spray drying.

The Peclet number (*Pe*) defined as the ratio of the evaporation rate to the solute diffusion rate is a conventional way of describing spray drying. Rapid evaporation and slow diffusion of the solute to the interior of the drying particles interior imply a large *Pe* and give hollow particles, while a homogeneous distribution of components is expected at small *Pe* (i.e., slow evaporation and/or rapid solute diffusion).

Drying chambers can be of different shapes, determined by the spray pattern. Nozzle chambers are taller than disk chambers, the latter have larger diameter. As mentioned above, co-current and counter-current flow arrangements are possible, or even mixed flow could be arranged. Counter-current spray drying is used for products with higher bulk density, minimizing production of hollow particles.

Downstream of spray drying, fluid beds for second-stage drying or fine agglomeration are typically installed. In this way, solids after spray drying with 5–8% residual moisture content undergo final drying under milder conditions with lower energy requirements. An interesting alternative for drying of sticky products with a high moisture content is integration of a fluidized bed dryer into a drying tower (Figure 2.20). The powder from the spray tower falls directly into a fluidized bed zone where drying takes place in a cloud of fine particles.

In centrifugal atomization, a wet product is accelerated to velocities 100–200 m/s, and the thin fluid film enters the dryer through an opening in the atomization disk. The rotational speed is used to alter the average particle size.

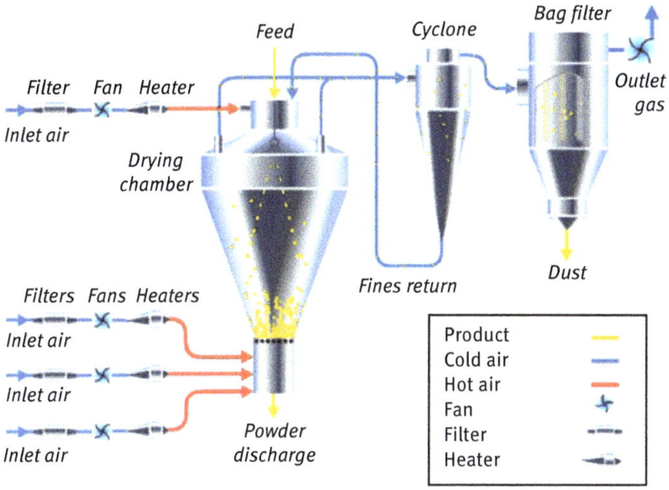

Figure 2.20: Spray dryer with an integrated fluidized bed [15].

One-fluid nozzles are used when a narrow particle size distribution is needed. Diameter of the nozzle opening, and the atomization pressure along with the materials' properties (viscosity) and the content of the solids are the parameters influencing the average drop size.

In a two-fluid nozzle less prone to clogging compared to one-fluid nozzles, the wet product is introduced under minor or no pressure. In this option with less control problems, higher energy requirements and smaller sizes of the dry products, as well as a broader particle size distribution, are obtained.

Size of particles is important in defining the size of a drying tower, as larger particles dry slower. As a consequence, the residence time is longer, influencing solubility in a negative way.

Conduction is the method of heat transfer in contact dryers, with steam and hot water acting as the heating media depending on the operating temperature. It is apparently clear that such heat transfer methods per se are more expensive when only the costs of a dryer itself are considered. At the same time with contact dryers, which overall are more energy efficient, there is a possibility to avoid expensive air treatment units, use small amounts of a purge gas and utilize less-expensive and smaller dust and vapor recovery systems.

Contact drying of various materials with varying viscosities, including pastes, is typically done with continuously operating roller dryers (Figure 2.21), allowing effective heat utilization.

Liquid is evaporated from the thin layers of the material by the heat from the heated rollers. The dried solid is peeled away by a scraper close to the end of the dryer, giving the final product in a form of a film, flakes or powder.

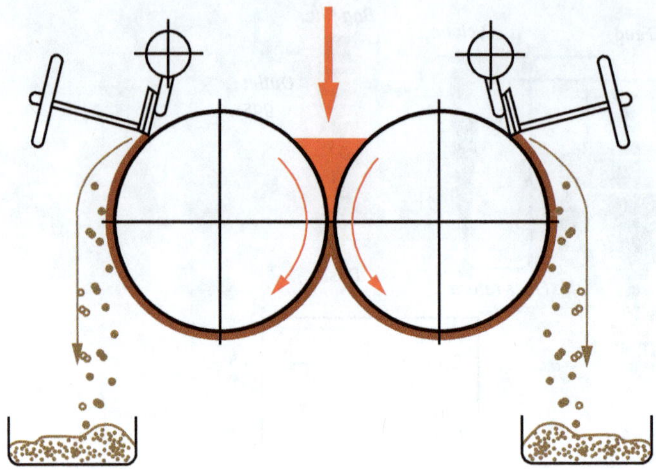

Figure 2.21: Double-roller spray dryer [16].

For drying of materials which can easily form dust, rotary dryers are used when the heat is transferred through the peripheral walls requiring only a small amount of air to carry off released moisture. Rotary dryers are used in drying, for example, coal and cement. A direct-heat rotary dryer is a horizontal rotating cylinder (Figure 2.22) with shell diameters of 0.5–6 m. The ratio of the dryer length to diameter depends on the drying type, ranging from 1 to 2 for batch rotary dryers to 4–10 for continuous drying. In the latter case, the cylinder may be slightly inclined for material flow control (Figure 2.22).

Material loading in a continuous dryer is ca. 10–15% of the cylinder volume. Gas flow in these rotary dryers may be co-current or counter-current. Co-current gas flow is preferred for heat-sensitive materials, while a counter-current operation has a better efficiency as the product temperature is higher than the exit gas temperature.

For temperature-sensitive products, vacuum drying or drying under reduced pressure is used as vaporization temperature is lower and time at the maximum temperature is shorter. Such type of drying with a more economical recovery of solvents and lower emissions of potentially poisonous compounds to the atmosphere is preferential for pharmaceutical products.

2.3.2 Drying of solid materials not related to agglomeration

Similar to the drying options discussed above the heat is transferred either from the drying agent (convection dryers) or from a heated surface (contact dryers). Heat transfer can be also arranged by radiant heating or dielectric drying and freeze drying (Figure 2.23).

The latter is a popular option to deal with sensitive products. First, a material is frozen and the pressure is reduced below 0.6 kPa, resulting in water removal from

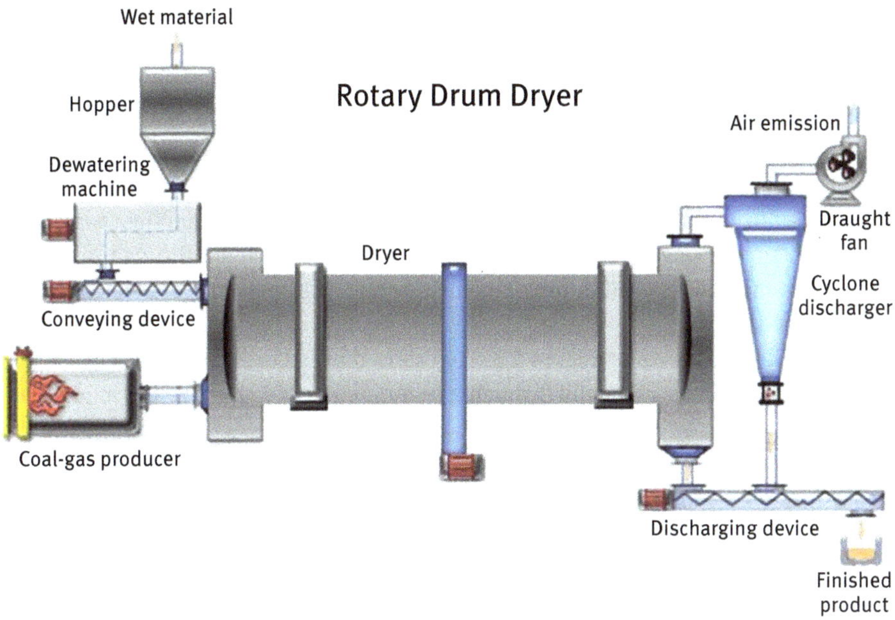

Figure 2.22: Direct-heat rotary dryer [17].

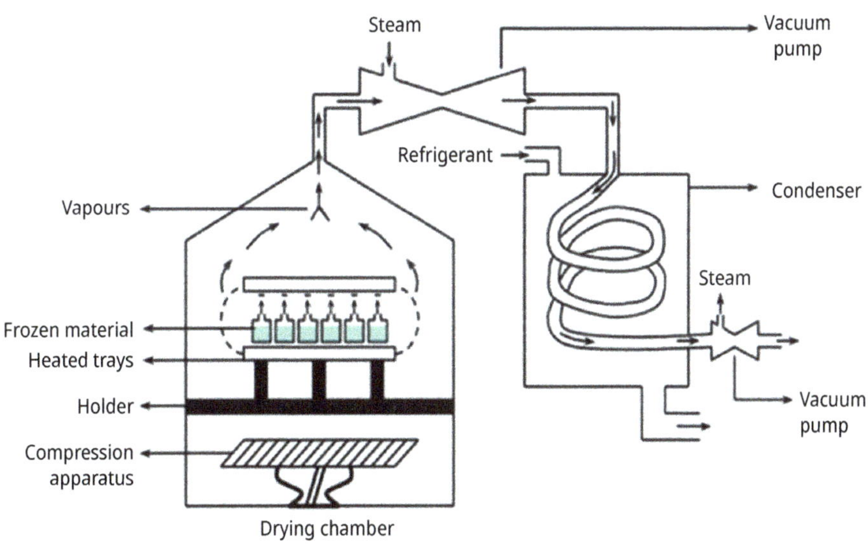

Figure 2.23: Freeze dryer [18].

the solid by ice sublimation directly to the gas phase. The freeze drying of pharmaceutical compounds in typically done at –5 to −35 °C allowing to avoid thermal inactivation of the active ingredient and evaporation of volatile compounds required for the final formulation. Obvious disadvantages of the freeze drying are its periodic nature, lengthy process time, high capital cost of equipment and high energy cost. Free dried compounds can be potentially hydroscopic thus susceptive to humidity.

In order to avoid mechanical stress during drying, flowing gas is applied. Tunnel dryers are applied to process large quantities continuously passing a material through a tunnel when air is moved either co-currently or counter-currently. With belt dryers, a belt passes through a drying chamber similar to a tunnel once or in multiple passages.

Convection drying by aeration is possible if air can be blown through a layer of solids. This mode of operation provides high drying rates even under relatively mild thermal conditions.

In the design shown in Figure 2.24, the solid moves horizontally in a chute and the drying agent flows vertically through a perforated floor to fluidize the solid.

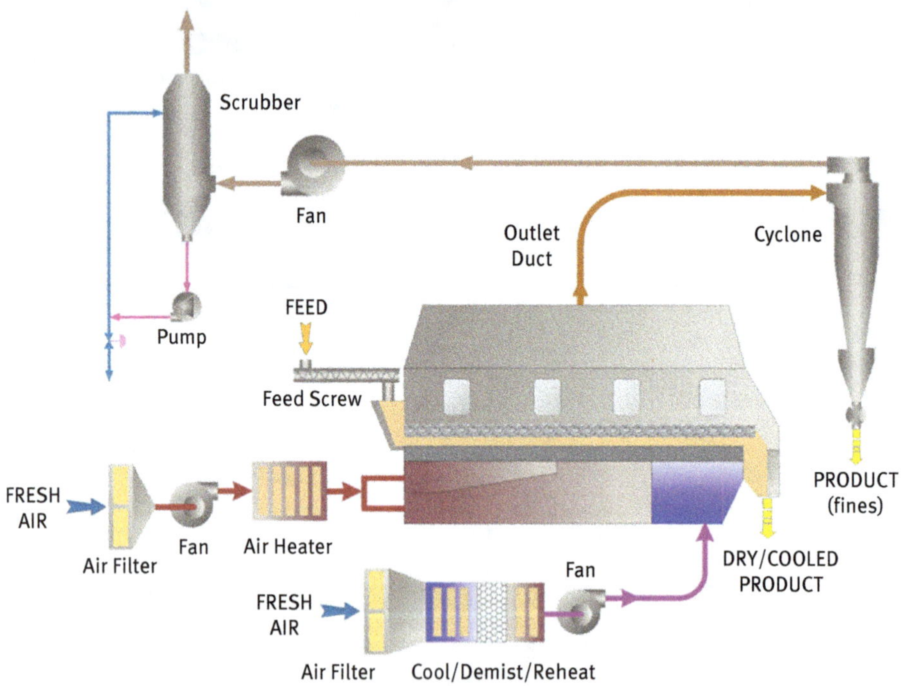

Figure 2.24: Fluidized bed dryer [19].

Drying of painted surfaces should be done in a very fast way, preventing the formation of an undesirable surface skin. Such short drying times can be arranged using infrared radiation, which is, however, expensive. Owing to the costs of this method, there should be very specific properties of the material requiring such type of drying. An example is the production of thin-walled ceramic products for avoiding formation of cracks.

2.4 Colloids

2.4.1 General properties

This term is used to define a two-component system in which the elements of the dispersed phase cannot be easily observed by optical microscopy. Another characteristic of colloidal particles is that their motion is affected by thermal forces, and they do not readily sediment in a suspended media. The size of colloidal particles is not very precisely defined ranging from nanometers to microns. Types and classification of dispersed systems are given, respectively, in Figure 2.25 and Table 2.1.

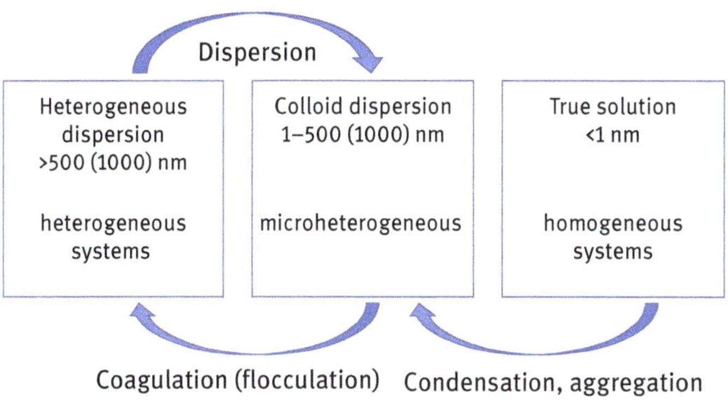

Figure 2.25: Types of dispersed systems.

Table 2.1: Classification of dispersed systems.

Medium/phase	Dispersed phase	Dispersed phase	Dispersed phase
	Gas	Liquid	Solid
Gas	None	Liquid aerosol (hair spray)	Solid aerosol
Liquid	Foam	Emulsion	Sol
Solid	Solid foam (aerogel)	Gel (agar)	Solid sol (gemstone)

Often, dispersions are classified as molecular dispersions or true solutions, colloidal dispersions and heterogeneous or coarse dispersions. In the case of molecular dispersions, the size is below 1 nm. In the case of colloidal dispersions, particles can be detected by electron microscopy and can be settled by centrifugation. Finally, in coarse dispersions (suspensions and emulsions) particles are visible under ordinary microscope, do not pass through filter paper or semipermeable membrane, do not diffuse and settle down under gravity.

Because of their small size (especially in the nanometer range), the surface area of colloids is sufficiently large for the surface effects to be extremely critical in defining stability of colloidal dispersions.

Colloids might have spherical as well as other shapes also.

Colloidally dispersed particles can have different types of size distribution, namely narrow (monodisperse) or wide (polydisperse) distribution. Bimodal and multimodal particle size distributions are also possible. For some applications involving concentrated dispersions (paper coatings, ceramics, catalysts), particle shape, size and size distribution are very important.

In this chapter, micellar solutions made by an action of surfactants will also be considered to be of colloidal dimension (Table 2.2).

Table 2.2: Types of colloidal sols.

Lyophilic	Associated	Lyophobic
Dispersed phase (large organic molecules with colloidal size)	Dispersed phase (micelles of organic molecules or ions – size below the colloidal range)	Dispersed phase (inorganic particles)
Molecules of dispersed phase are solvated. Formed spontaneously	Hydrophilic and lyophilic portions are solvated. Formed at concentration above CMC	Not formed spontaneously
The viscosity increases with an increase of the dispersed phase concentration	The viscosity increases with an increase in the micelle concentration	The viscosity does not greatly increase
Stable dispersion in the presence of electrolytes	CMC decreases with electrolytes	Unstable dispersion in the presence of electrolytes

Note: CMC, critical micelle concentration.

Particulate colloids are classified as lyophobic (hydrophobic when dispersed in water) or lyophilic (hydrophilic). Some examples of lyophilic sols include sols of gum, gelatin, starch, proteins and certain polymers (rubber) in organic solvents.

In solvent-repelling or lyophobic colloids, the particles resist solvation and dispersion in the solvent; therefore, the concentration of particles is usually relatively low. Lyophobic dispersions are not thermodynamically stable, but they have, however, some sort of kinetic stability. Such colloids are easily precipitated by heating, shaking or addition of small amounts of electrolytes. After precipitation, reverting to the sol by simple mixing with the dispersion medium is difficult; thus, these sols are irreversible.

Understanding the stability of colloidal dispersions is important for practical applications of colloidal systems. Such systems include emulsions and microemulsions, cremes, cosmetics and cleansing agents, paints and pigments, paper coatings, printing inks and pastes, various adhesives and thickening agents. There are many other examples of colloids related to paper industry, pharmaceuticals, materials science and consumer products.

Colloidal interactions, which define the stability of colloids, depend on the structure of surfaces, which is quantified by surface charge density and can be influenced by parameters such as pH, salt concentrations and presence of organic materials.

2.4.2 Colloidal stability

Stabilization is needed to prevent colloids from aggregating and forming sediments (Figure 2.26).

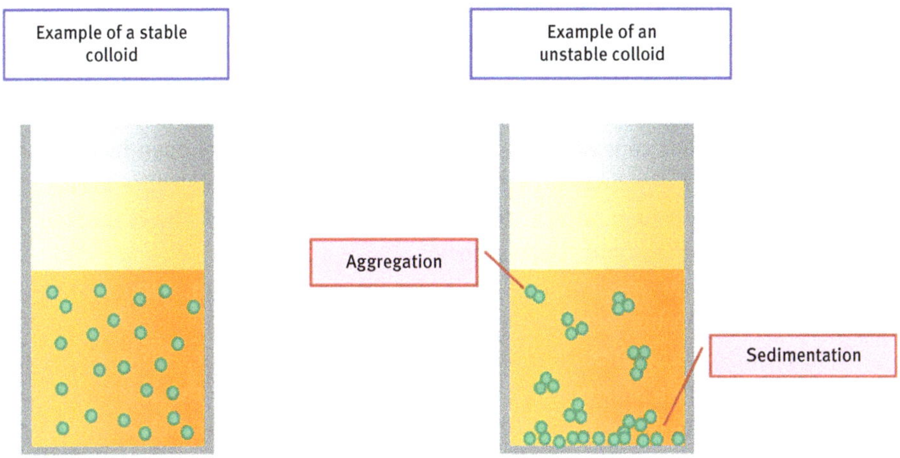

Figure 2.26: Examples of stable and unstable colloids [20].

Stability is mainly achieved by either steric stabilization (i.e., surrounding each particle with a protective solvent sheath, preventing adherence due to Brownian movement) or electrostatic stabilization (i.e., providing the particles with an electric charge through addition of small amounts of electrolytes).

Different paths are possible to go from a stable system to phase separation (Figure 2.27).

Sedimentation, or creaming, stems from differences in the density between the dispersed and continuous phases. Sedimentation gives two separate layers of dispersion with different dispersed phase concentrations. In one of these layers, concentration of the dispersed phase is larger, which can eventually result in aggregation. The latter means that two or several dispersed species touch each other without almost any changes in the overall surface area. For suspensions by aggregation, two kinds of phenomena occur: coagulation and flocculation. The latter means formation of a loose network of particles, while in the case of coagulation, compact aggregates are

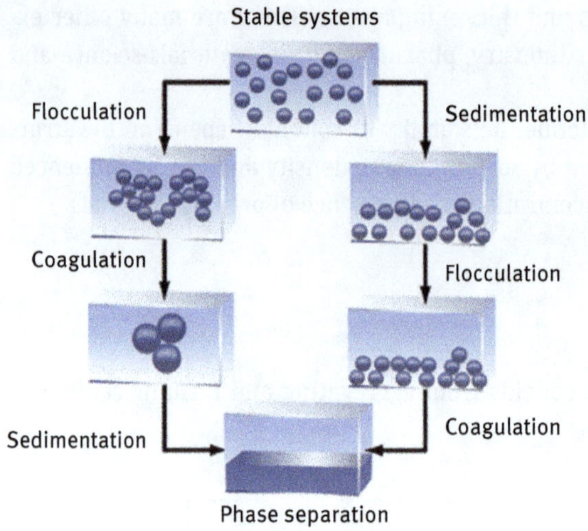

Figure 2.27: Pathways to phase separation.

formed. Flocs are thus considered to be colloidal aggregates very open in structure and are capable of being broken by shear. When a critical concentration of flocs is reached, flocs can sediment or form gels. The latter exhibit some features inherent to solids, such as yield stress and viscoelasticity.

In aggregation, the species retain their identity, and at the same time move as a single unit, thereby losing their kinetic independence.

Contrary to agglomeration, in coalescence, two or more particles, droplets or bubbles are combined to a single larger unit. The surface area as a result is decreased and the initial species do not preserve their identity. Some colloids such as emulsions can be unstable for aggregation, and at the same time can resist coalescence. On the contrary, suspensions could be kinetically stable against aggregation, and be prone to sedimentation. Understanding kinetic stability is important as colloidal dispersions being thermodynamically unstable can still have reasonable stability with time. Dispersions of lyophobic colloids may be stable for weeks, months or even years.

Lyophilic sols and association colloids are typically stable, and addition of moderate amounts of electrolytes does not lead to coagulation, contrary to lyophobic sols. For the latter case, the so-called salting, in other words agglomeration and precipitation, is possible. This can be done by, for example, addition of large amounts of electrolytes. Efficiency of anions to precipitate is related to ion hydration and their ability to separate water molecules from colloidal particles. Anions can be ranked in a decreasing order of precipitating power: citrate > tartrate > sulfate > acetate > chloride > nitrate > bromide > iodide. Alternatively, addition of a less polar solvent (acetone and alcohols) can influence the stability of colloids in a negative way.

Coacervation or formation of a layer rich in the colloidal aggregates can be done by mixing negatively and positively charged hydrophilic colloids. For instance, small amounts of hydrophilic or hydrophobic colloid added to a hydrophobic colloid of an opposite charge can result in sensitization (coagulation) of the particles. Thus, negatively charged colloidal silica particles can be flocculated by addition of a positively charged polymer.

Stability of the dispersion depends in general on the forces and interactions in a colloidal system.

Forces influencing a particle should be considered first. An expression for gravity is

$$F_g = mg = \frac{4\pi R^3}{3} \Delta\rho g \tag{2.16}$$

where $\Delta\rho$ is the difference between density of the colloidal particle (ρ_2) and the medium (ρ_1). When $\Delta\rho = \rho_2 - \rho_1$ is positive, sedimentation occurs, and the opposite case of negative $\Delta\rho$ results in creaming. Among other forces influencing a colloidal particle, the drag force defined by the Stokes law and Brownian movements (thermal motion in the liquid acting on the particle) are worth mentioning. In the case of dilute suspensions of rigid noninteracting particles, the hydrodynamic force according to the Stokes law is

$$F_{drag} = 6\pi\eta R v \tag{2.17}$$

where η is the fluid viscosity and v is the particle velocity. For more concentrated suspensions with the solid fraction ϕ below 0.2, hydrodynamic interactions between the particles diminish the rate of sedimentation v', as these particles do not sediment independently

$$v' = \mathrm{v}(1 - 6.55\ \phi) \tag{2.18}$$

For concentrated suspensions, the sedimentation velocity becomes as complex function of the solid fraction.

In general, for multiparticle colloidal systems, interactions between the particles, which can be either repulsive or attractive, should be considered. The main cause of repulsive forces keeping the particles apart is the electrostatic repulsion between similarly charged objects. Such repulsion thus occurs when surfaces of particles or droplets have an electric charge. The particles are surrounded by counterions creating electrostatic repulsions.

Attractive forces working in the opposite direction of destabilization are mainly van der Waals dispersion forces. Dispersion forces arise from the ubiquitous quantum mechanical effects caused by fluctuations in the electron clouds surrounding atoms.

For two spheres of radius R in vacuum, separated by a distance h, substantially smaller than the radius ($h < 10$–20 nm), the attractive energy V_A can be approximated by

$$V_A = -\frac{AR}{12h} \tag{2.19}$$

where the Hamaker constant A depends on the density and polarizability of atoms in the dispersed species. Polymers have low Hamaker constants (~ 4×10^{-20} J), while such materials as metals possess relatively large Hamaker constants (~ 3×10^{-19} J). In order to account for the medium other than vacuum, where the attraction is less pronounced, an effective Hamaker constant is introduced:

$$A = \left(\sqrt{A_2} - \sqrt{A_1}\right)^2 \tag{2.20}$$

with A_2 and A_1 representing the Hamaker constants for the dispersed species and the medium, respectively. Chemical similarity between the particles and the medium results in the lowest attraction between particles.

In addition to dispersion forces, surface forces are important. They arise from the proximity of colloidal surfaces in a colloidal dispersion. Surfaces in colloidal dispersions can be quite complex, bearing surface charges, having adsorbed ions, nanoparticles, surfactants, polymers, or even being covered with surface-grafted polymers. Surface forces can act in favor of stabilization or alternatively destabilization of colloidal particles.

Colloids in the absence of a stabilizing force will simply aggregate, leading eventually to phase separation. Stability of colloids is predominantly related to the repulsive forces created when for two charged surfaces their electric double layers overlap. Thus, in what follows, surface charge of colloidal particles will be considered since suspending particles in liquids lead to surface charging of the surface. The presence of dissociated chemical groups on the surface, adsorbed species and composition of the liquid result in a complex electrostatic layer near the particle surface (Figure 2.28).

As an example, Figure 2.28 features a double layer of positively charged counterions surrounding a surface with negative charges. In general, the surface can be negatively or positively charged because of dissociated functional groups, presence of hydroxyls, protons, etc. The innermost layer of ions is adsorbed to the surface, resulting in electroneutrality. The remaining part of the counterions, with density reflected by shading, is located in the solution. Other counterions such as additional electrolytes can also be present in the liquid.

An important characteristic of the double layer is the electrostatic potential ψ, which depends on the distance r. The Stern layer is related to immobile ions defining the region with a linear potential decay. The potential at the outer limit of this region, beyond which the liquid around a moving particle is no longer trapped to move with the particle, is defined as the zeta potential. The latter is determined by electrophoretic mobility measurements.

The surface charge is dependent on the solution pH. The value of pH at which the surface is neutral, the so-called point of zero charge (pzc), corresponds to the association of colloids, which lack electrostatic stabilization. More broadly defined colloids are quite unstable due to agglomeration for the zeta potential between +5 and –5 mV. Approximate criteria for colloidal stability are given in Table 2.3.

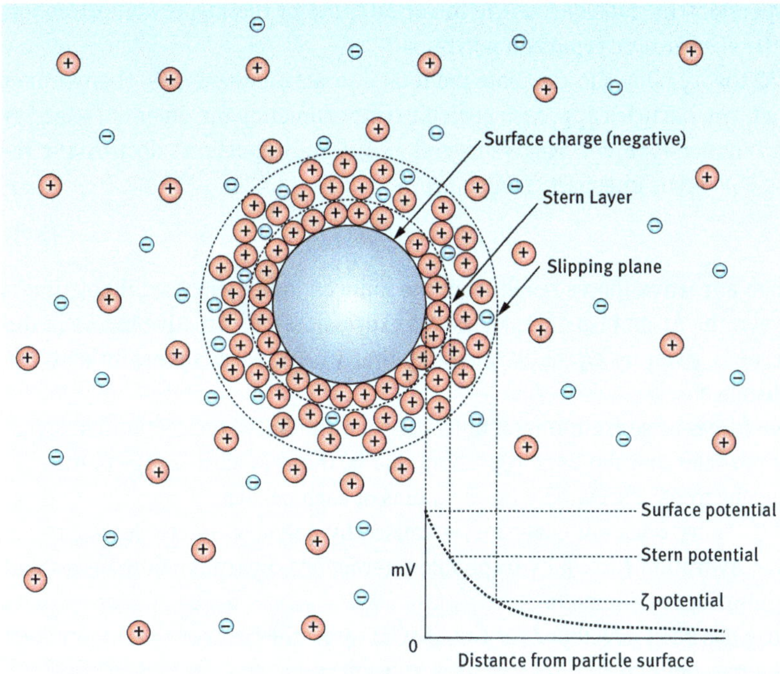

Figure 2.28: Schematic of a double layer in solution at the surface of a colloidal particle [21].

Table 2.3: Stability criteria based on zeta potential.

Stability characteristics	Zeta potential (mV)
Maximum agglomeration and precipitation	+3 to zero
Excellent agglomeration and precipitation	−1 to −4
Fair agglomeration and precipitation	−5 to −10
Agglomeration threshold (agglomerates of 2–10 particles)	−11 to −20
Plateau of slight stability (few agglomerates)	−21 to −30
Moderate stability (no agglomerates)	−31 to −40
Good stability	−41 to −50
Very good stability	−51 to −60
Excellent stability	−61 to −80
Maximum stability – for solids	−81 to −100
Maximum stability – for emulsions	−81 to −125

In the simplest example of colloid stability, dispersed species would be stabilized entirely by the repulsive forces created when two charged surfaces approach each other and their electric double layers overlap.

The theory proposed by Derjaguin and Landau in 1941 and, independently, Verwey and Overbeek in 1948 (thus the name DLVO) explained the complex behavior of colloidal dispersions by considering the linear addition of the dispersion attraction potential to the electrostatic repulsion potential.

The DLVO theory allows to calculate the total interaction energy, V_T, changes taking place when two particles approach each other by estimating the potential energies of attraction (London–van der Waals dispersion, V_A) and repulsion (electrostatic including Born, V_R) versus interparticle distance

$$V_T = V_R + V_A \tag{2.21}$$

The long-range attractive forces result from the induced dipole–induced dipole interactions between the assembles of particles. An expression for attractive forces in the case of spheres is given by eq. (2.19), meaning that V_A decreases inversely with the separation distance.

Repulsive forces arise from interactions between similarly charged double layers surrounding particles and the Born repulsion, that is, the very short-range mutual repulsion of the electrons associated with the atoms of each particle.

The DLVO theory does not cover forces caused by polymers (steric forces, bridging): repulsive hydration forces, hydrophobic interactions, hydrodynamic forces and other nonequilibrium phenomena.

The theory has been developed for two special cases: the first considering interactions between parallel plates of infinite area and thickness, and the second considering interactions between two spheres (Figure 2.29).

The first case seems to be less realistic from the practical viewpoint. However, force $F(h)$ acting between two colloidal particles having a surface separation h can be related to the free energy of two plates $V(h)$ per unit area by the Derjaguin approximation

$$F(h) = 2\pi R_{\text{eff}} V(h) \tag{2.22}$$

with the effective radius is given by $R_{\text{eff}} = R_+R_-/(R_+ + R_-)$, where R_+ and R_- are the radii of the two particles (Figure 2.29). For parallel plates, the interaction energy per area is thus proportional to h^{-2}. Derjaguin approximation can be used when the size of the particles is larger than the range of the forces involved, which is typical for colloidal systems except small nanoparticles.

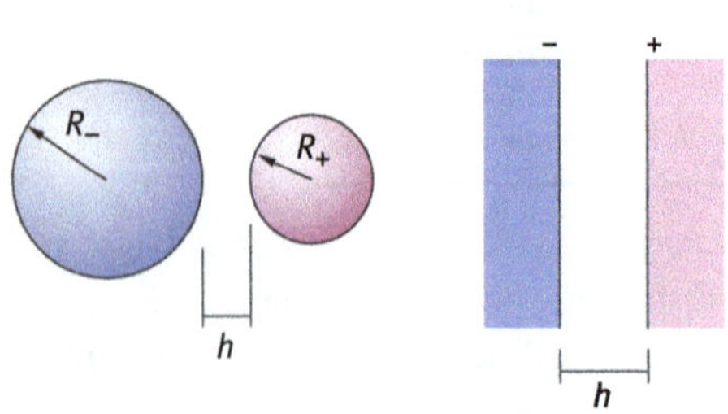

Figure 2.29: Interactions between colloidal particles [22].

The interaction free energy for the case of two parallel plates can be approximated as

$$V_{\mathrm{DL}}(h) = \frac{2\sigma_+ \sigma_-}{\varepsilon_0 \varepsilon k} e^{-kh} \tag{2.23}$$

and decreases exponentially with the separation distance h. In eq. (2.23), σ_+ and σ_- are the surface charge densities per unit area of the right and left surfaces, ϵ_0 is the permittivity of vacuum, ϵ is the dielectric constant of water and κ is the inverse Debye length. This latter term is expressed by

$$k^{-1} = \sqrt{\frac{k_{\mathrm{B}} T \varepsilon_0 \varepsilon}{2q^2 N_{\mathrm{A}} I}} \tag{2.24}$$

where q is the elementary charge, N_{A} is the Avogadro's number, I is the ionic strength, k_{B} is the Boltzmann constant and T is the absolute temperature. Equation (2.24) implies that Debye length is proportional to the square root of ionic strength.

Figure 2.30 displays how the overall potential energy of interactions (V_{T}) depends on interparticle distances, showing that with an increase of the separation distance the combined curve exhibits a primary minimum, followed by the primary maximum and the secondary minimum. Flocculation of a reversible type occurs in case of a sufficiently deep secondary minimum.

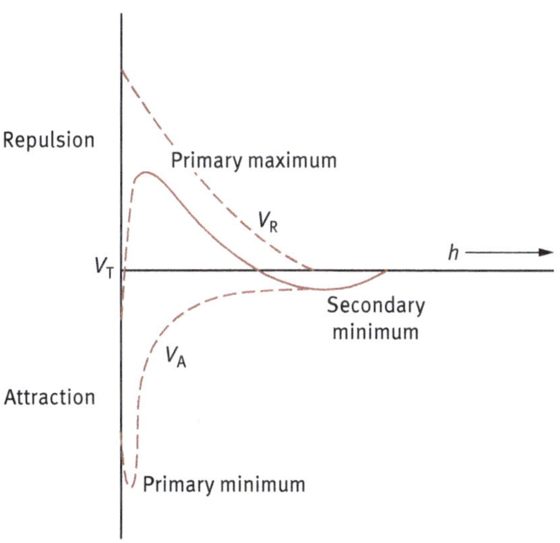

Figure 2.30: Potential energies of interaction between two colloidal particles as a function of their distance of separation h, for electrical double layers due to surface charge (V_{R}), London–van der Waals dispersion forces (V_{A}) and the total interaction (V_{T}) [23].

Despite its apparent simplicity, the DLVO theory can account for a rich colloidal behavior because both barrier height and distance from the surface are influenced by electrostatic repulsion, which depends on electrolyte concentration.

From the viewpoint of colloidal stability, the DLVO theory suggests that colloids are stable if repulsion V_R (V_{DL}) exceeds the van der Waals attraction V_A between the particles. The system becomes unstable when the van der Waals attraction overcomes the electrostatic repulsion.

The DLVO theory is a very useful starting point in understanding the behavior not only of such complex colloidal systems as suspensions but also emulsions and foams. It is clear from the discussion above that the DLVO theory is useful for cases when electrical effects have an influence on colloidal stability. This happens, for example, upon addition of inorganic ions or surfactants. According to the DLVO theory, the dispersion stability is increased with an increase in the particle size and the surface potential and with a decrease in the effective Hamaker constant, temperature and ionic strength of the dispersing liquid. The latter in practice implies that the dispersion becomes unstable when the concentration of salts added to the dispersion exceeds a certain critical value.

Some practical consequences of the DLVO theory are considered below.

Electrostatic stabilization can be achieved in the presence of electric charges on the particle surface. Such charges can be introduced by isomorphous substitutions in case of clays, silicates or zeolites and changes in the surface charge density by adjusting pH or adsorption of ions.

In the domain of pH close to pzc, dispersions are unstable being prone to coagulation. Subtle changes in pH close to pzc can influence significant stability of colloidal dispersions.

Colloidal systems are also very sensitive to multivalent cations according to the Schulze–Hardy rule. Upon addition of bivalent or trivalent counterions, the surface charges are neutralized, leading to coagulation of dispersion. An influence of multivalent cations can be explained by several reasons: changes of pH bringing it closer to pzc; formation of larger complexes of multivalent hydrated cations decreasing stability and more pronounced specific adsorption in the Stern layer compared with the monovalent ones.

From the practical viewpoint, it is interesting that an excess of multivalent counterions can recharge the surface, thus by increasing amounts of multivalent ions, a dispersion can pass from stabilization to destabilization and again to stabilization.

Despite practical usefulness of the DLVO theory, it should be mentioned that this theory should be used with caution when other effects are operational, such as steric stabilization forming some sort of organic envelope around the particle.

Long-chain molecules (macromolecules) adhering to the particle surface cause such steric stabilization bringing an advantage of insensitivity to dissolved salts. Destabilization, which is in general reversible, can be done by changing polymer–solvent interactions via changing the solvent or temperature even in a narrow temperature range.

Such changes in the solvent or temperature can act in both directions, favoring or disfavoring steric stabilization.

Steric stabilization requires threshold concentrations of polymers. Amounts needed for stabilization decrease with the polymer molecular mass. The prerequisite of steric stabilization, widely used industrially, is that macromolecules, typically amphipathic block or graft copolymers, should be bound to the surface. Steric stabilization has several advantages compared to electrostatic stabilization. The latter results in an increase of viscosity at high concentration of solid materials, forming gels. Moreover, electrostatic stabilization is less effective in nonaqueous media, while steric stabilization can also be used in organic solvents. Additional advantages of sterically stabilized dispersions are their relative insensitivity to electrolytes and reversibility of flocculation as mentioned above.

Polymers influence the stability of colloid dispersions in different ways. Usually, low concentrations of macromolecules decrease the stability of colloidal dispersions (sensibilization), whereas higher concentrations stabilize the colloidal state, in many cases in extreme ways.

Sensibilization at low polymer concentration is related to bridging the particles by macromolecules or by charge neutralization when charged polymer ions displace the counterions from the surface. The latter option requires a careful adjustment of parameters to prevent charge reversal and restabilization. An easier option is thus to adsorb macromolecules on the surface with a larger part of the polymer being free to form bridges.

Higher polymer concentrations allow stabilization by steric reasons in nonaqueous solvents. In this case, strongly adsorbed polymer molecules, such as terminally anchored block copolymer chains with a hydrophobic portion capable of strong adsorption, have also a hydrophilic part of the molecule that extends outward from the surfaces. The extent of the long-range repulsive force between particles is determined by the size of adsorbed macromolecules. The range of the available polymers is restricted by a potential overlap of the polymer layers on the respective particles; subsequently, steric stabilizers are typically low-molecular-weight polymers.

Illustration of different cases with adsorbing and nonadsorbing polymers is given in Figure 2.31.

Another option to influence the stability of dispersions is to add surfactants whose counterions are strongly adsorbed on the particle surface. For surfactants bearing a counterionic charge, this results in rapid destabilization of the colloidal system because of charge neutralization. After compensation of, for example, positive surface charges by surfactant anions, further adsorption of anionic surfactants causes the adsorption of protons by surface groups increasing the density of the positive surface charges. The effect of a particular surfactant depends on pH, which determines whether the surface of the oxide is positively or negatively charged.

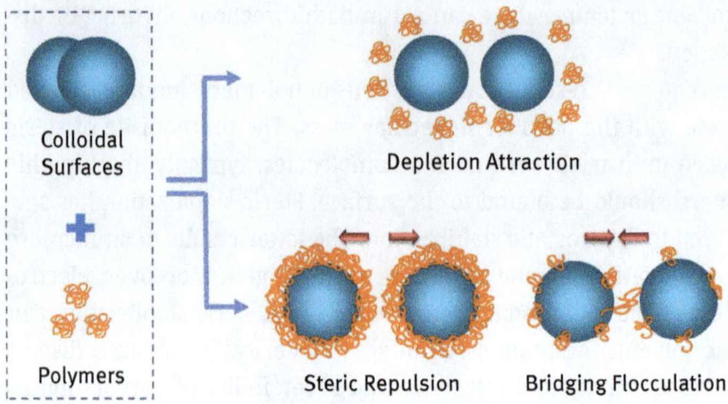

Figure 2.31: Schematic of polymer-induced interactions between colloidal surfaces. Depending on whether adsorption of the polymers exists on the surfaces, the induced interaction can be either depletion attraction for unabsorbing polymers, or steric repulsion/bridging flocculation for adsorbing cases (reproduced from [24] with permission from the Royal Society of Chemistry).

From a practical point of view, it is important to know the aggregation kinetics, especially for thermodynamically unstable dispersions, which is the case for many lyophobic dispersions. The aggregates as already mentioned kinetically behave as single units.

When sedimentation is slow, the overall aggregation rate r_a is given by

$$r_a = k_d n^2 + k_s n \tag{2.25}$$

accounting for diffusion-related collisions (expressed with the rate constant k_d) as well as shear-induced collision processes with the corresponding rate constant k_s. In eq. (2.25), n is the number concentration of dispersed species. The rate constants depend on physicochemical properties of the colloidal systems comprising particles, droplets or bubbles and are influenced by possible chemical reactions, phases composition, their size, size distribution, shape, surface properties, dipole moments and dielectric constant to name a few. Therefore, these constants are not really constant and can change during aggregation.

For small particles, droplets or bubbles, the energy barrier to aggregation is removed by adding excess electrolyte. For a monodisperse suspension of spheres, an expression for the number of aggregates with time, in this case of rapid coagulation, is

$$n = \frac{n_0}{1 + k_d n_0 t} \tag{2.26}$$

where n_0 is the initial concentration of particles at the beginning of coagulation. For the diffusion-controlled coagulation, the rate constant is defined as

$$k_d = \frac{4ak_B T}{3\mu} \tag{2.27}$$

where a is the collision efficiency ranging between zero and unity and μ is viscosity. It follows from eq. (2.27) that the diffusion-controlled rate constant depends linearly on temperature and inversely on viscosity.

Mutual repulsion exists when there is an energy barrier to aggregation, and the rate of coagulation decreases. Fuchs proposed to incorporate in the kinetic equations a term accounting for existence of such energy. As a result of theoretical analysis, the stability ratio W was introduced, relating the rates of fast and slow coagulation. The reciprocal ratio ($1/W$) corresponds to the fraction of encounters leading to attachment. For electrostatic energy barrier, W linearly decreases with salt concentration increase at low values of ionic strength.

For dispersions with larger concentrations of particles, mutual interactions between them become significant. This can influence the rheological properties of colloidal systems. Dilute colloidal systems behave like Newtonian fluids as the particles move independently of each other. Newtonian fluids represent the simplest models of fluids accounting for viscosity. Rheology of dispersions will be discussed separately in more detail in Section 2.6. While the concept of a Newtonian fluid is an idealized one, many common liquids and gases under ordinary conditions can be approximated as Newtonian for practical calculations. With increasing particle concentration, there is a gradual transition from Newtonian to plastic or pseudoplastic behavior.

Viscosity dependence of colloidal silica dispersions on parameters such as concentration, addition of electrolytes or polymers is widely used in practical applications. Another parameter influencing the flow behavior is pH. Changes of pH thereby can be used to diminish undesirable thickening or thinning of dispersions.

2.5 Emulsions

2.5.1 Basics

Emulsions are disperse systems consisting of two or more mutually insoluble or sparingly soluble liquids. The continuous phase is usually present in excess, while the other liquid is located in the dispersed phase.

If the external phase consists of water and the internal phase of an organic liquid, for example, mineral oil, the term “oil-in-water” (O/W) emulsion is used (Figure 2.32, left). On the other hand, if water is finely dispersed in a nonaqueous liquid, a “water-in-oil” (W/O) emulsion is produced (Figure 2.32, right).

More complicated emulsions (Figure 2.33) such as $W_1/O/W_2$ emulsions can be made by dispersing a W/O emulsion in water.

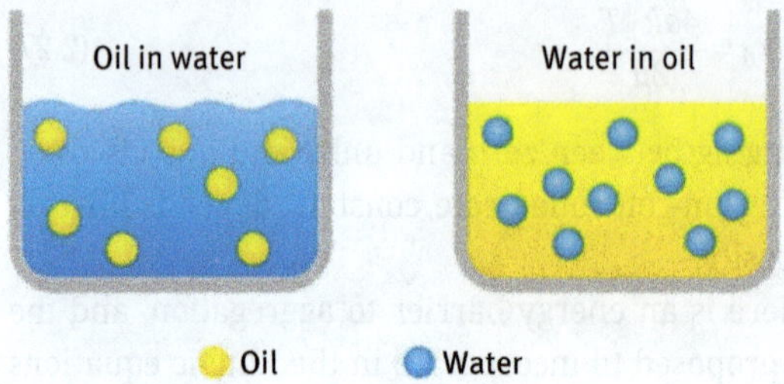

Figure 2.32: Types of emulsions.

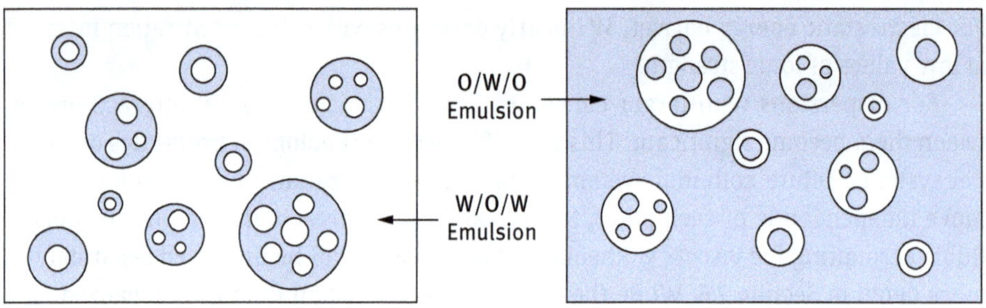

Figure 2.33: Multiple emulsions [25].

Examples of emulsions and suspensions related to personal care products are presented in Table 2.4.

Table 2.4: Some multiphase products in personal care.

Emulsions	Cosmetic and skin care creams	W/O, O/W, W/O/W
Suspensions	Microemulsion hair dyes	W/O
	Exfoliating scrubs	S/W
	Facial masks	S/W
	Lipsticks and lip balms/glosses	S/O

Note: W, water; O, oil; S, solid.

When two immiscible liquids are simply mixed together and vigorously shaken, the resulting emulsion might not be stable as some kind of a stabilizing agent, an emulsifier, is needed.

The presence of an emulsifier along with the volume ratio of oil to water and the method of emulsion preparation are the parameters defining if a particular emulsion can be prepared.

Stabilizing agents for emulsions can be surfactants, which act by reducing the interfacial tension and thus diminishing the work required to form an emulsion with a specific droplet size/interfacial area. A surfactant, which is a chemical compound that combines oil-soluble and water-soluble properties, when adsorbed around droplet acts as a physical barrier. Addition of a small amount of an emulsifier substantially decreases the interfacial tension and the interface energy forming emulsions which are thermodynamically unstable, but kinetically stabilized.

Thermodynamic instability is related to a small, but non-negligible value of the interfacial energy, which can be further diminished if the droplets coalesce. At the same time, the coalescence kinetics is substantially impeded in the presence of an emulsifying agent.

In the case of an O/W emulsion, the hydrophobic hydrocarbon chains of the emulsifying agent are located in the internal dispersed phase (Figure 2.34). For W/O emulsions, these hydrocarbon chains point outward into the external continuous phase (Figure 2.34).

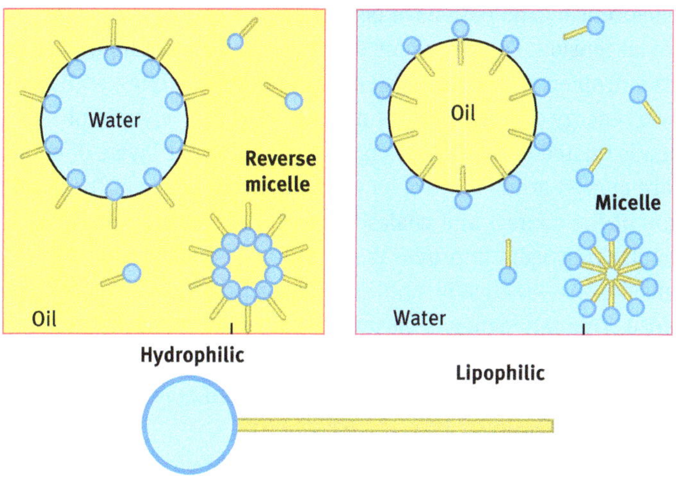

Figure 2.34: Envelopment of droplets with a surface film of emulsifier in O/W and W/O emulsions.

The emulsion type is thus determined by the phase where an emulsifier is placed. Emulsifying agents that are preferentially oil-soluble form W/O emulsions, while water-soluble emulsifiers generate O/W emulsions. This rule of thumb often referred to as the Bancroft's rule implies that the continuous phase corresponds to the liquid in which the surfactant is most soluble. Often mixtures of emulsifying agents are needed for particular applications as will be discussed below. Such behavior results from formation of a complex of several surfactants at the interface, lowering the interfacial tension and forming a stronger interfacial film.

Increasing polarity of the oil phase requires more hydrophilic emulsifiers. Alternatively, more lipophilic emulsifiers are needed for more nonpolar oil phases.

From empirical observations, it is also known that soaps of monovalent metal cations tend to produce O/W emulsions, while W/O emulsions are formed when polyvalent metal cations are applied.

O/W emulsions have a creamy consistency with viscosity similar to viscosity of an aqueous solution. Opposite to this W/O emulsions are of an oily character with buttery consistency as a result of a liquid crystalline gel structure.

In general, unless a special preparation procedure is applied, the sizes in the emulsions are broadly distributed. Emulsions with a mean particle diameter of 10^{-6} to 10^{-8} m have a typical appearance of colloidal solutions. Emulsions with diameter smaller than 10^{-8} m are highly dispersed, approaching micelles. Based on the size of liquid droplets, emulsions are either kinetically stable macroemulsions (above 0.2 mm) or thermodynamically stable microemulsions (0.01–0.2 mm).

In addition to the size of the droplets, the electrical charge of the droplets plays an important role in the emulsification and stabilization processes. It should be noted that the equations for the zeta potential of dispersions do not apply for emulsions.

There are several theoretical explanations of emulsification. The first is related to lowering of the interfacial tension. Otherwise for two pure immiscible liquids, high interfacial tension of the droplets drives the system generated by mixing with high energy input to the lowest energy configuration or in other words to separate phases.

Another explanation is related to the formation of monomolecular layers of emulsifying agents, which are curved around a droplet of the internal phase of the emulsion. Finally, prevention of the contact and coalescence of the dispersed phase by a protective film of an emulsifying agent improves the stability of emulsions.

Overall, since emulsions are kinetically stable, emulsion stability is determined by the rate of coalescence. This rate depends on several factors. Besides the physical nature of the interfacial surfactant film, viscosity of the continuous phase is important because stability of many emulsions is higher in concentrated media. Addition of thickening agents can increase viscosity.

Other parameters influencing stability are size distribution of droplets and the phase volume ratio. A uniform size distribution and a smaller volume of the dispersed phase allows generation of a more stable emulsion. On the contrary, an increase of the dispersed phase volume decreases emulsion stability, resulting eventually in the phase inversion.

Phase inversion can also be influenced by temperature. For instance, for O/W emulsions, temperature increase makes the emulsifier more hydrophobic, which might lead to phase inversions.

Other options to invert O/W emulsions is to add, for instance, strong electrolytes to emulsions stabilized by ionic surfactants or to make the emulsifier more oil soluble.

The type of emulsion might also depend on the order of addition of the phases. Thus, when oil and emulsifier are added to the aqueous phase, a W/O system is produced. On the contrary, addition of emulsifier in the aqueous phase to an oil phase gives a O/W emulsion.

Temperature, besides having an impact on phase inversion, also influences emulsion stability. Temperature increase influences emulsification stability in a negative way by affecting the interfacial tension, surfactant solubility, liquid viscosity and phases of the interfacial film.

For O/W emulsions electrical or steric barrier is significant because the electric double layer thickness is much larger in water than in oil. For nonionic emulsifying agents, the charge may arise due to adsorption of ions from the aqueous phase or contact charging because of differences in dielectric constants between phases.

Oil field W/O emulsions could be stabilized by a viscoelastic protective film around the water droplets or by adsorption of biwetting and mutually interacting solids at the oil/water interface. In the latter case, solid-stabilized or so-called Pickering emulsions (Figure 2.35), a strong barrier to both aggregation and coalescence is ascribed to the presence of mechanically strong films around the dispersed droplets.

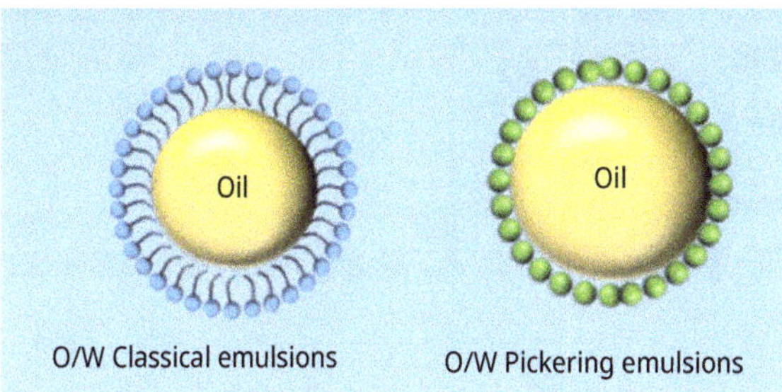

Figure 2.35: Comparison of Pickering and traditional surfactant-based emulsions [26].

In summary, emulsion stability is positively influenced by low interfacial tension, helping to maintain a large interfacial area easier; higher surface viscosity and mechanically stronger interfacial film preventing coalescence; larger electric double layer and more prominent steric repulsions preventing aggregation; smaller attraction; lower volume of the dispersed phase and droplet size; and finally smaller density difference between the phases.

When stability of emulsions is disrupted several options are possible (Figure 2.36) the dispersed droplets can be moved either upward (creaming typical for O/W emulsions) or downward (sedimentation occurring in W/O emulsion).

In general, the maximum possible stability of an emulsion is not always required; it should be merely stable under the conditions required for its performance and then disintegrate.

There are three main types of stabilizers for emulsions. Monomeric surfactants are low-molecular-weight amphiphilic molecules (ionic or nonionic) diffusing quickly to the interface and providing stability during formation of emulsions. Another option is to use polymers that have higher molecular weight. While adsorption is slower, there could be multiple attachment points resulting in stronger adsorption. Polymers provide significant steric stabilization. In some cases, addition of a polymer with a too high molecular mass, nonionic character or a change of the same sign as the particle can lead to flocculation.

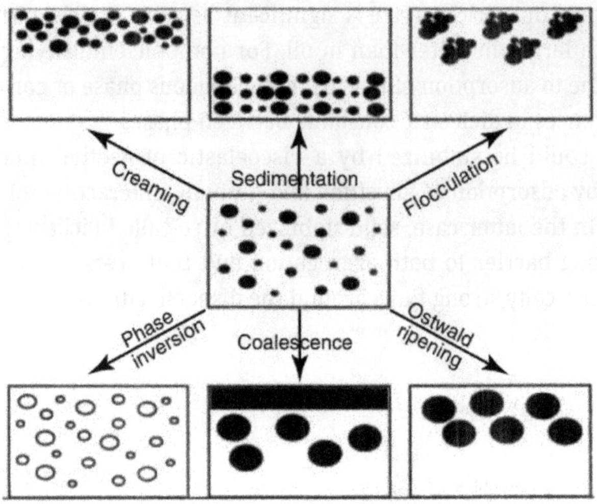

Figure 2.36: Schematic representation of the various breakdown processes in emulsions [27]. Reproduced with permission.

It is also possible to use particles (clays or activated carbon) for stabilization that are much smaller than the droplets. Hard protective coating on the droplet allows for emulsion stabilization.

2.5.2 Surfactants

Surfactants (surface-active agents) are wetting agents that lower the surface tension of a liquid and can also decrease the interfacial tension between two liquids. They are usually amphipathic organic compounds with both hydrophobic and hydrophilic groups being soluble in both organic solvents and water.

The chemical structure of the emulsifier should match the chemical structure of the internal phase. Emulsification of fats or aromatic compounds is thus done with fatty acid esters and emulsifiers, respectively, containing aromatic rings in the hydrophobic part.

A compromise is often made regarding solubility since an emulsifier should be sparingly soluble in the internal phase and as soluble as possible in the external phase.

The surface tension γ depends on the concentration of surfactant c:

$$\gamma - \gamma_0 = Bc \tag{2.28}$$

where γ is the value of surface tension in the absence of the surfactants, and on the structure. For instance, according to the Traube rule in a homologous series, each additional CH_2 group increases the surface tension reduction effect threefold.

Commercially produced surfactants are not pure chemicals but have instead a significant variation within each chemical type. The hydrophobic group exists as a mixture of chains of different lengths. The same is valid also for the polar head group. For example, polyethylene oxide (the major component of nonionic surfactants) has a distribution of ethylene oxide units. In practical applications, this means that commercial products bearing the same generic name could vary substantially in properties.

A nonpolar hydrophobic portion is typically a straight or branched hydro- or fluorocarbon with the chain length of 8–18 carbon atoms. The hydrophilic part can be nonionic, ionic (anionic and cationic) or amphoteric (zwitterionic) (Figure 2.37). The polar or ionic head group interacts strongly with water via dipole or ion–dipole interactions, while there are only weak interactions of the hydrocarbon chain with water.

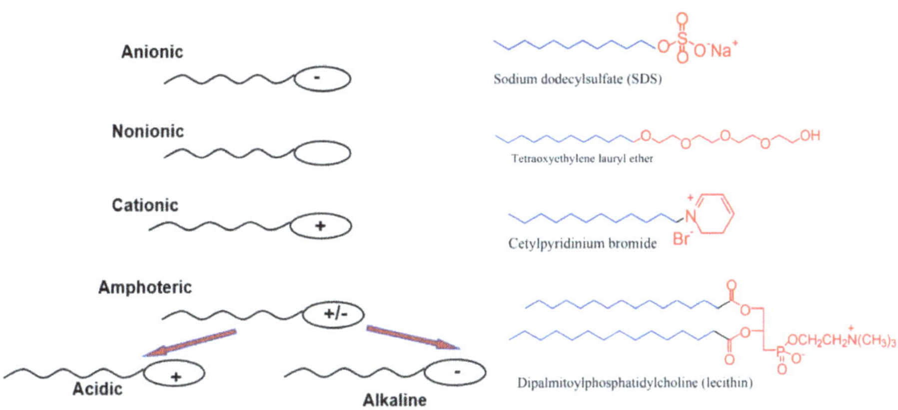

Figure 2.37: Surfactants of different types with examples.

Anionic surfactants are most widely used in laundry detergents due to their soil removal ability through emulsification and solubilization as well as cost-effectiveness. Such surfactants fully dissociate in water into ions. The negative charge is carried by the surfactant molecule, while the counterions are cations.

For optimum detergency, the hydrophobic chain is a linear alkyl group with a chain length ca. 12–16 carbon atoms. Linear chains are more effective and degradable than the branched ones. The most commonly used hydrophilic groups are carboxylates, sulfates, sulfonates and phosphates. Surfactants with these groups are the major components not only in laundry detergents but also in other household and personal care products. Some carboxylates are commonly known because commercial soaps are a mixture of fatty acids obtained from different sources (tall oil, coconut or palm oil, etc.). Low costs, easy biodegradability and low toxicity are clear advantages of such surfactants. An apparent drawback is their precipitation in the presence of metal ions, for example, Mg^{2+} or Ca^{2+}.

Naphthalene and alkyl naphthalene sulfonates are commonly used as dispersants, while phosphate surfactants are applied in the metal-working industry due to their anticorrosive properties.

Table 2.5: Anionic, cationic, and nonionic surfactants [28].

Class	Head group	Examples	Structures
Anionic			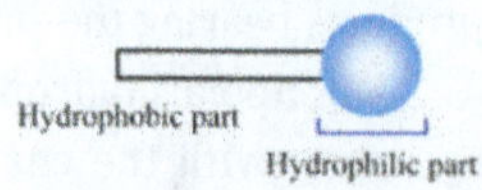
	Carboxylate	Sodium laurate (C_{12}Soap)	$C_{11}H_{23}COONa$
	Sulfonate	Sodium dodecylsulfonate	$C_{12}H_{25}SO_3NA$
		Sodium dodecylbenzene sulfonate (LAS; x+y=11)	
	Sulfate	Sodium dodecyl sulfate (SDS)	$C_{12}H_{25}SO_3Na$
		Sodium dodecylethoxylate sulfate (SLES)	$C_{12}H_{25}O(EO)_pSO_3Na$
	Phosphate	Sodium dodecylphosphate	$C_{12}H_{25}OPO_32Na$
Cationic			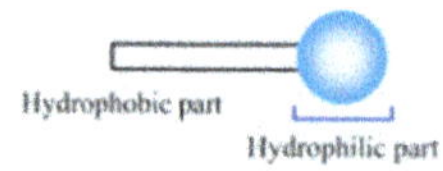
	Alkyl ammonium salt	$C_{12}N(H)a(Me)bCl$	
	a=3, b=0	Dodecyl ammonium chloride	$C_{12}H_{25}NH_3Cl$
	a=0, b=3	Dodecyltrimethyl ammonium chloride	$C_{12}H_{25}N(CH_3)_3Cl$
	Dialkyl ammonium salt	$R_2N(Me)_2Cl$	
	R=18	Dioctadecyldimethyl ammonium chloride	
	Alkyl pyridinium salt	$C_{12}NPyCl$	
		Dodecyl pyridinium Chloride	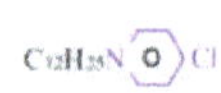
	Alkyl guanidinium salt	$C_{12}AmGCl$	
	m=0	Dodecylguanidinium Chloride	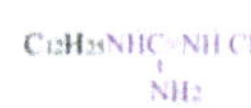
	m=4	Dodecylaminobutyl guanidinium chloride	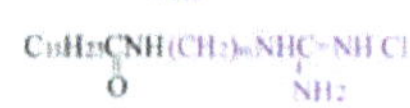

Table 2.5 (continued)

Class	Head group	Examples	Structures
Nonionic			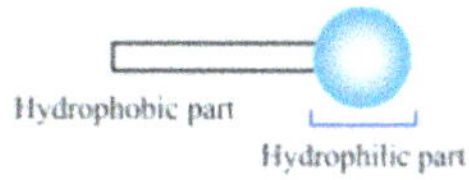
	Polyethoxylate type	Polyoxyethylene dodecylether ($C_{12}EOp$)	$C_{12}H_{25}O(CH_2CH_2O)pH$
		Polyoxyethylene Nonylphenylether (NPEOp)	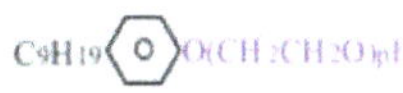
		Lauroyl methylester Ethoxylate ($C_{12}MEEp$)	
	Polyglycerin type	Polyglycerin Laurate	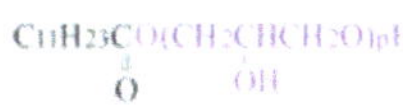
	Block copolymer type	Diblock ABTriblock ABA	PEOa-PPObPEOa-PPOb-PEOa
	A, polyoxyethylene; B, polypropylene.		

Source: Copyright Elsevier, reproduced with permission.

More expensive surfactants than anionic are cationic ones. For such water-soluble cationic surfactants, the head is positively charged. Cationic surfactants are quaternary ammonium compounds being mostly used for their disinfectant and preservative properties. A frequently used cationic surfactant is cetrimide bearing a tetradecyl trimethyl ammonium bromide group with a minimum amount of dodecyl and hexadecyl compounds. Cationic surfactants, generally stable under pH variations, are incompatible with most anionic surfactants. They can be used, however, together with nonionic surfactants. Cationic surfactants can adsorb at negatively charged surfaces and can be used as dispersants for inorganic pigments, antistatic agents for plastics, in hair conditioners and as anticaking agents for fertilizers.

Nonionic surfactants are resistant to water hardness deactivation, as they do not bear any electrical charge. The most common nonionic surfactants are based on ethylene oxide (Table 2.5) as their properties (the critical packing parameter and the hydrophilic–lipophilic balance (HLB) to be discussed below) can be tuned by the degree of polymerization of ethylene oxide units. Alcohol ethoxylates generally produced by ethoxylation of a fatty chain alcohol have a distribution of ethylene oxide chain length. The critical micelle concentration (CMC) of nonionic surfactants is about two orders of magnitude lower than for the corresponding anionic surfactants with the same alkyl chain length. Although alkyl phenol ethoxylates prepared by reacting ethylene oxide with an alkyl phenol (i.e., nonylphenol) are inexpensive, they suffer, however, from biodegradability and potential toxicity because of their degradation products. Utilization of nonylphenol ethoxylates in industry is justified, besides cost-efficiency, also by solubility both in aqueous and nonaqueous media as well as good emulsification and dispersion properties.

Polyglycerin fatty acid esters with higher hydrophilicity than polyethylene oxide because of the hydroxyl group of glycerin, as well as more structural diversity as the glycerol chain can be in the linear or cyclic form, have also found their application as emulsifiers. Another frequently used nonionic surfactants are triblock polymers, for example, PEO blocks polymerized with polypropylene blocks.

An interesting type of most commonly used nonionic surfactants is based on sugars (Figure 2.38) such as sucrose fatty acid esters and their ethoxylated derivatives. Higher hydrophilicity than the polyethylene oxide chain of nonionic surfactants is directly linked to a large number of hydroxyl groups.

Amphoteric or zwitterionic surfactants (Figure 2.37) contain both cationic and anionic groups that make them dependent on the solution pH. The positive charge is almost always represented by ammonium, whereas the source of the negative charge can vary. In acidic solutions, they behave like cationic surfactants bearing a positive charge. Under alkaline pH conditions, because of the negative charge, such surfactants exhibit anionic properties. These surfactants exhibit chemical stability in both

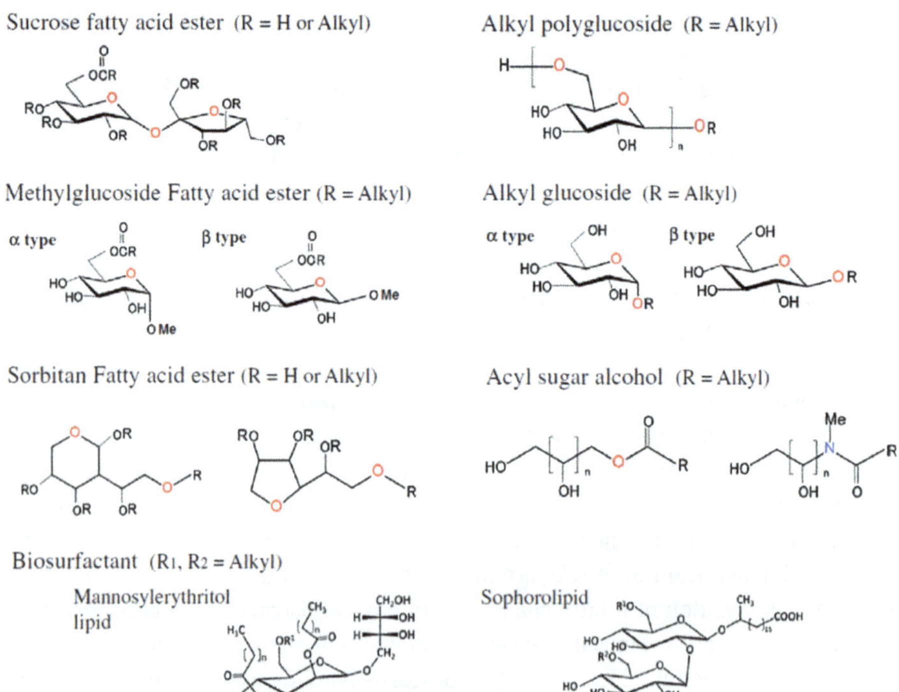

Figure 2.38: Molecular structure of sugar-based surfactants [28].

acidic and alkaline media and excellent compatibility with other surfactants. The surface activity of zwitterionic surfactants depends on the distance between the charged groups with a maximum at the isoelectric point.

Zwitterionic surfactants have excellent dermatological properties causing also low eye irritation. As a result, these surfactants are frequently used in shampoos and other personal care products including cosmetics and hand dishwashing liquids.

An interesting type of surfactants emerged in the late 1980s and early 1990s. Gemini surfactants are dimeric, meaning that there are two hydrophilic, mainly ionic, groups and two corresponding tails (not necessarily identical), linked by a spacer (Figure 2.39). Such spaces can have varying length and type (polar or nonpolar).

A larger total number of carbon atoms in the hydrophobic chains of the Gemini surfactants gives much better performance compared to conventional single-headed, single-tailed surfactants, by increasing surface activity. The Gemini surfactants have CMC values lower by an order of magnitude and enhanced solution properties such as hard water tolerance and superior wetting times.

A close packing of the chains of Gemini surfactants at various interfaces makes more coherent interfacial films resulting in superior foaming and emulsifying properties.

2.5.3 Selection of emulsifier

Selection of the emulsifier types depends on the system. Often, nonionic emulsifiers are used to prepare W/O emulsions while ionic emulsifiers have not proved effective for this purpose. Nonionic emulsifiers are also applied for O/W emulsions. Nonionic emulsifiers typically require larger quantities for making stable emulsions. Addition of anionic emulsifiers to nonionic ones resulting in mixed emulsifiers has been popular.

The Bancroft rule mentioned above explains how an emulsifier determines the type of emulsion (O/W or W/O). According to this rule, the emulsifier stabilizes the emulsion of the type (O/W vs W/O) which has the continuous phase where this emulsifier is most soluble.

A more quantitative approach for easy selection of nonionic emulsifiers is based on the semiempirical HLB concept reflecting emulsifying characteristics of an emulsifier, rather than its efficiency. For example, all emulsifiers with a high HLB will promote O/W emulsions with however a different degree of efficiency.

The HLB concept implies that an emulsifier contains both hydrophobic and hydrophilic groups, whose ratio affects emulsification, and in addition for any particular type of emulsion, there is an optimum HLB for its stability.

A dimensionless number ranging arbitrary for nonionic emulsifiers from 0 to 20 gives information on water and oil solubility. Numbers between 0 and 9 correspond to oil-soluble hydrophobic products, whereas values between 11 and 20 reflect water-soluble oleophobic compounds. Some examples for nonionic and ionic surfactants are

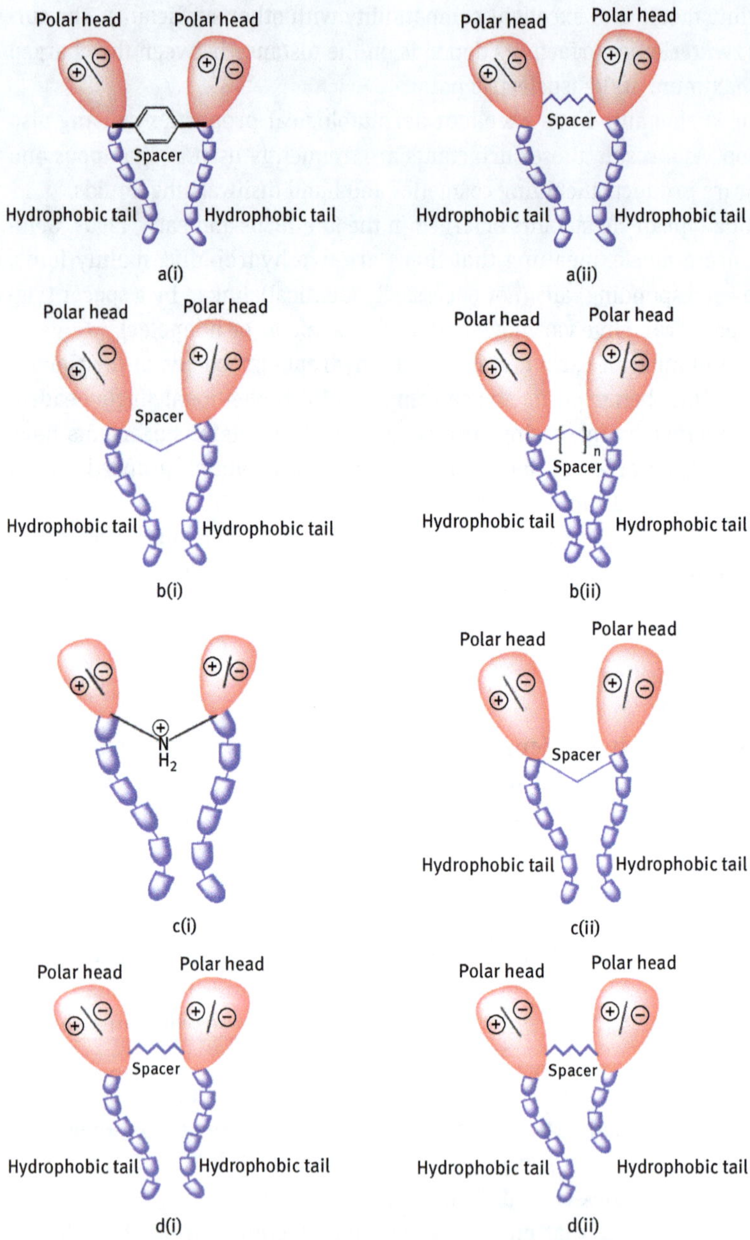

Figure 2.39: a(i) Gemini surfactant with rigid spacer; **a(ii)** Gemini surfactant with nonrigid flexible spacer; **b(i)** Gemini surfactant with short-chain spacer; **b(ii)** Gemini surfactant with long-chain spacer; **c(i)** Gemini surfactant with a polar spacer; **c(ii)** Gemini surfactant with a nonpolar spacer; **d(i)** Gemini surfactant with two identical hydrophobic tails; and **d (ii)** Gemini surfactant with two nonidentical hydrophobic chains (reproduced with permission from [29]).

Table 2.6: Values of HLB for various surfactants.

Surfactant	HLB
Oleic acid	1
Sorbitan tristearate (SPAN 65)	2
Sorbitan monooleate (SPAN 80)	4
Diethylene glycol monolaurate	6
Sorbitan monolaurate (SPAN 20)	9
Glycol monostearate	11
Polyoxyethylene (10) cetyl ether (BRIJ 56)	13
Polyoxyethylene sorbitan monooleate (TWEEN 80)	15
Sodium octadecanoate	18
Sodium dodecanoate	21
Sodium octanoate	23
Dioctyl sodium sulfosuccinate	32
Sodium heptadecyl sulfate	38
Sodium dodecyl sulfate	40
Sodium octyl sulfate	42

given in Table 2.6. In fact, the ionic surfactants have recently been assigned relative to HLB values exceeding 20 allowing to extend the range to 60.

Low HLB emulsifiers are lipophilic or soluble in oil and give rise to W/O emulsions. Hydrophilic emulsifiers being soluble in water have HLB exceeding 11. Substances with an HLB value of 10 are distributed between the two phases with the hydrophilic group (molecular mass M_H) projecting completely into water and the hydrophobic hydrocarbon group (molecular mass M_P) adsorbed in the nonaqueous phase. Ideally, HLB of a particular emulsifier of a molecular mass of M and a hydrophilic portion M_H is calculated according to the Griffin method proposed for nonionic surfactants in 1954:

$$\mathrm{HLB} = 20\frac{M_H}{M} \tag{2.29}$$

For O/W emulsions, the HLB range is 8–18, whereas for W/O emulsions, this range is 3–6. As a comparison, it can be mentioned that wetting agents are within the 7–9 range and detergents have HLB between 13 and 15.

Among empirical equations for calculation of HLB, an expression applied for polyoxyethylene alcohols ($C_nH_{2n+1}(OCH_2CH_2)_mOH$) can be mentioned:

$$\mathrm{HLB} = E/5 \tag{2.30}$$

where E is the mass percentage of ethylene oxide in the molecule.

Mixtures of surfactants (Figure 2.40) are of practical interest as use of mixed surfactants allows more surfactants to pack effectively at the oil–water interface giving lower interfacial tensions and therefore a more stable emulsion by steric stabilization. For two surfactant molecules with HLB values HLB_1 and HLB_2, respectively, the average HLB is calculated by knowing the weight fractions x_1 and x_2 of these surfactants:

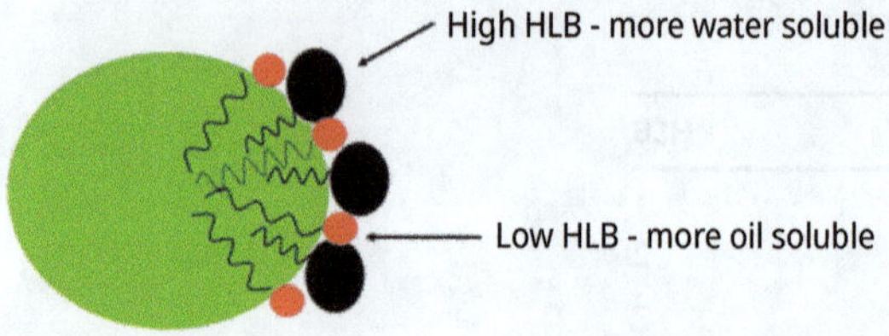

Figure 2.40: An illustration of paired emulsifiers.

$$HLB_{av} = x_1 HLB_1 + x_2 HLB_2 \tag{2.31}$$

The same approach can be extended for multicomponent mixtures of surfactants.

From the experimental viewpoint, selection of an emulsifier system is done by mixing emulsifiers with low and high HLB, investigating their ability to act as an emulsifier for the desired oil/water composition and further developing the best formulation by interpolation (Figure 2.41).

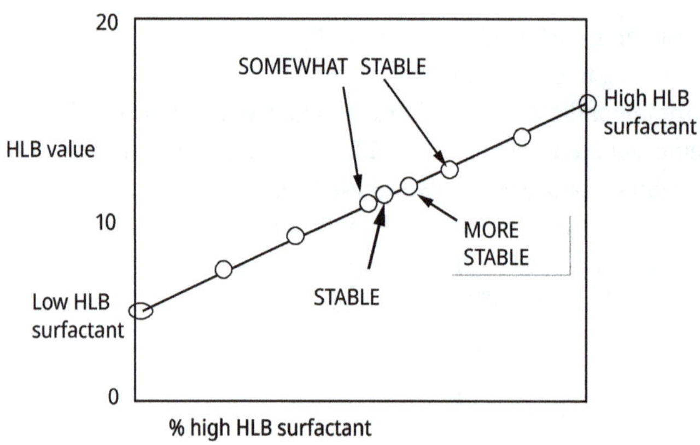

Figure 2.41: Selection of the best emulsifier composition [30]. Reproduced with permission.

In practice, for such mixed emulsifier systems, the high HLB surfactant is dissolved in the aqueous phase while the surfactant with a low HLB value is placed in the oil phase followed by making an emulsion of these two phases.

The optimum HLB number for producing the most stable emulsion depends on the nature of the oil. Stability as a function of the HLB number has a complex behavior (Figure 2.42).

HLB is a function of temperature, especially for the nonionic surfactants. As a result, at low *T*, a particular surfactant may be beneficial for stabilization of O/W emulsions, while increase of temperature might result in the opposite behavior with W/O emulsions being more stable. For practical applications, operation close to this phase inversion temperature (PIT) should be avoided, and emulsifying agents are chosen to

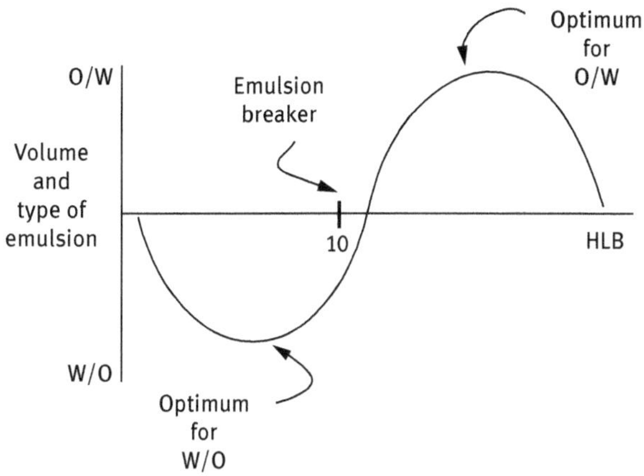

Figure 2.42: Dependence of the volume and emulsion type on HLB [31].

have storage and operation temperature far from the PIT. The measurement of the HLB value from the PIT being suitable for nonionic surfactants has an advantage of a direct measurement including the influence of the oil type and aqueous phase composition.

It should be stressed that the HLB concept is an empirical one not considering the structure and composition of the oil phase and the aqueous phase, including the presence of electrolytes. Moreover, relative positions of several functional groups present in a molecule are not taken into account. Further limitations include the inability to treat mixtures of interface active substances and lack of predictive power.

The main critics of the concept is related thus to the fact that HLB is considered to be a feature of a surfactant, while the emulsion stability is linked to the whole system, which depends, for example, on the concentration of electrolytes, oil type and temperature.

Besides taking into account the influence of an emulsifier on the interfacial tension between two phases, other important processes should be considered. This is because a decrease of the interfacial tension is necessary, but alone this is insufficient for emulsification.

One such processes is spreading of emulsifier molecules in a freshly prepared interface (Figure 2.43), called the Marangoni effect. This effect is related to the presence of a gradient in surface tension, resulting in spreading of the liquid. A liquid with a higher surface tension pulls stronger on the surrounding liquid, thereby in the case of a surface tension gradient the liquid flows away from regions of low surface tension.

After formation of a larger droplet from two emulsion droplets by Brownian motion or mechanical division, the point of rupture is initially free of emulsifier molecules. As a result, there is an interfacial tension gradient between this free unoccupied surface and areas on the interface where an emulsifier is present. Stabilization of this gradient can be achieved by interfacial adsorption (diffusion effects). This can happen by adsorp-

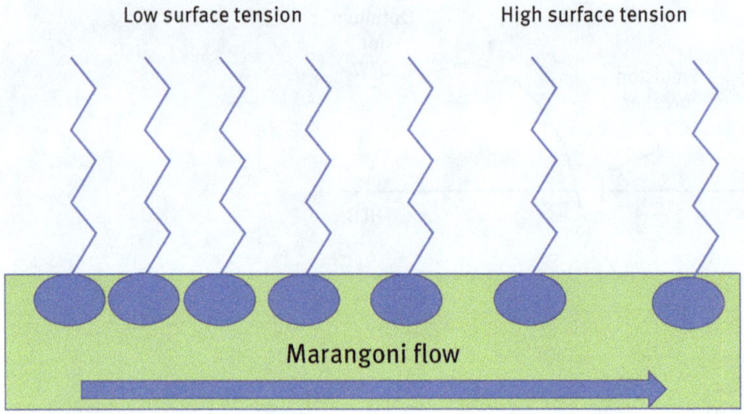

Figure 2.43: The Marangoni effect.

tion of additional emulsifier from the solution or from the interior of the droplet. Alternatively, the molecules present in the interface spread out (Marangoni effect).

In general, if the diffusion, spreading and adsorption rate of the emulsifier molecules are too low and a freshly formed droplet surface is not occupied rapidly enough by emulsifier molecules, adequate emulsification is not achieved and the droplets coalesce immediately after formation.

In the case of polymer emulsifiers, which are suitable for stabilizing emulsions because of the steric effects, potential untangling of the polymer molecules adsorbed at the interface should be considered.

Further factors influencing emulsion stability are temperature (increase in temperature deteriorates stability), density difference (preferably as low as possible), droplet size (should be small) and distribution (as uniform as possible), zeta potential (should be high) and optimum HLB values.

2.5.4 Micelles

Aggregation of surface-active agents in solution with formation of micelles is driven by reduction of a contact between the hydrocarbon chain and water. This decreases the free energy of the system. In the micelle (Figure 2.44), the surfactant hydrophobic groups are directed toward the interior while the polar head groups are directed outward toward the solvent. These micelles can be visualized as drops of oil in water while the reverse micelles are like drops of water in oil. Micelles can have different shapes, with the simplest ones being spherical.

Micelle formation becomes significant when the concentration of a surfactant reaches the CMC as illustrated in Figure 2.44. The latter depends on various factors such

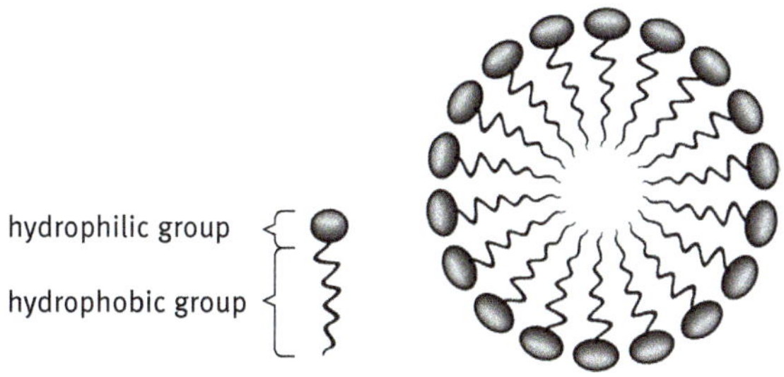

Figure 2.44: Micelle structure [32].

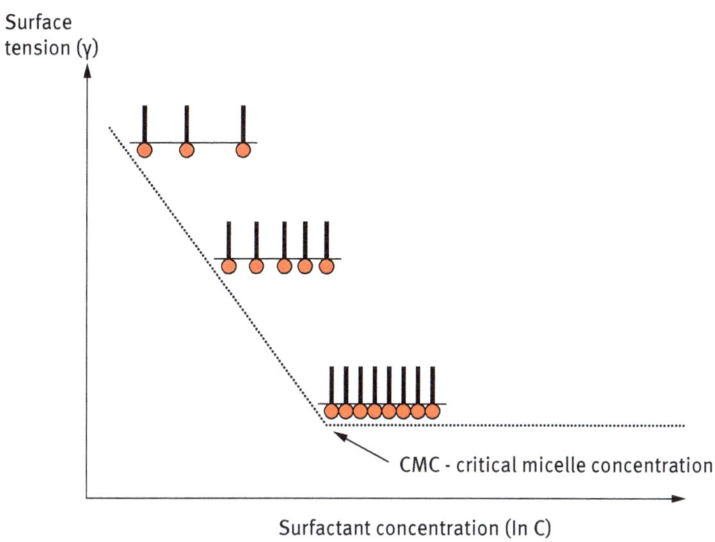

Figure 2.45: Dependence of surface tension on CMC [33].

as the surfactant nature, temperature, pressure, presence and type of additives. The latter either alter the solubility of single molecules or change the easiness of micelle formation. Dependence of CMC on both temperature and pressure is often rather weak even if there are exceptions. Influence of added substances is more complex as additives can be solubilized either in the micelle or in the intermicellar solution. For instance, long-chain polar molecules upon solubilization are incorporated in the micelle structure substantially lowering CMC. In case of ionic surfactants, addition of salts decreases repulsion of similarly charged head groups by some sort of shielding of the head-group charges in the structure of a micelle. As a consequence, introduction of salts decreases CMC.

Overall, lower CMC can be typically achieved by temperature decrease and addition of an electrolyte. The values of CMC show minor variation with regard to the nature of the charged head group. The main influence appears to come from the charge of the hydrophilic head group.

For a homologous series with an increase of the alkyl chain length, the micelles tend to form at lower concentration as solubility in water decreases

$$\text{In(CMC)} = \alpha - \beta\lambda \tag{2.32}$$

where α and β are constants. As a general rule, CMC decreases by a factor of 2 for ionic surfactants (without added salt) and by a factor of 3 for nonionic surfactants when an additional methylene group is added to the alkyl chain. The latter surfactants have lower CMC values than their corresponding ionic surfactants with the same alkyl chain length.

For nonionic surfactants, an increase in the hydrophilic group (polyethylene oxide) length results in CMC increase.

Influence of the bulkiness of the alkyl chain is more complex. With more bulky alkyl chains, the value of CMC increases except the case when benzene is present in the hydrophobic part of the alkyl chain, giving an opposite behavior. Introduction of the phenyl group also has a much smaller influence on hydrophobicity compared to an increase of the alkyl chain length with the same number of carbon atoms.

The physicochemical properties of surfactants (conductivity and surface tension) vary significantly above and below the CMC value, indicating a highly cooperative association with formation of micelles. For example, surface tension decreases when the concentration of surfactants decreases up to the point when CMC is reached (Figure 2.45). Above the CMC, the surface tension remains constant.

Solubility of micelle-forming surfactants shows a strong increase above a certain temperature (the Krafft point) because of high solubility of micelles, whereas single surfactant molecules have only limited solubility. The Krafft phenomenon is the temperature-dependent solubility of ionic surfactants. Along with HLB, the Krafft point for ionic surfactants (Figure 2.46) and the cloud point for nonionic surfactants are the characteristics applied in formulation development. Figure 2.46 illustrates that the Krafft point is the triple point of the surfactant monomer solubility curve, the CMC temperature curve and the phase transition line (T_c) of hydrated solids to micelles and/or liquid crystal. Subsequently, the Krafft point is often viewed as the melting point of a surfactant hydrated solid as below the Krafft point the surfactant exists as hydrated crystals having a turbid appearance at low temperature.

The Krafft point is affected by both the hydrophilic and lipophilic groups. When the hydrophilic group is identical, the Krafft point increases and the solubility to water decreases with increasing chain length. Furthermore, the Krafft point is lowered by branched chains and presence of double bonds. If the lipophilic group is identical, the solubility changes with the number and position of hydrophilic groups, and when the hydrophilic group is closer to the center of the alkyl chain, the solubility tends to increase. The Krafft point increases by addition of salting-out electrolytes.

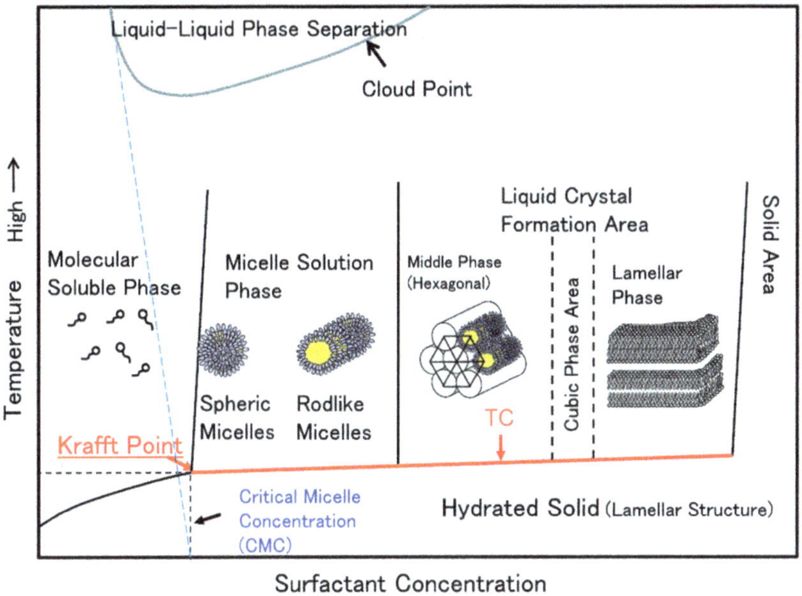

Figure 2.46: The concentration and temperature dependency of the phase conditions of a surfactant aqueous solution [34]. Reproduced with permission.

From the viewpoint of formulation development, crystallization of the surfactant should be minimized; therefore, it is important to have a lower Krafft point ensuring low temperature stability of a formulation.

Nonionic surfactant solutions have a tendency to cloud upon elevation of temperature when the temperature rises. The cloud point corresponds to a phase inversion that results from spinodal decomposition that renders the nonionic surfactant more hydrophobic. Nonionic surfactants with a higher cloud point have higher hydrophilic property; therefore, the cloud point is used as a descriptor of nonionic surfactants' solubility. An increase in the alkyl chain length and a decrease in the EO chain length of alcohol ethoxylates reduce the HLB and lower the cloud point.

Surfactants have a wide application range from mineral processing to personal care products and drug delivery systems. Their concentration should in general be above CMC.

2.5.5 Application of emulsifiers

Emulsions are typically prepared using one or more emulsifiers with different HLB values. The application areas are numerous and will be discussed in more detail in Chapters 3 and 4. Hydrophilic emulsifiers with a high HLB value favor the formation of O/W emulsions, while hydrophobic emulsifiers with a low HLB value can be used for preparing W/O emulsions.

In personal care products (cold cream, vanishing and deodorant cream, hand lotion), emulsions allow several advantages, including efficient cleansing and ease of application. Water- and oil-soluble ingredients can be used at the same time, which is important because besides emulsifiers many other ingredients (fragrances, pigments, moisturizers, thickeners, preservatives, pH adjusters, oils and waxes) are used in cosmetic emulsions. This complexity is needed to provide the desired sensory properties such as softness, creaminess, coolness and spreadability. Both types (W/O and O/W) of creams are used in skin care products. It might be difficult to make nonstick and easily spread W/O emulsions meeting all the requirements in terms of safety, smoothness, softness and spreadability. Besides good skin compatibility, stability during storage is important, resulting in a certain compromise. O/W emulsions are popular for providing an initial cooling effect due to evaporation of water even though W/O emulsions more closely resemble the skin physiological conditions. When emulsions are applied to wet skin, such as in hand and body lotions, O/W emulsions are used.

Skin and hair can be protected from UV radiation by a photoprotective emulsion bearing Gemini surfactants.

Shampoos are normally a "gelled" surfactant solution of well-defined association structures, for example, rod-shaped micelles.

Emulsions are also used in household cleaning and maintenance products. Because there is a significant variety of types of dirt and substrates from which such dirt should be removed, a single mechanism of dirt removal does not exist depending on the substrate, that is, glass, china, metal or textile. While a commonly perceived opinion by the general public that foaming is necessary for efficient cleaning, this is not the case; for example, nonionic detergents being efficient cleaning agents of liquid dirt do not foam. More information about laundry detergents will be given in Chapter 3. In household and industrial cleaning, the combination of chemical effects due to the surfactants and mechanical effects due to the shear added is needed to remove contaminants. Detergents typically comprise several surfactants with the aim to remove hydrophobic dirt. Besides the surfactants per se, commercial laundry detergents contain builders to remove hardness ions; antiredeposition agents; bleaching agents (e.g. to remove juice and wine stains), enzymes, breaking-up protein-based stains; fluorescent whitening agents and dyes, providing white appearance; foam boosters in hand dishwashing and shampoo formulations or alternatively antifoaming agents in mechanical cleaning formulations; and several other ingredients as will be explained in Chapter 3.

Metal-working fluids providing lubricity and cooling during metal-working and metal-cutting operations are traditionally based on petroleum oils with added surfactants to increase lubrication and wetting. O/W microemulsions can also be used for the same purpose, being able to switch to O/W macroemulsions when diluted with water during application.

Furniture, floor and automobile polishes also comprise oil or wax emulsions in water.

Other examples of emulsions are latexes or emulsion polymers used in paper coating, adhesive and latex paints.

Water-repellent finish is applied as an aqueous solution in waterproofing textile materials. Immersing fiber materials in such solution allows uniform wetting. Subsequent water removal by drying is substantially facilitated when the emulsion is broken at a lowest temperature.

While many readers have personal experience witnessing road paving for making asphalt, it is less known that asphaltic bitumen emulsions are used when the road surface should be hydrophobic repelling (Figure 2.47). Relatively low viscosity of ca. 50% internal phase emulsions allows application of very viscous asphalt. Because of low emulsion viscosity compared to hot applied bitumen, the bitumen emulsions have a good penetration and spreading capacity. From the viewpoint of rheology, the emulsion needs to be able to shear thinly during application, and then break relatively quickly. In most cases, no additional heat is required, allowing cost saving and application of cold materials at remote sites, ensuring workforce safety. A wide variety of emulsion types are available; nowadays, reduction of the interfacial tension and stabilization of the O/W asphalt emulsions can be done with the aid of natural naphthenic, synthetic anionic and commonly cationic surfactant. The road-building aggregate adsorbs the surfactants making these aggregates hydrophobic, and thus allowing preferential wetting by the liquid asphalt droplets. Positively charged aggregates (e.g., limestone and dolomite) require thus anionic surfactants, while cationic ones are applied to negatively charged aggregates, such as silicates.

Figure 2.47: Road pavement with bitumen emulsion [35].

Emulsions are being used increasingly as delivery systems for pharmaceutical treatments. There are several requirements for successful application of emulsions in pharma. Emulsifying agents, besides being stable, should be compatible with other ingredients, should be nontoxic and should not interfere with the efficacy of the active ingredient.

An important issue in drug delivery is the release kinetics. Emulsions can be formulated in a way, which allows a slow release either preventing overdosing, thereby reducing toxicity or prolonging the effect of the drug. On the other hand, retention of the active substance by emulsifiers should not be excessive; otherwise, the pharmaceutical compound might not be efficient because of very low concentration. Emulsions, which are pH sensitive, can be broken, for example, by a change in the pH in the intestine.

Other requirements for emulsions in pharmaceutical applications are related to their odor, taste and color.

One of the most useful applications of surfactants in pharmaceuticals is solubilization of poorly water-soluble drugs by surfactant micelles. This is done by incorporation of additional amphiphilic components, making thermodynamically stable isotropic solutions of a substance typically either insoluble or sparingly soluble. With nonpolar compounds, an increase of the alkyl chain length of the surfactant improves solubilization. When the alkyl chain length is the same, solubilization increases in the order: anionic < cationic < nonionic.

Among emulsifying agents, some carbohydrates (e.g., agar) or high-molecular-weight alcohols (e.g., stearyl alcohol) give O/W emulsions, while others result in W/O emulsions.

In some instances, like preparations for crop protection, emulsion concentrates are prepared and then put in contact with water on the spot before use. Such strategy is needed when the active substances undergo hydrolysis and are therefore sold in anhydrous formulations. For such compositions, intensive agitation is not required contrary to emulsifiable concentrates for cutting oils or cooling lubricants for metalworking.

Agrochemical formulations with spontaneous emulsification should thus be prepared with correct properties implying turbulence at the interface and thus low viscosity in the emulsifiable concentrate, presence of alcohols as solubilizers to improve mutual solubility of the phases and introduction of other additives. Among several emulsifiers applied in industrial formulations, besides the active ingredient present in ca. 95% (or even lower 80–90%), calcium or amine salts of alkyl aryl sulfonates are used as ionic surfactants, while ethylene oxide adducts of fatty alcohols, alkyl phenols, castor oil or sorbitan esters with different HLB values are utilized as the nonionic emulsifiers.

Temperature and the salt content of water applied in preparation of emulsions from the concentrates determine also the type of emulsifiers. More hydrophilic HLB with a higher fraction of ionic surfactants should be applied in warm, hard water contrary to colder soft water when the proportion of such emulsifiers can be diminished. As stabilizers for emulsifiable concentrates, epoxides such as epoxidized soya oil can be used.

As apparently clear from the description above, emulsion compositions besides surfactants contain emulsifying aids, such as thickeners (methylcellulose, sodium alginate, waxes, other proteins and polysaccharides) to increase the viscosity of the outer phase, solubilizers (aliphatic alcohol in crop protection or urea in textile or dye auxiliaries) to influence density and viscosity of both phases, protective colloids (carboxymethylcellulose in household laundry detergents) to diminish droplets coalescence, defoaming agents (e.g., a nonionic polyalkylene glycol ether with mixed ethylene and propylene chains) along with various preservatives to prevent fungal and bacterial growth. The latter are not required if the emulsifiers possess themselves such properties.

2.5.6 Emulsion technology

The order of ingredient addition and homogenization may have a large impact on the product properties. Emulsions are typically prepared by dissolving the emulsifying agent into the phase where it is most soluble. This is followed by adding the second phase and applying shear by efficient mixing. For O/W emulsions, such vigorous agitation can be crucial for making sufficiently small droplets. Thus after an initial mixing, a second mixing with very high applied mechanical shear forces might be required. The main processes during homogenization are given in Figure 2.48. Several process parameters are important for proper homogenization. Energy density (energy input per volume) defines the minimum achievable droplet size. The latter typically decreases with an increase of energy density, unless mixing is inefficient. Energy efficiency influences heat losses and manufacturing costs, while production capacity depends on the volume flow rates. Some limitations on which type of materials can be homogenized are imposed by the product rheology.

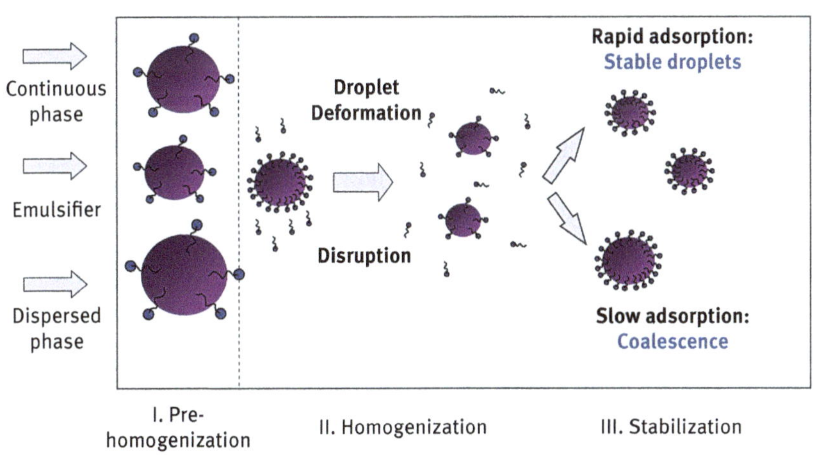

Figure 2.48: Physicochemical processes during homogenization [36].

The following devices can be used for preparation of industrial emulsions: vessels with high-speed stirrers, agitation or impact machines, centrifuges, colloid mills, metering pumps, vibrators, ultrasonic generators and homogenizers. Some of the devices used for homogenization are presented in Figure 2.49.

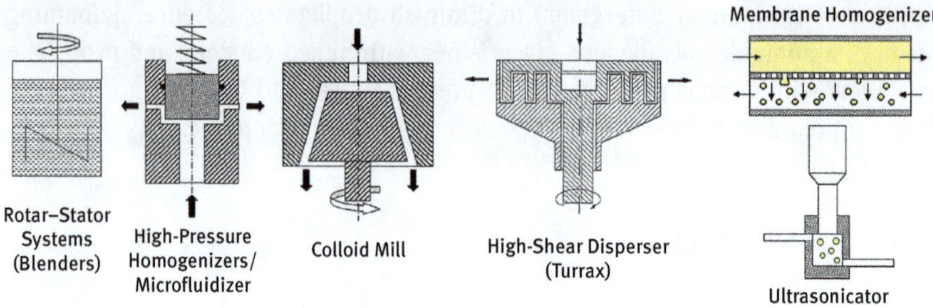

Figure 2.49: Devices for homogenization [36].

Flow profiles are typically complex and based on the geometry of homogenizer, they may be laminar (rotational, simple shear and elongation) or turbulent.

Stirred vessels have low power density distributed inhomogeneously with simultaneously present high-shear zones and dead zones. Droplets with mean diameters below 1 µm and with a narrow distribution can therefore be rarely produced. Moreover, long residence time can even lead to undesired by-products. The presence of air might lead to foaming. For this type of fairly cost-effective homogenization mainly used in making premixes, the impeller/blender geometry and rotational speed are of primary importance.

More efficient from the viewpoint of energy density are devices with small disruption zones, for example, colloid mills or ultrasound generators. In rotor–stator homogenizers such as colloid mills, the force for disruption comes from the energy of the rotor. The part rotating within a stator is a truncated cone, with a variety of surface profiles most often toothed. The conical design allows transportation of emulsions without applying external pressure.

Droplet disruption in colloid mills occurs in a flow channel between a rotor–stator assembly. The narrow gap to the stator provides the shearing action. Very high shear rates with speeds up to several thousand rpms are achieved to accomplish droplet breakup. The residence time can be adjusted with the size of the gap.

High shear dispersers or Ultra-Turrax Systems, being rotor–stator devices similar to colloid mills, are composed of coaxial intermeshing rings with radial openings. As shown in Figure 2.49, the fluid enters in the center and is accelerated by the rotor. Because of multiple acceleration and deceleration while passing from such type of disperser, high tangential forces along with the turbulent flow are generated, leading to droplet disruption.

In ultrasonic generators (batch or continuous), the emulsifying action of ultrasound is attributed to the cavitation forces generated in the liquid. One of the limitations with ultrasonic generators is related to the power of sound inducers, posing restrictions in terms of high throughput, which could be difficult to achieve. Transducers that are either piezo-electric or magnetorestrictive generate sound waves with associated pressure gradients, thereby causing cavitation, turbulent flow and high shear. Overall, this gives rise to deformation of droplets. Among the process parameters influencing the performance of ultrasonic generator, the frequency and sonication time should be mentioned. Size of cavitational bubbles decreases with increasing frequency, and a minimum sonication time is needed to achieve droplet disruption.

The presence of dissolved gas reduces the intensity of cavitational collapse due to dampening, despite more cavitational events. Temperature also influences droplet disruption.

High-pressure homogenizers are most common in the food industry (milk, cream, etc.). For such homogenizers, disruptive energy originates from relaxation of high-pressure buildup across the homogenization valve. Pressures can reach 50–500 bar and even up to 1,600 bar in microfluidizers. Scaling up of high-pressure homogenization is challenging because the flow conditions should not be changed with increasing nozzle throughput. A viable option is to use a numbering-up approach (i.e., have more nozzles) rather than scaling up.

In membrane homogenizers, the dispersed phase or emulsion is pushed through a microporous membrane into a continuous phase, allowing high energy efficiency, low droplet size (twofold or threefold the pore size) and a narrow particle size distribution. Performance of such homogenizers is influenced, besides temperature, by viscosity and flow rates, the membrane pore size, material and the pressure across the membrane.

Comparison of these homogenizers is provided in Figure 2.50, which shows the dependence of the droplet size on the energy input per volume E_V. This figure can be interpreted in the following way, when the coalescence rate is low, membrane homogenizers are efficient in obtaining small size of droplets with narrow distribution. If coalescence rate and viscosity are low, ultrasonic homogenizers can be applied at the expense of somewhat broader droplet distribution and energy density. Conical mills being better in terms of energy density than devices relying on ultrasound, at the same time, give larger droplet size with less uniform distribution and could be recommended for the cases of high coalescence rate or alternatively when coalescence is not that profound but viscosity is high.

Purely mechanical emulsification gives emulsions of low concentration (0.1%); therefore, small quantities of emulsifiers are typically added in most mechanical processes.

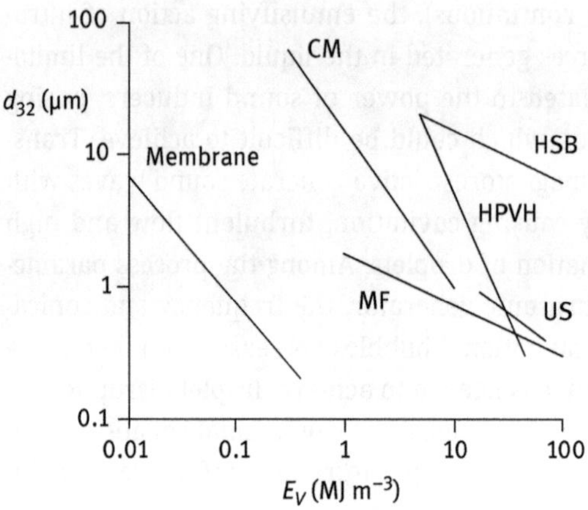

Figure 2.50: Comparison of various homogenizers in terms of droplet size versus energy density [36].

2.5.7 Microemulsions

Contrary to turbid and minimally stable (macro)emulsions with a broad distribution of droplets in the 0.05–100 μm range, microemulsions have structural diameters in the nanometer (10–50 nm) range and are clear, isotropic and thermodynamically stable.

Because of these differences, despite apparent similarities with (macro)emulsions, microemulsions are often considered as solutions of swollen micelles rather than dispersions of one liquid in another. Preparation of microemulsions generally requires much higher amounts of an emulsifier (15–30%) than for macroemulsions (1–5%). Microemulsions are thus thermodynamically stable fluid dispersions of two or more ordinarily immiscible liquids (hydrophilic and hydrophobic, i.e., water and oil), stabilized by at least one amphiphile. The latter could be a short-chain alcohol stabilizer such as 1-pentanol. Strictly speaking, such amphiphiles have very small molecular weights to be considered as a true surfactant. The selection of surfactants for producing microemulsions requires determination of the phase diagram for the ternary system oil–water–surfactant–cosurfactant to define regions where O/W or W/O microemulsions are formed.

Microemulsions typically contain discrete domain microstructure (Figure 2.51) with a broad distribution of sizes and different topologies.

Microemulsions are used in many cosmetic products because of long-term stability and product aesthetics including clarity. In sunscreen formulations, transparent aesthetics are combined with good sensory and water-resistant properties. Such formulations can be made with two layers: the first being an O/W microemulsion with nonionic surfactants and organic oils, while the second is a silicon oil phase with the active components needed to ensure desired UV protection. Finally, hair styling waxes with microemulsion structure are worth mentioning. The content of nonionic

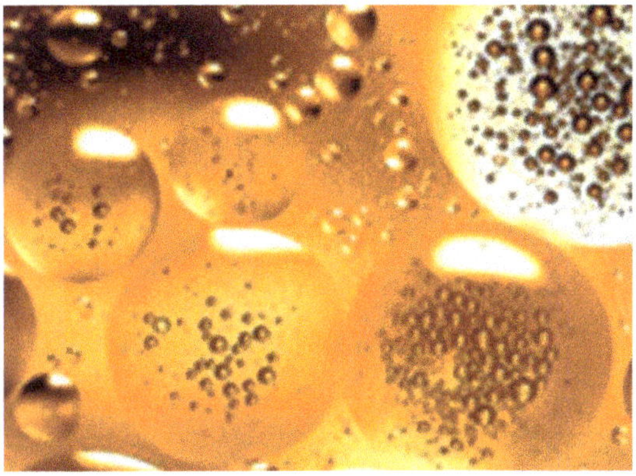

Figure 2.51: Microemulsion [37].

surfactants can be as high as 30% in addition to 10% oil. High viscosity, good spreadability and visible clarity are attributes of styling products appealing to even demanding customers.

2.6 Basics of rheology

In various materials discussed in this book and related in a broader context to formulation product technology, very often, certain rheological properties should be achieved to ensure ease of formulation, mixing and application, allowing long-term physical stability without or minimum phase separation (creaming or sedimentation). Such requirements can be found across the applications ranging from personal care products, cosmetics and pharmaceuticals to paints, printing ink and coatings to agrochemicals. These substances have a complex microstructure and, contrary to the flow of gases and simple liquids, their flow can be described by a single property, namely the shear viscosity.

Shear is an action of stress when forces are applied, causing two contiguous parts of a body to slide relative to each other in a direction parallel to their plane of contact. Figure 2.52 illustrates deformations produced between parallel plates when the upper plate, of surface area A, is moved in response to a force F. The displacement of a plane layer (dx) over the separation between layers (dy) is termed the shear (dx/dy) acting on the fluid.

Depending on a material, there are differences in the behavior when the shear stress is removed. While a solid will tend to return to its original shape, a plastic material may show only partial recovery.

For liquids, an ideal behavior, also known as Newtonian behavior can be characterized by a single coefficient of viscosity. The latter appears from the following considerations.

The force per area in the plane termed the shear stress (τ) is expressed as

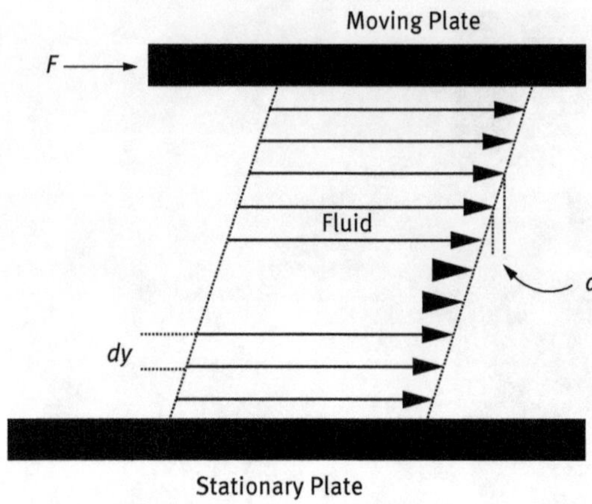

Figure 2.52: Illustration of deformations between parallel plates.

$$\tau = F/A \tag{2.33}$$

Shear stress can be viewed as friction to overcome when one layer slides with respect to another. According to Newton's law of viscosity, the shear stress is proportional to the derivative dV/dy termed the shear rate $\dot{\gamma}$:

$$F/A \propto dV/dy \tag{2.34}$$

giving dependence of shear stress on shear rate

$$\tau = \eta\dot{\gamma} \tag{2.35}$$

where η is the coefficient of viscosity, with units of mass/length · time. In SI units, τ is in Pa and γ in s^{-1}; therefore, η is expressed with Pa/s. The coefficient of viscosity, being the ratio of the shear stress to the shear rate, represents the internal resistance to flow of the fluid. Equation (2.35) is applied to ideal or Newtonian fluids (e.g., water or mineral oils) when their behavior can be characterized by a single coefficient of viscosity. For the large class of fluids relevant for formulation product technology, this does not hold, and viscosity depends on the shear rate:

$$\tau = \eta(\dot{\gamma})\dot{\gamma} \tag{2.36}$$

Such deviations from the ideal behavior are related to, for example, polydispersity, high concentration of the dispersed phase and mutual orienting and structure formation of the dispersed species in flow. Subsequently, dispersions, emulsions, and polymer solutions often exhibit flow properties distinctly different from Newtonian behavior (Figure 2.53), and their viscosity is not constant being a function of the shear rate. For polymers, a correlation exists between rheological properties and the average molecular mass. Rheology of polymers is sensitive to the structure of the polymer chain, side groups, presence of copolymers, side group content, etc.

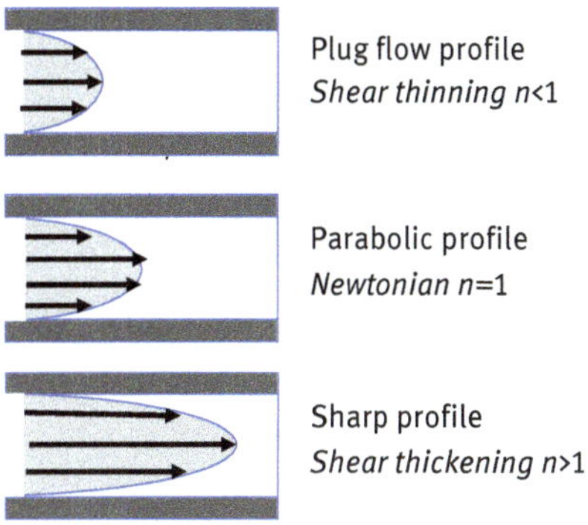

Figure 2.53: Flow profiles. Coefficient n appears in eq. (2.37).

Figure 2.53 illustrates a parabolic profile of the Newtonian liquid. Deviations for shear-thinning liquids from the ideal behavior are attributed to a low viscosity near the wall, where the shear rate is high and a high viscosity is in the middle. The velocity profile is more flat, resembling a plug flow. The opposite behavior with a sharp profile is observed for shear-thickening fluids.

Dependence of viscosity on the shear rate is shown in Figure 2.54. Shear viscosity decreases or increases with increasing shear rate, leading to shear thinning or shear thickening, correspondingly.

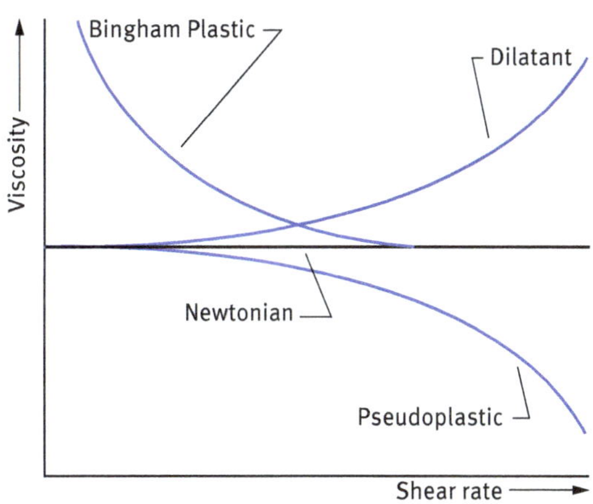

Figure 2.54: Typical curves of viscosity versus shear rate [38].

When interparticle forces cause structuring of particles and lead to some sort of particle locking, the shear viscosity is increased and shear thickening or dilatancy is observed. In this case, viscosity increases with increasing shear rate. An example could be concentrated suspensions of polyvinyl chloride particles in a plasticizer liquid. Alternatively, the shear viscosity decreases with increasing shear rate resulting in shear thinning or pseudoplasticity (Figure 2.55) when the particles, drops or bubbles are deformed and aligned. Examples of shear thinning or pseudoplastic systems include emulsions, foams and suspensions. A decrease of viscosity for bimodal or multimodal size distributions is often related to an optimized packing when small particles are located in the interstitial volume between the larger ones. Typically, such effect of viscosity reduction in systems with particles of different size is seen for the particle volume fraction exceeding 0.5. Various equations have been proposed in the literature to account for the effect of particle loading and size distribution on viscosity as will be mentioned below.

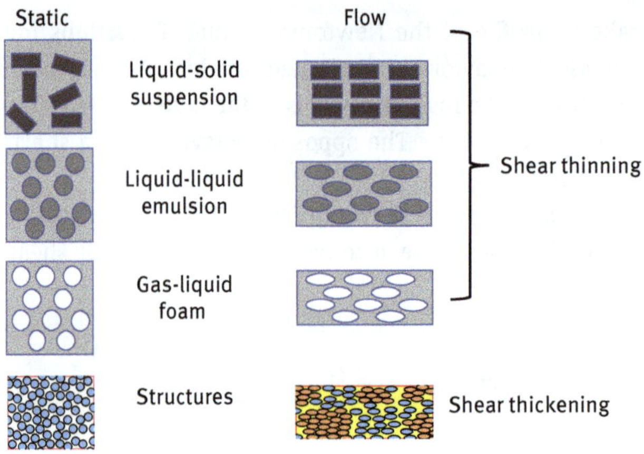

Figure 2.55: Flow of dispersions.

In some colloidal dispersions, a minimum stress to be applied is required before they start to flow. Figure 2.56 displays Bingham fluids for which the shear rate (flow) remains at zero until the yield stress (τ_Y) or a threshold shear stress is reached. Thereafter either Newtonian or pseudoplastic flow is observed. Bingham plastics are strictly speaking non-Newtonian as they require the yield stress different from zero; however, after the flow starts, the behavior of Bingham plastics is the same as that of Newtonian fluids meaning that the shear stress is linearly proportional to the shear rate. An example of this kind of behavior is toothpaste. The behavior of pseudoplastic Bingham fluids in terms of their viscosity dependence on the shear rate is similar to that of shear-thinning fluids. A familiar example to many readers is ketchup.

The rheograms for shear-thinning (pseudoplastic is sometimes used as synonymous) and shear-thickening (dilatant) fluid can often be described by an empirical power law function called the Ostwald–de Waele equation:

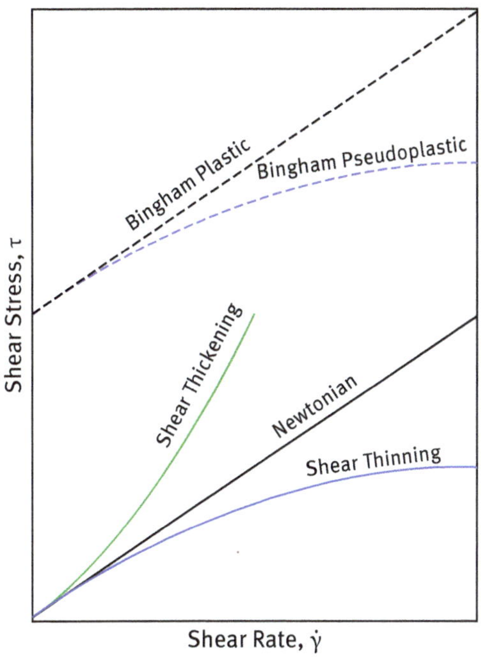

Figure 2.56: Rheogram shapes [39].

$$\tau = k\left(\frac{d\gamma}{dt}\right)^n \tag{2.37}$$

where k is the consistency index. For shear thinning $n < 1$, for Newtonian flow $n = 1$ and for dilatancy $n > 1$. The entire flow curve for pseudoplastic behavior can often be fitted using the Reiner–Philippoff equation:

$$\tau = -\dot{\gamma}\left(\eta_\infty + \frac{\eta_0 - \eta_\infty}{1 + \tau^2/A}\right) \tag{2.38}$$

where A is an adjustable parameter, η_0 is the viscosity at $\gamma \to 0$ and η_∞ is the viscosity at $\gamma \to \infty$.

In Bingham fluids mentioned above, a threshold shear stress is required for the shear rate to deviate from zero. Subsequently, the Bingham plastic fluids can be described with the following expression:

$$\tau = \tau_0 + \eta_p \dot{\gamma} \tag{2.39}$$

where τ_0 is the yield stress, η_p is the Bingham plastic viscosity, τ and $\dot{\gamma}$ are shear stress and shear rate.

The Herschel–Bulkley model of the power law type is another model with the addition of a yield stress:

$$\tau = \tau_0 + k(\dot{\gamma})^n \tag{2.40}$$

where k is the consistency constant and n is the flow index. In this model, viscosity is expressed as

$$\eta = \frac{\tau_0}{|\dot{\gamma}|} + k|\dot{\gamma}|^{n-1} \tag{2.41}$$

The Roberston–Stiff model is another option to describe the shear stress dependence on the shear rate with a constant C

$$\tau = k(C + \dot{\gamma})^n \tag{2.42}$$

While the value of the yield stress related to a structure formed by hard solid components does not depend on matrix viscosity, for concentrated suspensions the yield stress depends on solid phase concentration φ, nature of solid particles, their size and surface properties.

For rather diluted cases of low particle concentration (<ca. 0.05 or even lower), viscosity can be described by the Einstein relation

$$\eta_r = 1 + 2.5\varphi \tag{2.43}$$

At higher concentrations (<0.1), the hydrodynamic interactions between the particles should be considered, resulting in the expression proposed by Batchelor

$$\eta_r = 1 + 2.5\varphi + 6.2\varphi^2 \tag{2.44}$$

At even higher concentrations, multiparticle interactions should be considered and prediction of viscosity is often done using various empirical and semiempirical expressions. The data for relative viscosity appear to follow more of an exponential function of concentration according to the Mooney equation

$$\eta_r = e^{\frac{\xi\phi}{1-k\phi}} \tag{2.45}$$

where φ is the solid fraction, ζ is a coefficient determined by fitting and constant k is the self-crowding factor, being in the range between 1.35 and 1.91. The self-crowding factor is inversely proportional to the maximum concentration. The latter has a limit because of the exponential character of the curve. The self-crowding factor at maximum concentration does not allow the movement of neighboring particles as the suspension acts like a solid.

Krieger–Dougherty's semiempirical equation describes the concentration dependence of viscosity as follows:

$$\eta_r = \left(1 - \frac{\varphi}{\varphi_{\max}}\right)^{-K\phi_{\max}} \tag{2.46}$$

where $\varphi_{\max}$ is the maximum packing fraction or the volume fraction at which the zero shear viscosity diverges. This parameter is often taken as 0.64 corresponding to the value associated with random close packing. Parameter K in eq. (2.46) is equal to 2.5 for spherical particles while deviation from this value is related to the shape of the particles. Equation (2.46) reduces to the Einstein expression (eq. (2.43)) at low concentration.

An alternative approach of Brodnyan–Mooney was developed for ellipsoidal particles

$$\eta_r = \eta_o \exp\left[\frac{2.5\phi + 0.399(p-1)^{1.48}\phi}{1-k\phi}\right] \tag{2.47}$$

In eq. (2.47), k is the self-crowding factor.

The treatment above considered suspensions of hard spheres without colloidal or thermodynamic interactions. Repulsive surface forces stabilize dispersions and prevent aggregation, as the particles are kept away at a certain distance. Thus, electrostatic or steric repulsion results in an excluded volume inaccessible to other particles, which can be expressed as follows:

$$\phi_{\text{eff}} = \phi\left(\frac{a_{\text{eff}}}{a}\right)^3 \tag{2.48}$$

In eq. (2.48), a_{eff} is the effective particle radius defined as half the distance to which centers of two particles can approach each other under colloidal forces. Rheology of colloids, in many cases, is similar to hard sphere dispersions and is quantitatively described as a hard sphere system with the concept of effective radius. Equation (2.48) implies that the particle size is a parameter that influences φ_{eff}. An increase in the particle radius a at constant φ and a constant range of the repulsive colloidal interactions correspond to a decrease in φ_{eff}. This means that dispersions with the same φ but different a may exist in different phases having a strong impact on the shear rate-dependent viscosity.

As mentioned above, not only the size but also the particle size distribution can influence the phase behavior of colloidal dispersions. A significant viscosity decrease is typically observed by mixing particles with different sizes when the particle volume fraction φ exceeds 0.5. This effect is more pronounced with increasing φ and the ratio between large and small particles χ. When repulsive colloidal interactions become relevant, the viscosity passes through a minimum, increasing again when χ is increased at a constant total particle concentration and a fixed fraction of small particles.

Attractive particle interactions can lead to large more dense aggregates and rapid phase separation. Alternatively, loose aggregates attract water and increase the effective volume fraction φ_{eff}. This results in an increase in the zero shear viscosity. Upon increase in the shear rate, there is a gradual flocs breakdown and/or their alignment in the flow direction, and a decrease in viscosity.

Some equations have been developed specifically for emulsions as the equation of Taylor

$$\eta = \eta_c \left[1 + \left(\frac{\eta_c + 2.5\eta_d}{\eta_c + \eta_d} \right) \phi_d \right] \tag{2.49}$$

where η_c and η_d are the viscosities of the continuous phase and dispersed phase, respectively, mPa s; ϕ_d is the volume fraction of the dispersed phase.

Many suspensions, and also some polymer solutions, change in time (Figure 2.57a) because structures can be formed or broken by shearing. Thixotropic liquids (coating formulations and concentrated dispersions in the two-phase region) typically display a hysteresis loop when the shear stress is measured as a function of increasing and decreasing shear rate (Figure 2.57b). In anti-thixotropic fluids, shear stress is higher when shear rate is decreasing compared to the increasing branch.

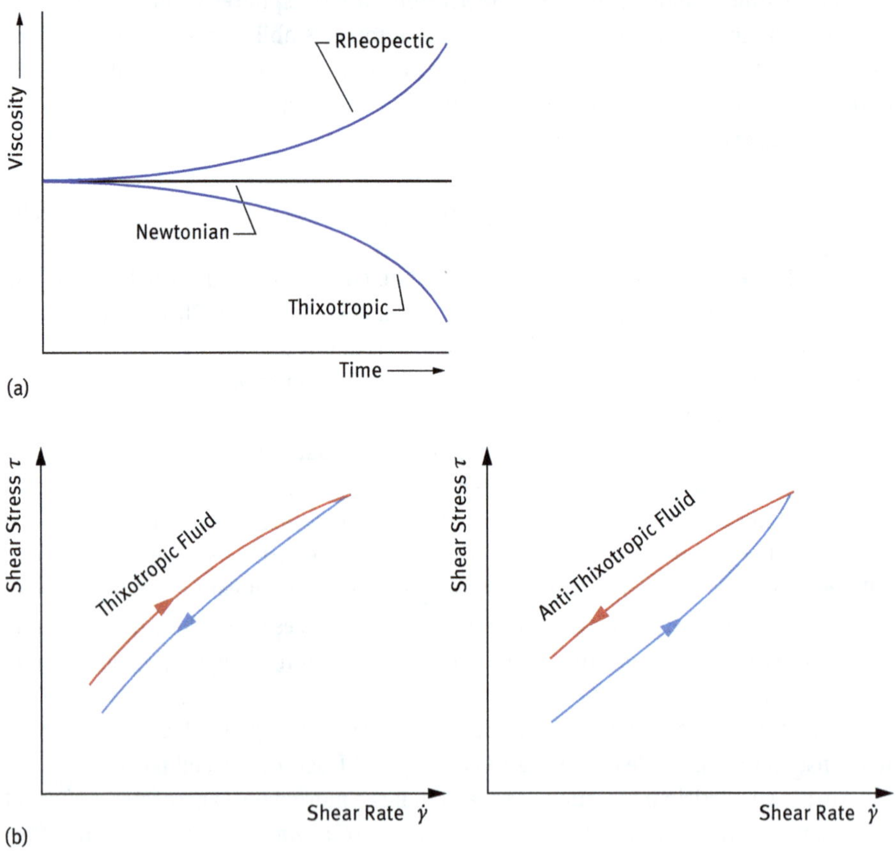

Figure 2.57: Time-dependent fluids: (**a**) dependence of viscosity on time [40] and (**b**) flow curve of thixotropic and anti-thixotropic materials [41].

Thixotropy is a time-dependent pseudoplastic (shear-thinning) property. Gels, thick or viscous liquids, upon application of shear (shaking, agitation, etc.) becomes less viscous (Figure 2.58).

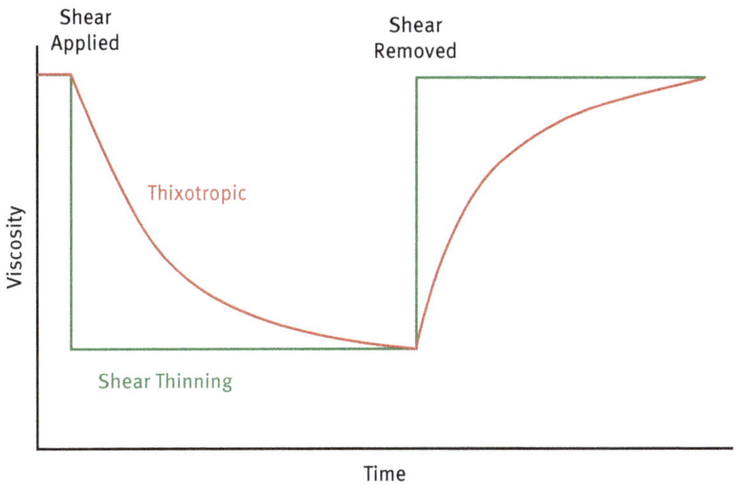

Figure 2.58: Time-dependent thixotropic behavior [42].

Some time is needed for the interparticle or intermolecular alignments to be altered after application of shear, as follows from the right part of Figure 2.58. After a fixed time, a more viscous state is reached. Longer periods of shear stress give lower viscosity.

An example of a pseudoplastic and thixotropic material is lipstick, which must flow during and immediately after application for a smooth finish. Thereafter, it quickly regains a high semi-solid viscosity; otherwise, the result might not be satisfactory (Figure 2.59).

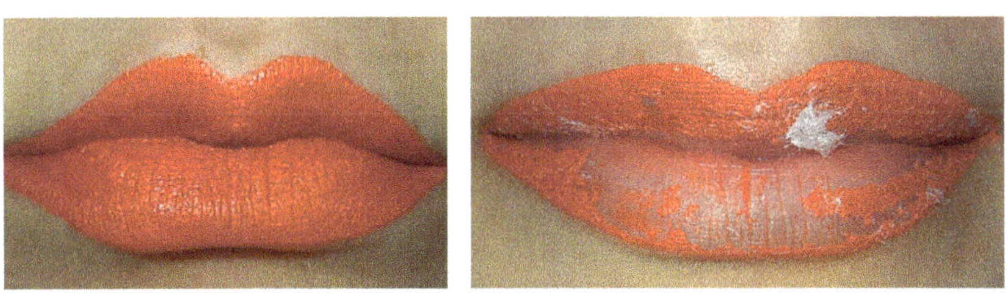

Figure 2.59: Comparison of different quality lipsticks [43].

Another example is paints, which must be plastic and thixotropic. They flow when brushed on and only immediately after brushing for smooth finishing. Shortly after brushing on, the paint should stop to flow.

In rheopectic liquids, which are not common, viscosity increases with time and after application of shear, by, for instance shaking, such rheopectic fluids (some lubricants, gypsum pastes and printer inks) thicken or solidify.

An overview of rheological behavior for several chemical systems is given in Table 2.7.

Table 2.7: Overview of rheological behavior.

Behavior	Specific features	Examples
Pseudoplastic or shear thinning: decrease of viscosity with shear rate increase	In paints, irregular particles can align, matching the induced flow and lowering viscosity.	Paints (pigment suspensions in liquids), polymer solutions
Dilatant or shear thickening: increase of viscosity with shear rate increase	Dense packing of particles and drying occur when walking on wet beach sand. Low shear results in the particles moving past each other. Under high shear, the particles wedge together and the fluid cannot fill the increased void volume.	Wet beach sand, starch suspensions and PVC plastisols
Pseudoplastic with yield stress (plastic): pseudoplastic or Newtonian flow begins only after passing the threshold shear stress (the yield stress)	In these materials, the interparticle network makes them resistant to positional changes. Only after overcoming these forces, flow can occur.	Toothpaste, lipstick, grease, oil-well drilling mud
Thixotropic: time-dependent pseudoplastic flow. At constant applied shear rate, viscosity decreases. Hysteresis in the flow curve	Time-dependent aligning to match the induced flow. After the shear rate is diminished, some time is needed for restoration of the original alignments.	Paint, quicksand
Rheopectic: time-dependent dilatant flow. At constant applied shear rate, viscosity increases. Hysteresis in the flow curve	Slow setting on standing. Fast setting after gentle agitation because of time-dependent particle interference under flow.	Clay suspensions
Rheomalaxic: time-dependent behavior with irreversible changes in viscosity after application of shear	Inversion to a higher or lower viscosity emulsion after shear without re-inversion after shear removal.	Emulsions which invert irreversibly when sheared

In various complex materials, rheological properties are not sufficient for intended applications. Cosmetic and personal care products need ease of product dispensing from its container. Moreover, they spread in a facile way and provide a good feeling when deposited on the skin. Good coating and lack of sag with even coating are needed for paints. Injection of a pharmaceutical compound through a small hole in a needle should be easy. There are also some requirements for agrochemical formulations regarding rheology, that is, flow from a large container with minimum agitation. Dispersion on dilution should also be easy when dealing with agrochemicals.

Such requirements can be achieved with application of rheology modifiers. These modifiers can be thickeners consisting of high-molecular-weight polymers. Polymers can also be hydrophobically modified by grafting of alkyl chains on the hydrophilic backbone of the polymer or cross-linked, making chemical gels. Gels produced because of repulsive interactions or attraction of finely divided particles can be used as rheology modifiers. Liquid crystalline structures of surfactants with hexagonal, cubic or lamellar phases can also be used for the same purpose.

2.7 Unit operations for specific chemical products

2.7.1 Extrusion

Extrusion is the most widely used technique of processing several types of materials such as polymers and plastics, ceramics, food products (pasta, breakfast cereals, ready-to-eat snacks, etc.), catalysts, some drug carriers or biomass briquettes. Behavior of materials during extrusion is mainly determined by rheology and can be tuned by modifying rheological properties. In what follows, two different types of materials will be considered, namely polymers, which are extruded as melts and catalysts, when extrusion is done with concentrated suspensions.

2.7.1.1 Extrusion of polymers

This is one of the methods for compounding polymers. In general, compounding is done when there is a need to alter properties of polymers or prevent degradation through introducing appropriate additives (antioxidants, UV and heat stabilizers, lubricants, pigments, dyes and flame retardants). Polymers can only be processed in the rubber state or when molten. Therefore, polymer extrusion is done by pushing a polymer melt across a metal die, which continuously shapes the melt into the desired form. A typical extrusion apparatus is shown in Figure 2.60. This figure illustrates that a rotating (an extrusion) screw conveys the polymer fed from hopper to the die. The polymer pellets, powder or flakes from the hopper fall through a feed throat (a hole) onto the extrusion screw placed inside the extrusion barrel. The screw pushes the polymer forward into a heated region of the barrel where the polymer melts because of external and frictional heating. The molten polymer moves forward until exiting through the die. The extrudate is immediately cooled and solidified typically in a water tank. Typically, feeding is done by gravity; therefore, for a sticky feed, forced feeding might be needed.

An extruder is a continuous pump without back-mixing; therefore, consistent feeding rate from the hopper into the screw is needed to ensure constant composition

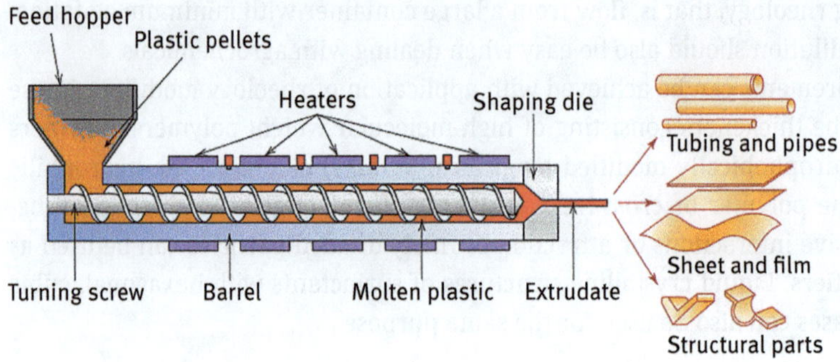

Figure 2.60: A typical extrusion apparatus for processing plastics [44].

and weight of extrudates. Physical and chemical characteristics of the feed (size and shape, and their distribution, solid density, friction on the metal surface) as well as the hopper and the feed throat design determine the feeding rate.

A typical single-screw extruder presented in Figure 2.60 has 20 or more turns with a pitch similar to the diameter, giving a long slender machine. These types of extruders are suitable for continuously processing a wide range of synthetic thermoplastic polymers into finished or semifinished products.

The pressure at the bottom of the hopper is very low, and the feeding rate is usually independent of the amount of feed in the hopper. Feeding is done under air, which is squeezed out of the feed as the polymer is pushed along the screw undergoing compaction. Initially, the feed can contain 30–70 vol% air on dry basis. Air flowing out of the screw through the feed throat back into the hopper acts in an opposite way to feeding by gravity. Another force opposing to gravity is the centrifugal force exerted by screw rotation.

Good feeding characteristics are obtained when the pellet size is small in comparison to the screw channel area; the feedstock has high bulk density and melting point and provides low friction (internal between the pellets and external on the hopper surface).

Compaction of the polymers, which are rigid at the feed temperature, happens along their travel in the screw. The pellets stay loose over the first 2–4 L/D (length-to-diameter ratio) of the screw and are quickly compacted over the following 2–3 L/D. Clearly, in the first several L/D of a screw, sticking of polymer pellets on the screw surface should be avoided. After 3–5 L/D from the hopper because of pellet compaction, some sort of a solid bed is formed, developing pressure. The solid bed after full compaction (5–7 L/D) is so strong that it cannot be easily compressed or sheared even if the pellets are not fused together.

Melting of the solid bed results in the formation of a thin melt film on the barrel surface. The screw operates above the polymer melting point because of heat conduction. Exception is the first several L/D of the screw, which in a steady-state operation is continuously cooled by a cold polymer feed.

High conveying rates of the solid bed are realized when several conditions are fulfilled. They include, but are not limited to, a large screw channel area and high rubbing force on the barrel. The barrel temperature and screw temperature should be significantly higher than the polymer melting point. Other requirements are highly polished screw surface, low friction coating on the screw surface and a low screw channel depth-to-width ratio.

When a melt film is formed on hot screw surfaces surrounding the solid bed, a large amount of heat is generated within the film. The thin melt film is scraped off the barrel surface and collected in a melt pool. Because of minor shearing of the melt film on the screw surface by the slow movement of the solid bed relative to the screw, this film is not scraped off the screw surface.

There is a gradual decrease of the solid bed width with a subsequent increase of the melt pool width as the solid bed melts along the screw. Melting of the solid bed happens primarily by heat conducted through the film on the barrel surface and is completed at ca. $L/D = 15$. Melting mainly through the barrel surface implies that it occurs on the surface of the solid bed, while the interior of the solid bed stays almost at the feed temperature along the screw. High screw speeds result in coexistence of the solid bed and the melt pool over most of the screw length. Completion of melting is required before the end of the screw; therefore, the last several L/D of the screw should not preferably contain any solids to achieve uniform melt temperature. The melt temperature exiting the die is the most important processing variable influencing processability and properties of the product. Polymer flow characteristics through the die are determined to a significant extent in the conveying section of the extruder.

For the cylindrical extrudate geometry, the ratio of the final diameter to that of the die capillary, D/D_0, varies from 1.12 at low shear rates to 2–4 times the extrudate diameter at high shear rates for such non-Newtonian fluids as polymer melts.

The pressure usually increases fast along the feeding section and the compression section, thereafter staying constant or decreasing based, for instance, on the screw and die design, and the operating conditions. Safety precautions against extremely high head pressures are needed because of potential damages if the melt stream is blocked. When the head pressure reaches a pre-set level, the screw stops automatically.

Other types of extruders are available for special purposes. For example, the Buss Ko-Kneader is used for a wide range of polymers, especially in the case of temperature-sensitive materials such as PVC and thermosetting plastics. In this extruder, stationary pins are attached to the barrel interrupting screw flights, thereby generating high shear in the narrow gap between pins and screw flights. This allows very good dispersive mixing and an excellent control over the melt temperature. The Buss kneader parts are illustrated in Figure 2.61.

When special forms such as foams used in thermal insulation or packaging are required, they can be made by introducing gas-generating blowing agents into the polymer melt. Low-boiling solvents can be used, releasing, for instance, nitrogen in case of azo compounds or carbon dioxide when isocyanates are applied. To achieve a

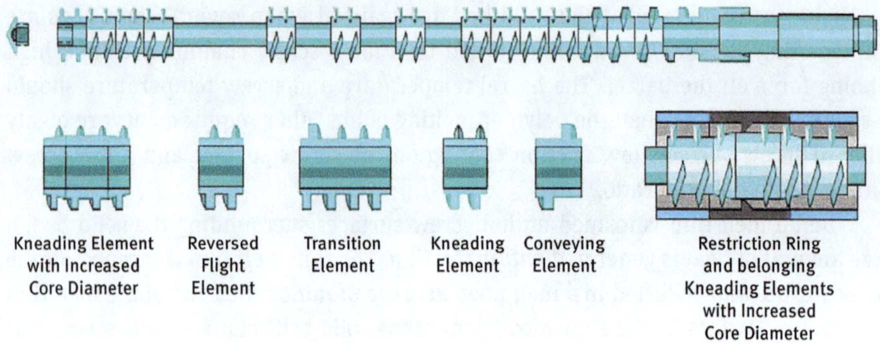

Figure 2.61: Buss kneader elements [45].

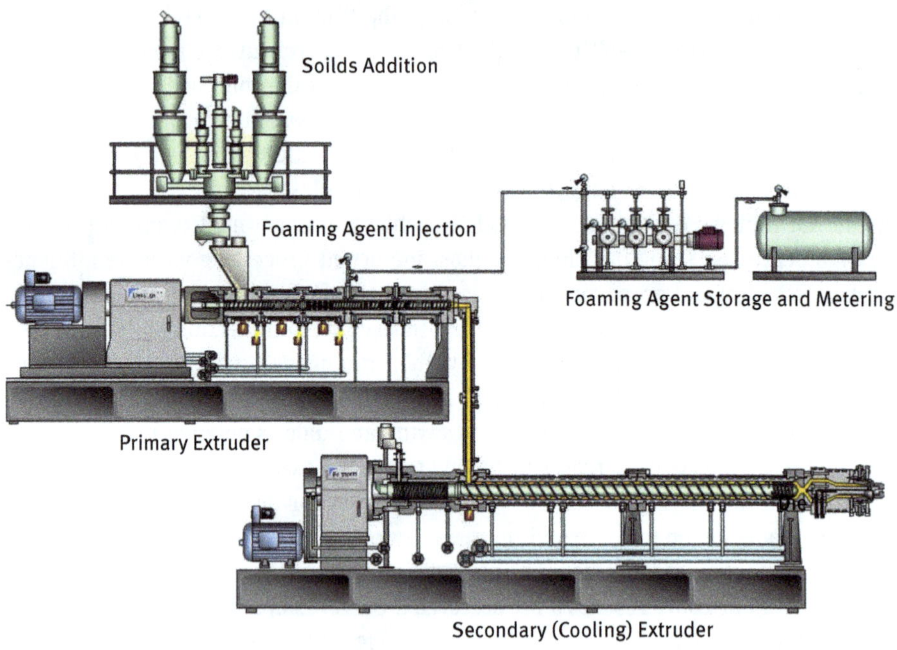

Figure 2.62: Foam extrusion line [46].

correct degree of the foaming agent decomposition, a precise temperature control is needed. Otherwise, there could be a premature blowing agent release. Good mixing and cooling of the melt just prior to the entry of the die is thus needed, which can be organized by a tandem arrangement of two extruders (Figure 2.62), where the blowing agent is completely dissolved in the first extruder, while optimum cooling is arranged in the second extruder by operating it at a slow speed.

In the case of chemical blowing agents, structural foam components can be manufactured using conventional injection molding equipment. An alternative is the addition of gases, which is more complicated from a practical viewpoint. Chemical blowing agents instead allow their simple mixing with the granulate.

2.7.1.2 Extrusion of catalysts

Extrusion is the most economical and commonly applied shaping technique for catalysts and supports. It is different from extrusion of polymers, which are melted. The polymer melt is essentially homogeneous, and the properties can be regulated by extruder temperature. The extrusion of ceramic masses for catalyst preparation requires application of highly concentrated dispersions, whose behavior is determined by rheological characteristics. Moreover, the pore structure and mechanical stability of extrudates are determined by the properties of the paste and extrusion conditions.

The extrusion process is thus influenced by the paste composition as will be specified below. It is also important that no large solid particles are present for the extrusion of complex shapes. The operating variables also include mixing time, aging and extrusion temperature.

While elevation of temperature is positive, improving relative deformations, from the viewpoint of technology, the process becomes much more energy intensive and unnecessarily complex. Moreover, temperature fluctuations can influence mechanical and rheological properties of the suspension, primarily the optimal water content. Therefore, extrusion of highly concentrated suspensions is carried out at ambient temperature.

The paste should be plastic enough to be able to pass through the die and obtain the desired shape. After extrusion, the extrudate should not have any visible or microdefects having adequate properties for further processing (transport, drying, calcination, etc.). The coagulation structure of the paste, on the one hand, should be strong enough ensuring adequate rheological properties and retention of the shape after leaving the die, and on the other hand too strong coagulation structure implies higher energy needed for extrusion.

An important parameter influencing properties of the extrudates is the size of the particles in the suspension, which should be small enough enhancing plasticity. An energy-intensive operation, ball milling, is typically needed to reduce the size of the solid phase.

In the case of catalysts and catalyst support, a certain pore structure should be developed, allowing transport of reactants to the active sites. Typically, extrudates contain large transport pores of 300–600 nm in addition to mesopores (10–25 nm) and are different from materials prepared by tableting, as the latter has mainly monomodal distribution of mesopores.

A need to process concentrated dispersions leading after extrusion to a product with not only certain porosity but also catalytic activity means that only a restricted number of additives or binders can be used that are not detrimental for the required catalytic properties.

During extrusion a wet paste from a hopper at the top is forced through a die and the emerging ribbon passing through holes in the die plate is cut to the desired length using a suitable device (Figure 2.63). The pressure, which is developed in the screw extruder as the paste moves toward the die, is affected by the screw geometry and the paste rheology.

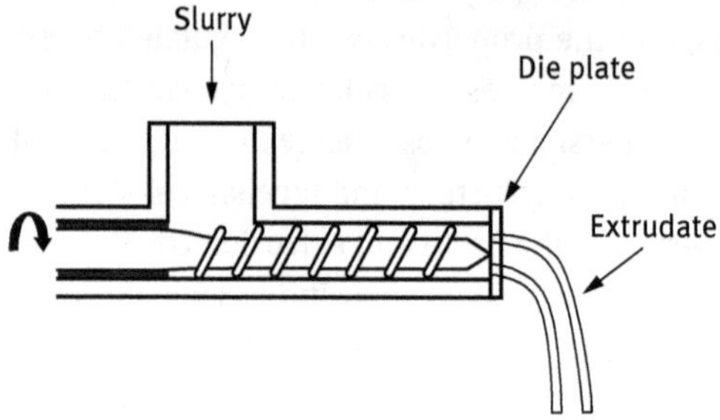

Figure 2.63: Principle of extrusion.

Unlike polymers, which are melted during extrusion, this does not happen with the catalyst pastes. Moreover, usually the catalyst powders obtained after the thermal treatments behave like sand, not possessing the required moldability and plasticity even after water addition. If the viscosity of the pastes is too low, it can result in unstable extrudates. On the contrary, in the case of highly viscous pastes the extruder can be blocked.

In order to improve the flow and rheological properties, various additives are used in the formulation of pastes, including clays and starch for better rheological behavior; binders (e.g., alumina) to keep the active particles together; peptizing agents (diluted acids) for de-agglomeration; combustible porogens for porosity increase (carbon black, starch, etc.) as well as plasticizers, lubricants and water (typically 20 and 40 wt%). It is also important that no large solid particles are present for extrusion of complex shapes.

When the amount of water is much lower than the solid pore volume, it might be difficult to process efficiently such dry pastes. Settling of the pastes after extruder can be on the contrary troublesome for a significant water content (Figure 2.64). The amount of water depends on the surface, lyophilic and adsorption properties of the solid phase and its specific surface area. For example, the optimum amount of water in mass% for an alumina support using 20% nitric acid solution as a peptizing agent is 22–23%, while somewhat higher amount of water (28–29 mass%) is required when polyvinyl alcohol (PVA) is used instead of nitric acid. A catalyst for oxidation of SO_2 to SO_3 based on vanadia on silica when processed with sulfuric acid as a peptizing agent needs as high as 37–38 mass% of water.

Porogenes (carbon black, starch and sawdust) are added to the paste to create porosity, which is generated during calcination. Besides porogenes, plasticizers and lu-

Figure 2.64: Extrusion of too wet paste (courtesy of Dr. Z. Vajglova).

bricants are also burned away during calcination, contributing to porosity generation. Such porosity while beneficial for mass transfer can deteriorate mechanical stability.

Modification of shaping masses, when the particle sizes are in the (sub)micrometer range, from the colloidal viewpoint is controlled by zeta-potential. These systems used in extrusion could be modified, for instance, by addition of peptizing agents (e.g., nitric, formic or acetic acids) influencing colloid–chemical interactions between particles and assuring deviation of pH from the point zero charge. Otherwise, the liquid phase might start to agglomerate. Corrosion behavior of peptizing agents (e.g., nitric acid) should be considered when selecting them. In addition, utilization of nitric acid during extrusion might influence acidity and can give NO_x during calcination. To mitigate these emissions, a selective catalytic reduction of NO_x should be installed, obviously increasing the catalyst production costs. As a consequence, application of electrolytes as peptizing agents acids is justified when other methods to regulate rheological properties of suspension pastes are not successful.

Plasticizers or their mixtures applied to improve the paste's rheological behavior are selected from a wide variety of plasticizers, including clays, starch, sugars, cellulose derivatives, polyethylene glycol and PVA. The amount of the plasticizers should not be too high (up to 3–5 mass%), to allow desired changes in rheological properties. Higher concentration might not be beneficial from the rheological viewpoint.

Selection of rheology improvers should also consider potential adhesion of the suspension paste to the extruder material, which should be minimized.

Formation of shaped catalysts implies cross-linking between particles in the suspension which can be achieved either by condensation of surface hydroxyls, resulting in oxygen bridges, or by sintering. In the absence of suitable surface hydroxyl groups in many catalysts or catalyst precursors, addition of inorganic binders allows interparticle cross-linking. Binders, such as hydroxide forms of alumina, silica-alumina, clays or silica sols, applied together with the catalytic phase result in cross-linking at sufficiently low temperatures, avoiding thereby damage of the catalytically active material. Binders required during extrusion make the extrudates strong enough, preventing a potential collapse (Figure 2.65).

Typically, inorganic binders such as alumina, silica sols and clays are utilized because organic ones will be burned away at the calcination step at 500–600 °C, which also results in the burnout of organic additives like plasticizers. When alumina is added in the form of boehmite or pseudoboehmite, it is transformed into transition alumina during calcination. Gelation of silica sols and agglomeration of delaminated fragments of clays are responsible for stability of extrudates when these binders are used.

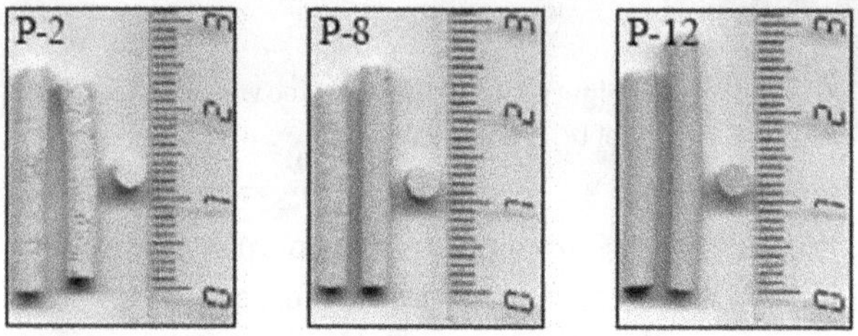

Figure 2.65: Extrudates with increasing amount of a binder. Higher binder content in the series P12 > P8 > P2 results in the absence of cracks and smoother surfaces [47].

Introduction of binders can influence catalytic activity sometimes in a negative way. In such case, a compromise should be made between the rheological properties and mechanical stability, on the one hand, and catalytic behavior on the other hand. For example, introduction of aluminosilicates for extrusion of monolithic blocks based on titania improves the rheological properties of the suspension paste at the expense of the catalytic properties.

A special care should be taken on the surface properties of binders, which can themselves be catalytically active. In production of traditional ceramics, this is a rare case, while in production of catalysts by extrusion, there are several examples when the main active component and the binder influence each other which has a consequence in terms of the final catalyst's physicochemical properties. Figure 2.66 shows temperature-programmed desorption of ammonia from fresh forms of H-ZSM-5 zeolite and the material with aluminum phosphate as an AlPO binder.

There are evident changes in acidity after addition of a binder, resulting in another distribution of sites in terms of acidic strength. Such example of the binder influence on the acidic and thereby catalytic properties can be interpreted by the phosphor migration from the binder to the zeolite crystal structure.

Application of binders can result in nonuniformity of the active component distribution. In implies that there could be zones with a high concentration of the active phase, and consequently, higher rate maybe local overheating and appearance of zones which are controlled by mass transfer rather than kinetics.

Binders can also influence the flow behavior, altering the velocity profile in the paste and thereby deteriorated extrudate properties due to inadequate adhesion during extrusion. The desired product during extrusion (high shear rates) is achieved for pseudoplastic pastes with low viscosity, because after leaving the die at a low shear rate, the viscosity is increased, leading to stable extrudates. Most catalyst pastes are, however,

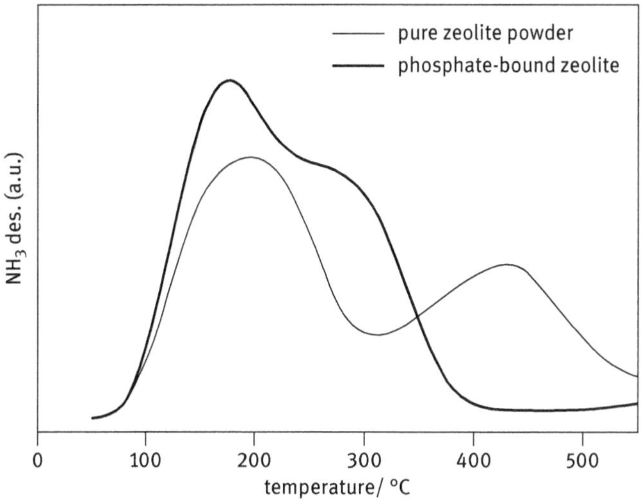

Figure 2.66: Ammonia TPD of H-ZSM-5 powder after calcination and $AlPO_4$-bound extrudate containing the same zeolite (reproduced with permission from [48]).

prone to shear deformation. The flow behavior of molding masses is well described by Ostwald–de Waele equation (2.37). The flow behavior index n in eq. (2.37) indicates pseudoplastic fluids. For simple shapes, this index is at most 0.7, while for preparation of honeycomb monoliths, it is below 0.3.

The value of the flow behavior index is related to the velocity profile. There is a radial profile of flow velocity in the die channel as the flow velocity decreases with an increasing distance from the flow axis. If the velocity gradient is significant, the defects will be generated after a granule leaves a die channel because of the velocity equalization. Subsequently, efficient molding will not be possible. A small flow behavior index corresponds to less significant velocity gradients and an ideal plug flow is achieved when the index approaches zero.

Molding pastes have non-Newtonian flow behavior, and during their movement through dye channels, different velocity profiles are observed (Figure 2.67), determining the tensile strength of shaped catalysts. Obviously, in the case of substantial velocity profiles in the radial direction, produced grains will have macrodefects, originating from improper "layering" of the material.

Figure 2.67: Flow deformation profile during extrusion (reproduced with permission from [49]. Copyright (2016) The American Chemical Society).

An important parameter of suspensions used in catalyst extrusion is viscosity, which can be expressed by the Einstein equation (2.43), which is valid for low concentrations. At higher concentration, the data for relative viscosity appear more to be an exponential function of concentration according to the Mooney equation (2.45).

An important parameter is the material of the extruder as well as of the die. Adhesion of the paste to the die should be minimal.

Special shapes (trilobates, rings, hollow cylinders, monoliths or honeycombs) can be obtained using appropriate dies (Figure 2.68).

Figure 2.68: Shapes of extruded catalysts.

Internal cavities can be formed when, for instance, a two-piece aluminum extrusion die set is used. In a set with separate parts, the piece at the right of Figure 2.69 is used for forming the internal cavity.

Figure 2.69: A two-piece aluminum extrusion die set to form extrudates with an internal cavity [50].

After extrusion, the formed extrudates are either broken due to gravity, giving a non-uniformly sized material, or cut off by, for example, a rotating blade, leading to very similar extrudates.

The quality of the extrudates depends not only on extrusion per se but also on downstream drying and calcination. These steps require special attention in case of larger structures such as extruded monoliths. Such monoliths are, for instance, used in selective catalytic reduction of NO_x at power and waste incineration plants. The active phase V_2O_5 is deposited on titania (anatase) support. To improve the mechanical strength, silicoaluminates or glass fiber are added as mechanical promoters (3.5 wt%). The suspension for extrusion contains clays (ca. 6.5 wt%), water and small amounts of methyl hydroxyethyl cellulose and polyethylene glycol as lubricants. Drying of the extruded monolith must be slow enough to prevent ruptures and cracks. Calcination is performed at ca. 500 ± 600 °C.

Screw extruders with short screw length-to-barrel diameter ratio are used for catalysts requiring less viscous pastes. For more viscous pastes, press extruders with a rotating pressing cylinder are utilized due to shorter distance at which the pressing force is applied. Figure 2.70 shows that this distance is just between two cylinders, the pressing and the screening ones. Because of lower shear deformation, the flow behavior of the paste is less critical than in screw extruders. The latter allow, however, the production of extrudates with more complex shapes.

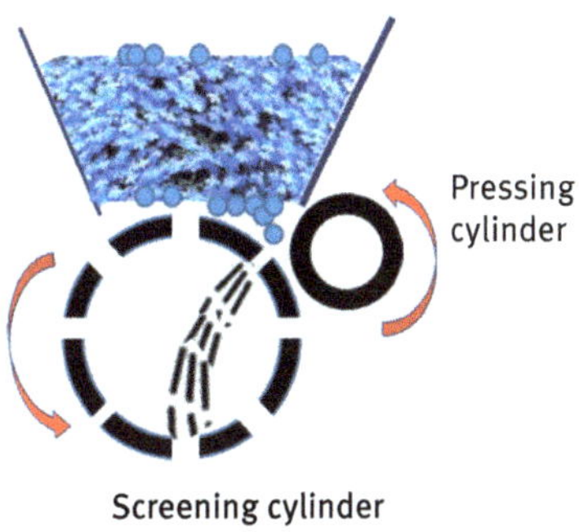

Figure 2.70: Extruder press.

Compared to other preparation methods, extrusion affords high throughput at relatively low costs, giving a variety of possible extrudate shapes. The downside of the method is a nonuniform shape of extrudates and lower abrasion resistance compared to pellets.

2.7.2 Molding

Molding, which will be discussed here, includes injection, compression and blow variants. It has a cyclic nature; thus, the products are made in batches. The specific examples will be related to polymers.

In injection molding, a batch of molten polymer is injected under pressure into a steel hollow mold (Figure 2.71). High viscosity of polymers prevents their direct flow into a mold (cast); thus, a plunger is needed to inject the melt in the mold cavity. Thus, two parts are essential for an injection-molding machine: an injection unit and a clamp unit housing the mold. The function of the former unit is related to melt generation, mixing and pressurization, and pushing the melt to the clamp unit. The latter is needed to hold the mold, shape the product, open and close the mold and finally eject the finished product.

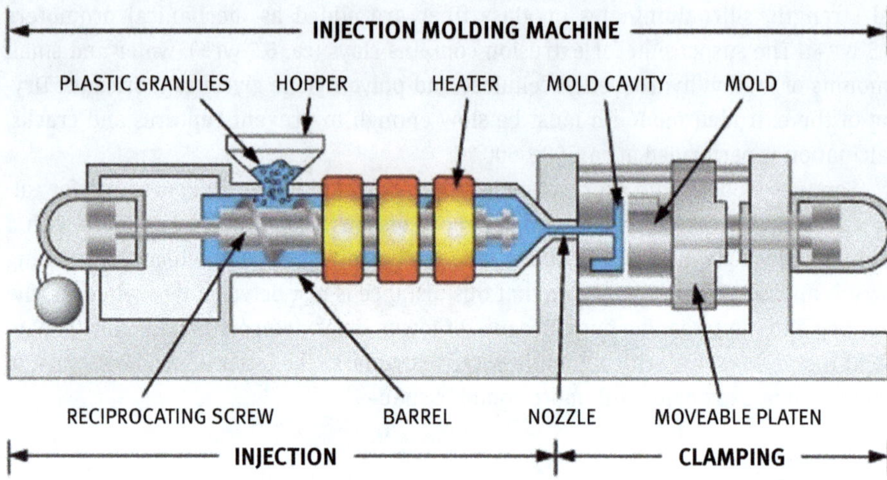

Figure 2.71: Schematic of an injection-molding machine [51].

A reciprocating screw most often used has many similarities to an extruder screw in terms of the sections (feed, compression and metering) being different in a reciprocating (back and forth) action. Such screw is much shorter than an extruder screw, having typical length-to-diameter ratios between 12:1 and 20:1. As a result in injection molding, there is less mechanical action added during melting, requiring therefore more thermal energy.

The polymer after the screw forms a pool at the mold entrance moving, thereafter, being injected into the mold. The injected melt is maintained under pressure during the so-called hold time when additional melt is injected to counterbalance polymer shrinkage during solidification. After solidification and opening the mold, the product is removed either by gravity or by ejector pins. Polishing to a very high gloss of the custom-made molds, adopted to a particular shape, is required; otherwise, every detail of the mold will be visible in the product.

When the melt is cooled in the mold, pressure is maintained during the early stages of cooling to counteract contraction. Thereafter, the pressure is released, the press and the mold are opened and the mold is removed. This is followed by closing the mold again and repeating the cycle.

Because a large part of the overall cycle time is taken by cooling, the rates of cooling are important for the economics of the injection molding.

Injection molding is suitable for mass production and is the most important process for making moldings from thermoplastics, elastomers and thermosets, allowing reproducible manufacture of complex molding geometries in a single stage.

Compression molding is a simple application and economical version of molding, which is mainly applied for molding thermoset polymers and elastomers. This method gives a relatively low amount of produced waste, and can also be used in the cases when the raw material (ultra-high-molecular-mass polyethylene) cannot be processed in extruders. An illustration of the principles of compression molding in a simplified form is provided in Figure 2.72. As can be seen from this figure, two parts (fixed and moving) are present. First, the cold plastic is placed in the cavity. Thereafter, it is melted by heating the mold under slight pressure (compression). When melting is done, the mold is closed and pressure is applied causing the material to flow, cross-link and harden. The next step after hardening is to open the mold and remove the molded product. Similar to injection molding, compression molding has a cyclic character.

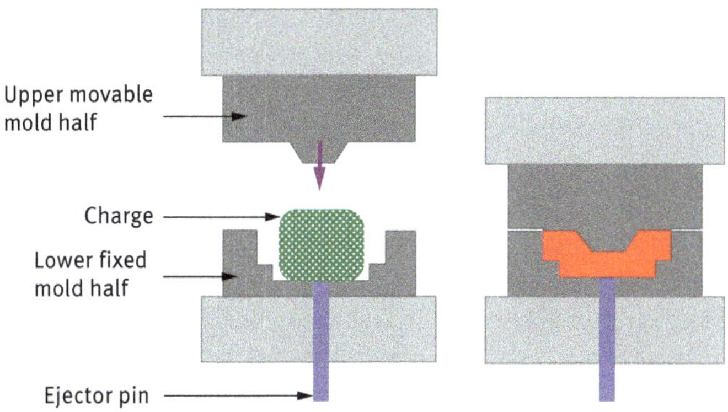

Figure 2.72: Schematic illustration of compression molding with mold open (left) and mold closed (right) [52].

An apparent difference between compression and injection molding is the mold condition. While in injection molding, the molten polymer will freeze after injection to the mold, which is cold, in the case of compression molding, the opposite happens and the mold is heated to allow the so-called thermoset curing. Curing takes longer than freezing; therefore, the molding cycles are longer (1–2 min vs 20–60 s for compression and injection molding, respectively).

Besides longer cycle time, other disadvantages of compression molding are related to geometrical restrictions and inability to reprocess rejected parts.

A natural symbiosis of compression and injection molding is transfer molding in which the charge is melted in a separate pot, which is a part of the heated mold. Thereafter, it is transferred into the mold cavity. This technology allows cycles shorter than in compression.

The most important factors from a processing viewpoint in plastic molding are degree of curing and filler orientation. Variation of temperature with time and geometrical position can lead to nonuniform degree of curing throughout the molding, with some layers fully cured and others still uncured. As a consequence, within the same molding piece, different regions may be found, leading to bubbles and cracks and poor behavior (internal stresses, surface gloss, stiffness and insufficient electrical insulation properties).

When processing thermoset molding compounds, the shear deformations are lower than the tensile deformations; therefore, the filler and reinforcing additives are oriented mainly perpendicular to the flow direction.

Molding shrinkage (or the difference between the mold dimensions and dimensions of the corresponding molding) is influenced significantly by the molding compound per se (content of inorganics and length of fibers), orientation of the filler as well as processing conditions (the mold temperature and pressure). Higher molding temperature, low curing, pressure and inorganic content increase molding shrinkage. A way to diminish or substantially compensate for molding shrinkage is to add suitable fillers, such as thermoplastic powders.

Longer fibers in impact-resistant compounds, local differences in the curing reaction because of mold temperature variations as well as inhomogeneous cooling of the molding result in more prominent distortion. Curing conditions might not be important for the mechanical strength unless there is no strong under or overcuring. However, for best possible insulation characteristics, the molding mass should be perfectly cured after drying in an oven prior to processing.

Two reactive monomers can be polymerized in a closed mold using a reaction injection molding (RIM) process for making polymeric articles. In this way, some of the apparent drawbacks of injection molding, especially with an increased size of molding articles, can be circumvented, namely a need to make a homogenized melt in the injection part and generate sufficient clamping pressure. The latter is required to keep the mold closed during filling and packing stages of the injection molding cycle. A schematic diagram of the RIM process is presented in Figure 2.73.

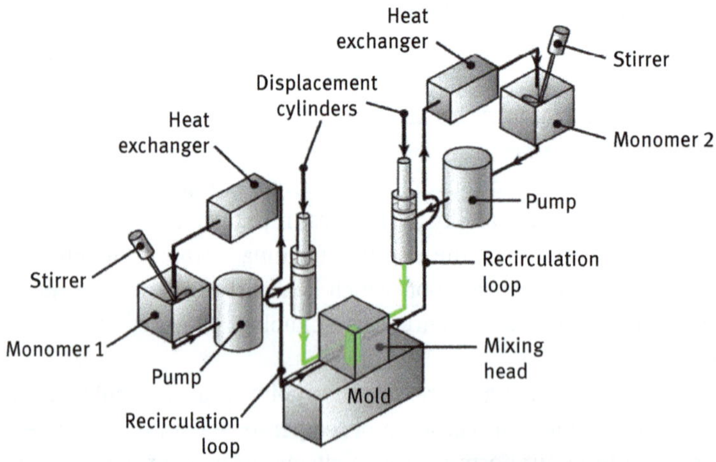

Figure 2.73: Schematic diagram of the reaction injection molding (RIM) process [53].

With this method, for example, polyurethanes can be made from polyols and diisocyanates. A proper selection of monomers and additives (e.g., foaming agents) results in materials with the desired properties ranging from flexible elastomers to rigid foams. Apparently, the rate of polymerization should be high enough and in balance with the cycle times of 2–4 min.

While relatively low injection pressures and an ease of polymer homogenization are clear advantages of RIM, for design and scale-up of this technology, heat generation during polymerization should be accounted for.

Some other polymer processing versions of injection molding are also realized. In resin transfer molding, which is a hybrid of transfer and RIM, reinforcement is incorporated during molding to create a composite. Another related process is structural RIM, where preform and mold preparation are similar to that of RIM, while some changes in mold release and reinforcement sizing are made to optimize their chemical characteristics. After the mold has been closed, the resin is introduced in a fast way into the mold, thereafter also quickly reacting to cure fully in few seconds without a need for postcure. High reactivity of liquids required for structural RIM implies thorough mixing before entering the mold and therefore very rapid, high-pressure impingement mixing.

Blow molding is a version of molding for producing bottles and other hollow objects in a more economical way than injection molding and operates at low inflation pressures and allows to create parts with narrow openings and wide bodies. Combination of high production rates, automatic polymer feeding systems and part removal as well as short cycles improves overall molding economics.

In extrusion blow molding, the resin is used directly from the extruder, which operates continuously, producing parisons at a single crosshead die. The mold moves to the die and closes on a length of parison, sealing one end. After separation of a parison from the die by cutting, the mold moves away from the die, allowing continuation of the next parison extrusions. Upon arriving of the mold at the blowing/cooling station, a blow pin is inserted into the open end of the parison sealing it. Thereafter, compressed air is fed through the blow pin into the parison, inflating it against the inside of the cooled mold surfaces. After sufficient solidification of the product, it is recovered by opening the mold. Figure 2.74 illustrates the sequence of steps in extrusion blow molding.

Continuous extrusion blow molding is arranged by running the extruder continuously. Extruder output is matched by mounting multiple molds, which seal and blow the parison on a rotating wheel. In essence, when one mold is closing to capture the parison, the mold ahead is in position for blowing of the part, while other molds are either closed for cooling or open for ejection of the product and preparing for capturing another parison. Apparently, the rotational speed of the wheel and the number of mounted molds should correspond to the extrusion rate.

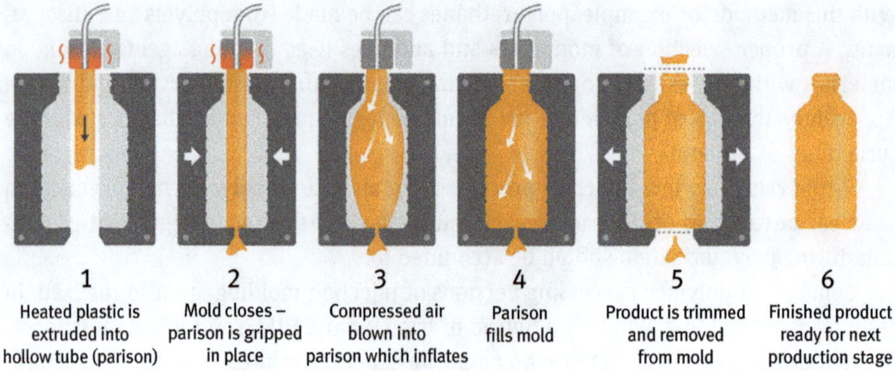

Figure 2.74: Sequence of steps in extrusion blow molding [54].

In injection blow molding used currently for production of polyethylene terephthalate (PET) bottles for carbonated drinks, an injection-molded preform is used instead of a directly extruded parison (Figure 2.75).

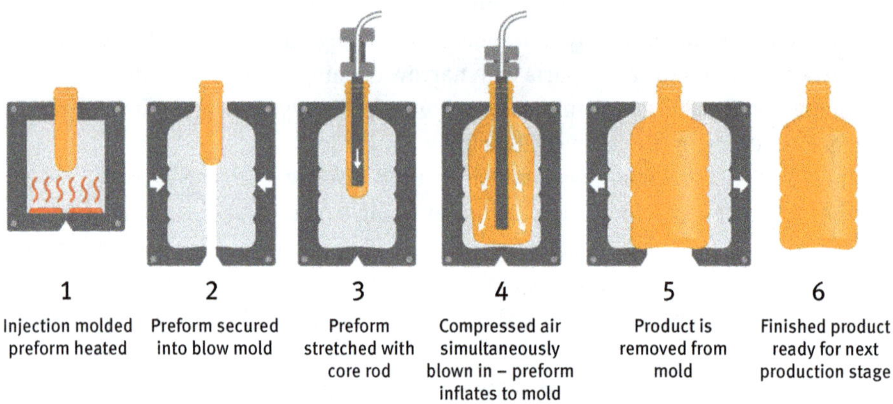

Figure 2.75: Stages in injection blow molding [55].

Injection blow molding consists, first, of extrusion of the tube and quenching it rapidly. Heating the tube to just above its glass transition temperature (90–100 °C) is followed by axial stretching, pushing down the blow pin and simultaneous blowing, resulting in radial expansion. After rapid crystallization and solidification of PET, the bottle is ejected.

2.7.3 Calendering

Continuous manufacture of polymer sheets or films is done using calendars that consist of four horizontal, counter-rotating steel rolls (Figure 2.76).

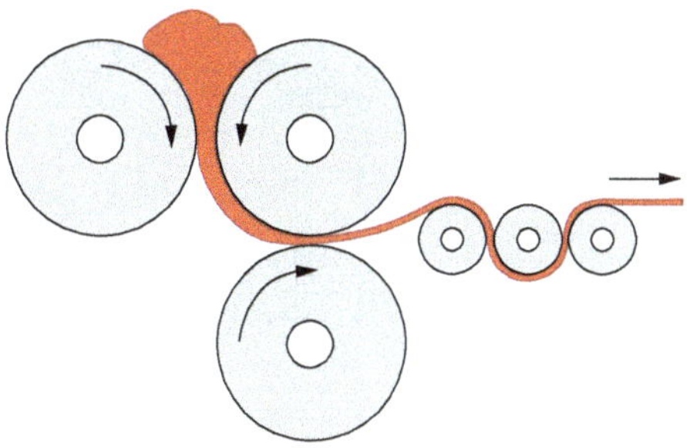

Figure 2.76: Schematic diagram of calendering system [56].

A uniform polymer melt coming from a melt-conveying operation (e.g., extrusion) passes through rolls, with decreasing gaps between subsequent rolls. The temperature of the rolls also gets lower. The thickness of the sheet is diminished along the process, allowing finally to make few millimeter-thick and meter-wide sheets with a proper surface finish applied by the final roller. This operation mode does not require utilization of an expensive and complicated die needed for an alternative method of direct sheet extrusion.

2.7.4 Fiber spinning

Polymer fibers, important, for example, as composite reinforcement and in clothing can be made by several spinning methods, including melt, dry solution and wet solution spinning (Figures 2.77 and 2.78). In melt spinning (Figure 2.77a), a molten polymer is forced through a single die and pressed through a spinneret (multiple orifice plate) consisting of 50–60 holes with diameter 0.12 mm. It is not the hole size which determines the final fiber diameter (denier), but rather pumping and winding rate. This method is applied for a range for materials, including polyester, olefin fibers or nylon.

When viscosity at the spinneret is not between 1,000 and 2,000 poise, but is higher because of too viscous polymer melt, melt spinning requiring a dangerously high temperature cannot be used. Instead, the polymer melt is spun in solution. Dry solution spinning is a method for making among others, cellulose acetate, spandex or polyacrylonitrile fibers. The notion of dry solution, which sounds unusual, implies that first a polymer is dissolved in an appropriate solvent. This is followed by pumping the solution through a spinneret (Figure 2.68b), followed by exposure to air with subsequent solidification because of solvent evaporation.

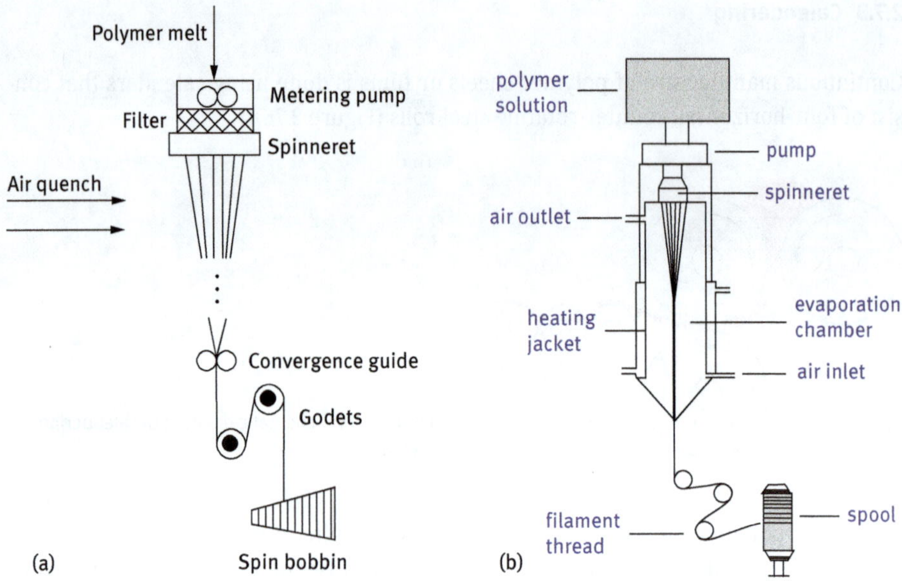

Figure 2.77: Polymer spinning: (**a**) melt [57] and (**b**) dry spinning [58].

In wet solvent spinning, which is otherwise similar to dry solution spinning, the solvent is leached out by another liquid (Figure 2.78a). The presence of a chemical bath, into which the spinneret is submerged, causes fiber precipitation and then solidification, as it emerges. This method is applied for making different fibers, including rayon (cellulosic), aramid and spandex. In dry jet-wet spinning (Figure 2.78b), which is a modification of wet and dry solvent spinning methods, the solution is extruded into air and drawn thereafter is submerged into a liquid bath.

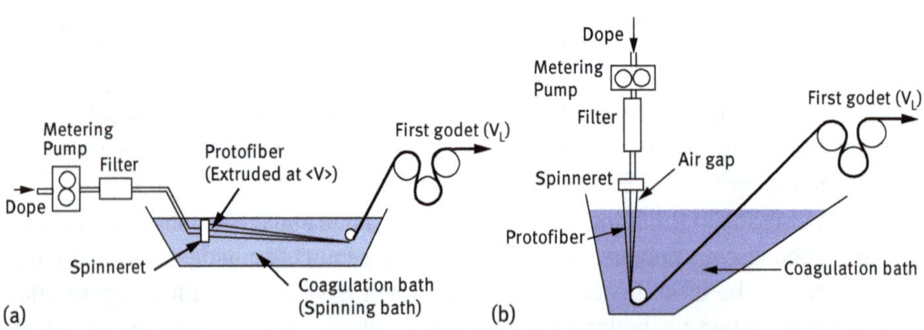

Figure 2.78: (**a**) Wet spinning and (**b**) dry jet-wet spinning [59].

2.7.5 Unit operations in ceramics processing

The casting of ceramics involves pouring of a ceramic slurry material into a mold, followed by formation of a solidified product along with the yield strength development. Different types of transformations can take place in ceramic casting, including physical, thermal and chemical transformations, or their combinations.

In slip casting, applied for objects with hollow centers, an aqueous slurry of fine clay (a slip) is used to make casting with a porous gypsum mold. The slurry particles coagulate near the mold surface, thereby forming the cast. The process rate can be increased by applying pressure, vacuum or centrifugal force. In drain casting (Figure 2.79), the molds are first filled with the slurry. After initial solidification, the excess slurry is drained from the mold.

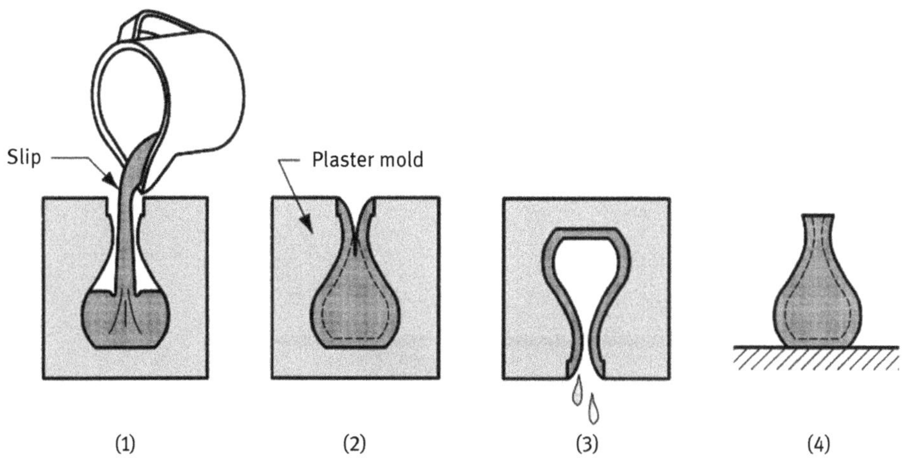

Figure 2.79: Schematic illustration of the drain slip-casting process [44]. The sequence of steps are (1) slip is poured into mold cavity, (2) water is absorbed into plaster mold to form a firm layer, (3) excess slip is poured out and (4) part is removed from mold and trimmed.

In solid casting, the whole amount of slurry in the mold is used and the entire slip remaining in the mold must be dried prior to the cast removal. After removal of the cast, it is subjected to post-casting treatment such as drying.

Slurry preparation, mold filling, draining and partial drying while in the mold are important steps in solid casting. Slip-casting slurries contain typically milled powders or granules of alumina (40–50 vol%), water (50–60 vol%) and some processing additives (binders, ammonium citrate as well as sodium citrate and sodium carboxymethylcellulose). Low viscosity is required if, for instance, screening of the slurry prior to filling of the mold is done. Minimization of particle settling or deflocculation is ensured by introducing deflocculants in the slip slurry.

An alternative to slip casting is tape casting when a slurry, containing besides solvents (water), deflocculated powders or granules (maximum particle size of 1–5 μm and specific surface area of 2–5 m^2/g), binders and/or plasticizers, can be processed into thin and flexible sheets (Figure 2.80). Wetting agents in a slurry are used to promote spreading of the slurry on the substrate while homogenizers are introduced to enhance surface quality.

Casting machines have the following characteristics: length up to 25 m, several meters in width, speed up to 1,500 mm/min, Teflon or cellulose acetate as carrier material and tape thickness in the range of 25–150 μm. Height of the blade, carrier speed and the drying shrinkage have a direct influence on the tape thickness.

The dried tape, being flexible due to a high binder content, has a very smooth surface. As a result, electronic materials can be printed on the tape surface, being integrated in the ceramic base after firing. Examples of tape-cast ceramics utilized for electronic materials include photovoltaic cells, sensors, electronic conductors, resistors and capacitors, as well as solid electrolytes for batteries.

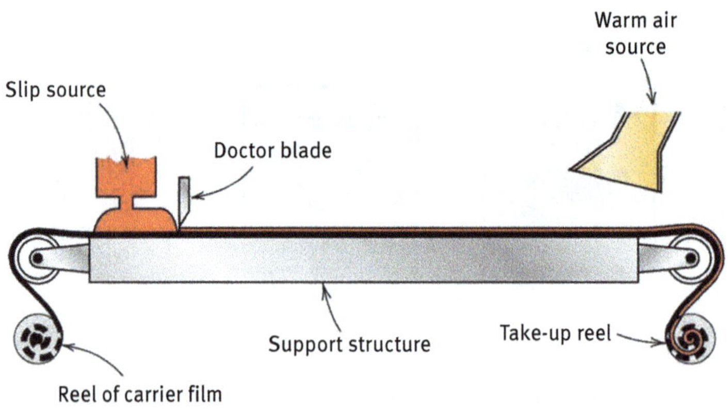

Figure 2.80: Schematic diagram of a typical tape-casting process [44].

Chemical homogeneity and particle size are critical parameters for fabrication of electronic components ensuring smoother tape surface and dimensional stability, respectively, during drying typically under air. As drying proceeds, a viscoelastic tape becomes more elastic. The drying rate varies with the solvent concentration on the surface and temperature. Short drying times can be achieved by using special solvent blends, which dissolve additives, allowing adsorption of the latter on ceramics. These blends should have low viscosity, boiling point and a low heat of evaporation. With aqueous systems, longer drying times and higher drying temperatures are required. Shrinkage during drying occurs primarily along the dimension of the tape thickness, resulting in the final thickness of ca. 50% the initial blade height.

Centrifugal casting is used for forming advanced ceramics of high cast density, making casting parts possessing axial symmetry. In this method, the molted materials (steel or sand) are poured without a gating system into a cylindrical mold spinning about its axis of symmetry. Centrifugal casting technology is widely used for manufacturing of iron pipes, bushings, wheels and other objects with axial symmetry. Figure 2.81 illustrates a centrifugal casting machine with metal as a casting material.

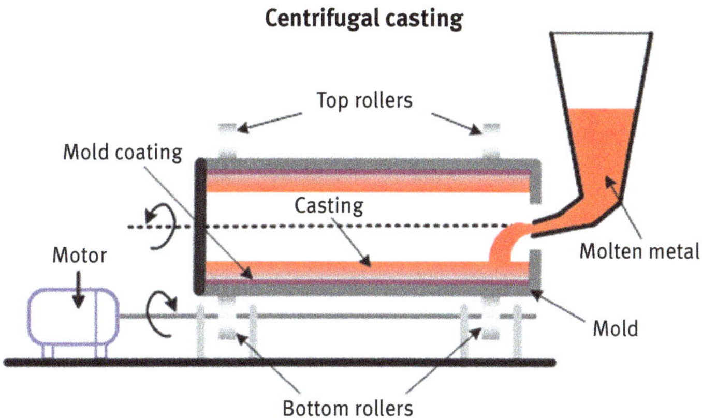

Figure 2.81: Centrifugal casting [60].

During centrifuging the liquid is displaced to the top of a centrifuge cell and the mold is kept rotating until solidification of the cast. The rotation speed can be varied in a broad range influencing the casting rate.

Firing involves heat treatment of the green ceramic in a high-temperature furnace, known as a kiln (Figure 2.82), to develop the desired microstructure and properties. Green body is a term used to describe an object whose main constituent is a weakly bound ceramic material not being related to a particular color. Firing or sintering in a furnace is needed to produce strong, dense objects, as due to extremely high melting temperature and viscosity of ceramics, their casting is difficult. Kiln design varies depending upon the ceramic to be fired, allowing to operate in either an intermittent or continuous fashion.

Temperatures in excess of 1,700 °C can be achieved by combusting natural gas. Processes during firing may include drying, decomposition (thermolysis) of organic additives (waxes, polyethylene glycol) as well as carbonates and sulfates (calcination), prior to densification on sintering. Thermolysis of binders depends on the nature of the binder, composition and flow rate of the gas in a kiln.

A high sintered density with a minimal grain growth is achieved for densely packed and homogeneously distributed fine particles present in the material for sintering. Grain growth inhibitors such as magnesium oxide are added to inhibit exaggerated grain growth.

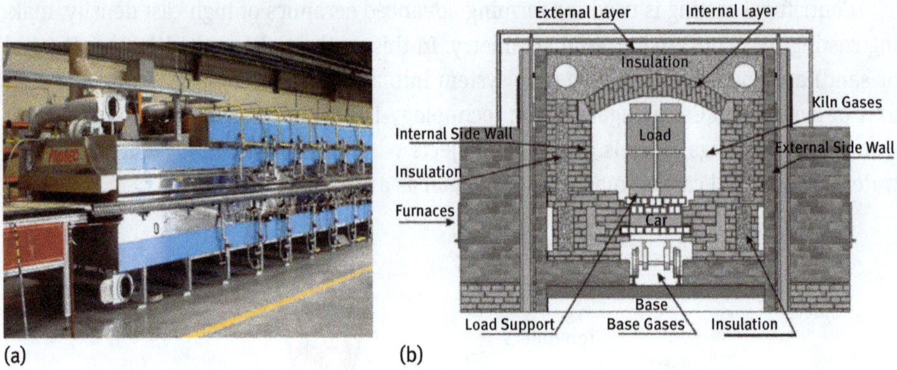

Figure 2.82: Kiln: (**a**) visual appearance of a continuous kiln [61] and (**b**) schematics of a continuous kiln [62].

2.8 Filtration

In the case of cake filtration (Figure 2.83), a relatively thin filter medium is used and separation is achieved upstream of the medium.

The particles should be either larger than the pores of the medium or should form bridges to cover the pores. A cake is formed from successive layers of solid deposits, apparently increasing the pressure drop for the case of dead-end filtration (Figure 2.83a). In cross-flow filtration (Figure 2.83b), the slurry moves tangentially to the filter medium, continuously shearing off the cake. Another option to the so-called deep bed filtration (Figure 2.83c) with the filter medium is a deep bed with pore sizes much larger than the particles to be removed.

An increase of the cake height during deposition of solids apparently results in an increase of the total cake resistance. In an incompressible cake with a linear increase of resistance with height, the filter cake porosity remains constant while the cake volume increases. Filtering at constant pressure as implemented in vacuum filtration results in decline of filtration rates as the filter cake grows in size. Too thick filter cakes lead to prolonged filter cycles because of low filtering, dewatering and washing rates. Steps in a filtering cycle include, besides filtration per se, dewatering, washing, filter cake discharge, cleaning, reassembly and filling of the filter.

In continuous large-scale vacuum filters, the suspension is introduced to the filter at atmospheric pressure. Vacuum applied on the filtrate side of the medium creates the driving force for filtration. The rotary vacuum drum filter is illustrated in Figure 2.84.

As the drum rotates, being partially submerged in the slurry, solids trapped on the drum surface are washed and dried. The cake discharge occurs at the end of the rotational cycle. The drum surface covered with a cloth filter medium can be precoated with a filter aid to improve filtration and increase cake permeability.

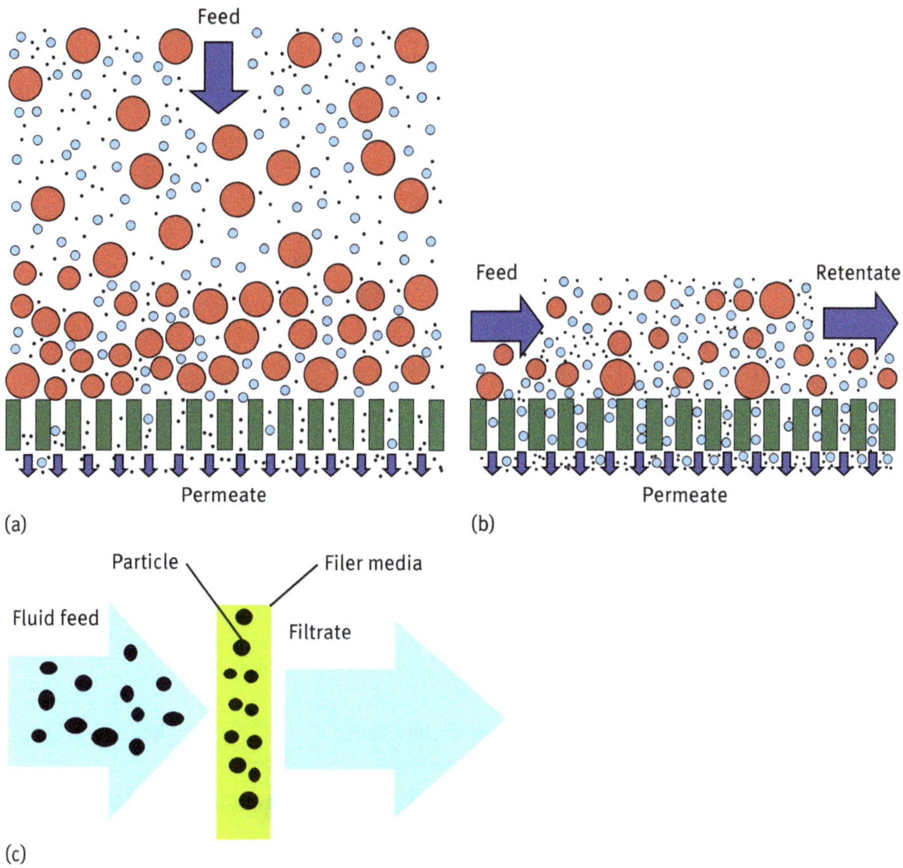

Figure 2.83: Schematics of filtration: (**a**) direct dead-end cake filtration (reproduced with permission from [63]), (**b**) cross-flow filtration (reproduced with permission from [63]) and (**c**) deep filtration [64].

Horizontal filters, such as the horizontal belt filter, presented in Figure 2.85 allow settling by gravity before the vacuum is applied. Horizontal belt filters with a simple design and low maintenance costs have difficulties in handling very fast filtering materials on a large scale.

The filters described above operate in a continuous mode. Nutsche filters, which are basically vessels divided into two compartments with vacuum applied to the lower compartment, can operate batch-wise. This might be necessary in case of stringent washing requirements or if there is a need to keep batches separated. The Nutsche (Figure 2.86) filters can handle batches of 25 m^3 and a cake volume of 10 m^3 and are thus able to work with an entire charge of slurry. Sufficient holding volume is required for fast charging and emptying of the vessel. The difficulties of operation with such filters arise when cakes are slow to form, sticky and the product deteriorates during long downtime. The operational sequence starts with filtration per se

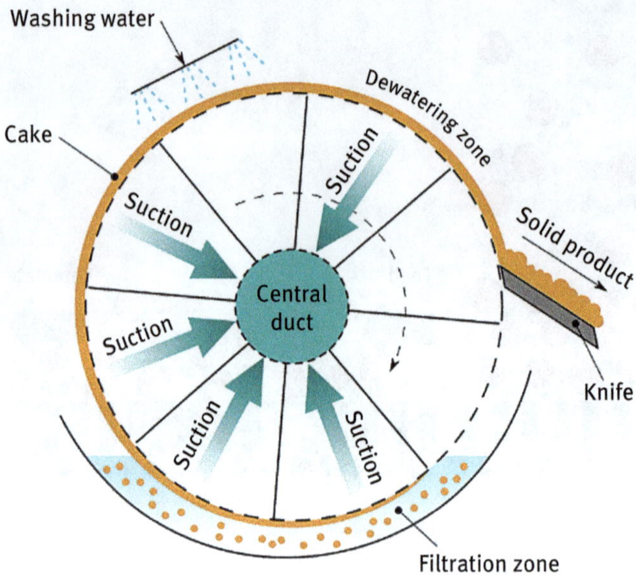

Figure 2.84: Schematic of a rotating vacuum filter [65].

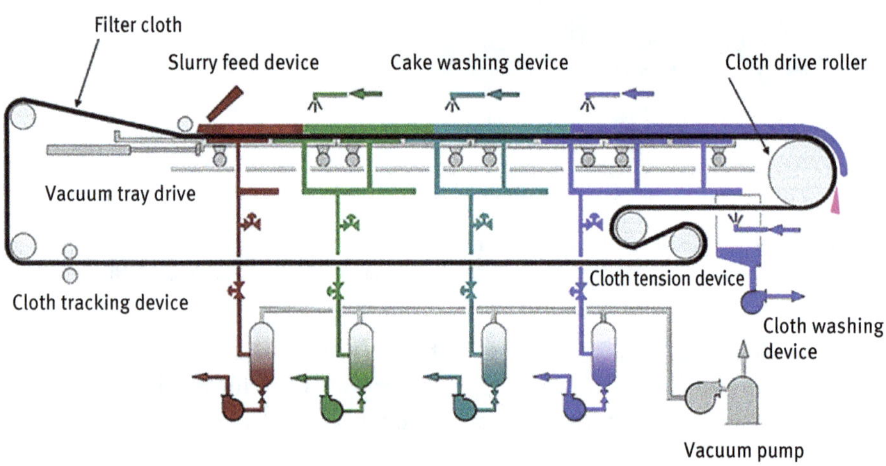

Figure 2.85: Schematic of a horizontal belt filter [66].

when the filter is charged with slurry and the pressure is applied. In the washing stage, the wash liquid is introduced over the cake, displacing the mother liquor. In the drying stage, air or gas purges the cake until the desired drying level. The final step is the cake discharge, and in some instances washing the cloth or woven mesh screen with water to remove any cake residue.

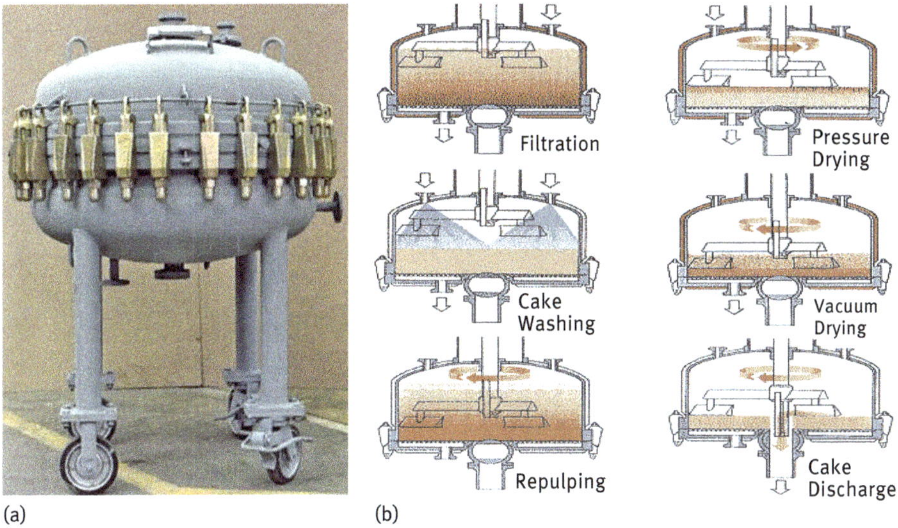

Figure 2.86: Nutsche filter: (**a**) photo [67] and (**b**) schematics of operation [68].

References

[1] https://depts.washington.edu/solgel/documents/class_docs/MSE502/Ch_3_Section_3.2.1–3.2.5.3.pdf, accessed on 23.12.2017.
[2] https://153160-us.all.biz/forced-circulation-crystallizer-g209154, accessed on 23.12.2017.
[3] http://www.thermopedia.com/content/5103/CrustF2.gif, accessed on 23.12.2017.
[4] http://www.thermopedia.com/content/5103/CrustF1.gif, accessed on 23.12.2017.
[5] http://meroli.web.cern.ch/meroli/Lecture_silicon_floatzone_czochralski.html, accessed on 23.12.2017.
[6] http://www.tankonyvtar.hu/hu/tartalom/tamop412A/2011-0016_01_the_theory_and_practise_of_pharmaceutical_technology/images/image385.jpg, accessed on 23.12.2017.
[7] http://bfdfilm.com/ball-mills-introduction-and-types/, accessed on 23.12.2017.
[8] https://www.linkedin.com/pulse/ball-mill-highly-efficient-grinding-milling-machine-evita-lee/, accessed on 11.02.2023.
[9] http://cdn.chemengonline.com/wp-content/uploads/2017/12/12.jpg, accessed on 23.12. 2017.
[10] https://fertilizerproductionlines.com/pan-granulator/
[11] http://ietd.iipnetwork.org/content/high-pressure-roller-press, accessed on 23.12.2017.
[12] http://www.tankonyvtar.hu/hu/tartalom/tamop412A/2011-0016_01_the_theory_and_practise_of_pharmaceutical_technology/ch24.html, accessed on 23.12.2017.
[13] N. Al-Zoubi, S. Gharaibeh, A. Aljaberi, I. Nikolakakis, Spray drying for direct compression of pharmaceuticals, Processes, 2021, 9, 267.
[14] https://www.eurotherm.com/spray-drying, accessed on 23.12.2017.
[15] http://ru6img.allhaving.com/upload/3633/o/13_1_fluidized-bed-spray-dryer_05.gif, accessed on 23.12.2017.
[16] http://www.katsuragi.co.jp/en/products/01/drumdryer.html, accessed on 23.12.2017.
[17] http://www.ft-dryer.com/products/Rotary-drum-dryer.html, accessed on 23.12.2017.

[18] https://electricalworkbook.com/freeze-dryer/, accessed on 12.02.2023.

[19] https://www.gea.com/en/products/fluid-bed-dryer.jsp, accessed on 23.12.2017.

[20] https://en.wikipedia.org/wiki/File:Colloidal_Stability.png, accessed on 23.12.2017.

[21] https://upload.wikimedia.org/wikipedia/commons/thumb/0/05/Diagram_of_zeta_potential_and_slipping_planeV2.svg/600px-Diagram_of_zeta_potential_and_slipping_planeV2.svg.png, accessed on 23.12.2017.

[22] http://www.colloid.ch/grouppage/pdfs/Overview_DLVO_Theory1.pdf, accessed on 23.12.2017.

[23] https://pocketdentistry.com/wp-content/uploads/285/B9780443101441000052_gr4.jpg, accessed on 23.12.2017.

[24] X. Gong, Z. Wang, T. Ngai, Direct measurements of particle-surface interactions in aqueous solutions with total internal reflection microscopy, Chem. Commun., 2014, 50, 6556.

[25] https://clinicalgate.com/emulsions-and-creams/, accessed on 23.12.2017.

[26] W. Deng, L. Wu, Y. Li, S. Chen, Pickering emulsions stabilized by polysaccharides particles and their applications: A review, Food Sci. Technol., Campinas, 2022, 42, e24722.

[27] T. W. Tandros, Emulsion Formation, Stability and Rheology, in *Emulsion Formation and Stability*, 1st Edition. Ed. T. F. Tadros. Wiley-VCH Verlag GmbH, Weinheim, 2013, pp.1–75.

[28] M. Miyake, Y. Yamashita, Molecular structure and phase behaviour of surfactants, in *Cosmetic Science and Technology: Theoretical Principles and Applications*, Elsevier, Amsterdam, 2017, pp. 389–414, http://dx.doi.org/10.1016/B978-0-12-802005-0.00024-0.

[29] M. H. Mondal, S. Malik, A. Roy, R. Saha, B. Saha, Modernization of surfactant chemistry in the age of Gemini and bio-surfactants: A review, RSC Adv., 2015, 5, 92707.

[30] R. Y. Lochhead, Basic physical science for the formulation of cosmetic products, in *Cosmetic Science and Technology: Theoretical Principles and Applications*, Elsevier, Amsterdam, 2017, pp. 39–76, http://dx.doi.org/10.1016/B978-0-12-802005-0.00003-3.

[31] http://soft-matter.seas.harvard.edu/index.php/File:HLBSchema.png, accessed on 23.12.2017.

[32] http://eng.thesaurus.rusnano.com/upload/iblock/383/micelle1.jpg, accessed on 23.12.2017.

[33] http://slideplayer.com/6214458/20/images/93/Critical+micelle+concentration+%28CMC%29.jpg, accessed on 23.12.2017.

[34] Y. Nakama, Surfactants, in *Cosmetic Science and Technology: Theoretical Principles and Applications*, Elsevier, Amsterdam, 2017, pp.231–244, http://dx.doi.org/10.1016/B978-0-12-802005-0.00015-X.

[35] http://www.nuroil.com/bitumen-emulsions.aspx, accessed on 23.12.2017.

[36] http://people.umass.edu/mcclemen/FoodEmulsions192008/Presentations(PDF)/(5) Emulsion_Formation.pdf, accessed on 23.12.2017.

[37] http://sylvatex.com/wp-content/uploads/renewables-micro-emulsion-spash-blending.jpg, accessed on 23.12.2017.

[38] http://www.thermopedia.com/content/5408/762NNFFig1.gif, accessed on 23.12.2017.

[39] http://www.rheosense.com/applications/viscosity/newtonian-non-newtonian, accessed on 23.12.2017.

[40] http://www.thermopedia.com/content/5408/763NNFFig2.gif, accessed on 23.12.2017.

[41] http://polymerdatabase.com/polymer%20physics/Viscosity3.html, accessed on 23.12.2017.

[42] http://coatings.mst.edu/media/research/coatings/images/Thixotropy_reduced.jpg, accessed on 23.12.2017.

[43] http://www.ladiesproject.ru/public/stoykaya-matovaya-pomada-me-now-generation-ii-ottenok-no-09, accessed on 23.12.2017.

[44] http://slideplayer.com/slide/4235773/

[45] http://www.intacps.com/products/spare-parts-for-extrusion-machinery/buss-kneader-parts/kneader/, accessed on 23.12.2017.

[46] S. Lee, R. Smith, S. Costeux, J. Alcott, M. Barger, D. Beaudoin, H. Clayton, S. Donati, J. Duffy, R. Fox, D. Frankowski, K. Giza, L. Hood, T. Hu, C. Kruse, T. Morgan, C. Shmidt, W. Stobby, C. Vo, Zero ozone-

depleting foaming agent technology for extruded polystyrene foam. SPE FOAMS 2009 Conference, Iselin, NJ, US. https://www.researchgate.net/profile/Stephane_Costeux/publication/260985730_Zero_ozone-depleting_foaming_agent_technology_for_extruded_polystyrene_foam/links/00b495336ee3bc0349000000/Zero-ozone-depleting-foaming-agent-technology-for-extruded-polystyrene-foam.pdf.

[47] M. I. Pariente, F. Martínez, J. Á. Botas, J. A. Melero, Extrusion of Fe2O3/SBA-15 mesoporous material for application as heterogeneous Fenton-like catalyst, AIMS Environ. Sci., 2015, 2, 154.

[48] J. Freiding, F.-C. Patcas, B. Kraushaar-Czarnetzki, Extrusion of zeolites: Properties of catalysts with a novel aluminium phosphate sintermatrix, Appl. Catal. A Gen., 2007, 328, 210.

[49] S. Y. Devyatkov, A. A. Zinnurova, A. Aho, D. Kronlund, J. Peltonen, N. V. Kuzichkin, N. V. Lisitsyn, D. Y. Murzin, Shaping of sulphated zirconia catalysts by extrusion: Understanding the role of binders, Ind. Eng. Chem. Res., 2016, 55, 6595.

[50] http://upload.wikimedia.org/wikipedia/commons/thumb/2/21/Al_extrusion_die_set.jpg/440px-Al_extrusion_die_set.jpg, accessed on 23.12.2017.

[51] http://www.visionplastics.net/what-is-plastic-injection-molding/, accessed on 23.12.2017.

[52] http://www.argomold.com/html/17-Different-ways-to-molding-a-part.asp, accessed on 23.12.2017.

[53] http://slideplayer.com/slide/4401990/, accessed on 23.12.2017.

[54] http://robinsonpackaging.com/plastics/blow-moulding/, accessed on 23.12.2017.

[55] http://robinsonpackaging.com/plastics/injection-stretch-blow-moulding/, accessed on 23.12.2017.

[56] https://glossary.periodni.com/glossary.php?en=calendering, accessed on 23.12.2017.

[57] https://www.hindawi.com/journals/mse/2011/138143/fig1/, accessed on 23.12.2017.

[58] http://www.tikp.co.uk/knowledge/technology/fibre-and-filament-production/dry-spinning/, accessed on 23.12.2017.

[59] http://nptel.ac.in/courses/116102010/18, accessed on 23.12.2017.

[60] http://www.substech.com/dokuwiki/doku.php?id=centrifugal_casting, accessed on 23.12.2017.

[61] http://www.prosec.es/descargas/drying-and-firing-processe.pdf, accessed on 23.12.2017.

[62] http://www.scielo.br/img/revistas/jbsmse/v31n4/a03fig02.gif, accessed on 23.12.2017.

[63] http://www.sciencedirect.com/science/article/pii/S0376738817306403#f0005, accessed on 23.12.2017.

[64] https://upload.wikimedia.org/wikipedia/commons/6/61/Filtration_solute.png, accessed on 23.12.2017.

[65] https://en.wikipedia.org/wiki/Rotary_vacuum-drum_filter#/media/File:Rotary_vacuum-drum_filter.svg, accessed on 23.12.2017.

[66] https://www.tsk-g.co.jp/en/tech/industry/horizontal-belt-filter.html, accessed on 23.12.2017.

[67] http://trends.directindustry.com/comber-process-technology-srl/project-34050–131310.html, accessed on 23.12.2017.

[68] http://www.solidliquid-separation.com/pressurefilters/nutsche/nutsche.htm, accessed on 23.12.2017.

3 Performance chemicals

3.1 Plastics and polymer composites

3.1.1 Thermoplastics and polymer blends

Thermoplastics (Figure 3.1), which depending on molecular structure, can be amorphous or semicrystalline, consist of macromolecules formed by addition polymerization (poly(vinyl)chloride, polycarbonate polyurethane) or polymer transformation (poly(vinyl alcohol). These materials are rigid and brittle below glass transition temperature, exhibiting elastic behavior above the softening temperature due to chain enlargements. Transition from solid to melt is reversible; however, complications might arise due to molecules alignment upon melting. Such orientation although beneficial from the viewpoint of mechanical properties in the direction of orientation, can lead to distortion of the finished product when temperature is elevated. For relaxation to occur, the melt is kept long enough at the maximum possible temperature possible.

Figure 3.1: Some examples of thermoplastics [1].

Strong dependence of polymer viscosity on temperature and shear velocity and poor thermal conductivity impose a few restrictions on processing of polymers. A required rapid input of energy cannot be achieved by conductive heating that is too slow. High-molecular melts exhibit laminar flow; therefore convection is of minor importance. Consequently, melting by dissipation in screw machines, combined with conductive heating via the contact surfaces is the preferred method of polymer melting. Heat removal from melts is done by thermal conduction, which can lead to problems when thick-walled moldings are used.

https://doi.org/10.1515/9783110797961-003

In polymer composites where a polymer is the major or continuous phase and a filler (metal, ceramic or another polymer) is the second phase, thermosetting and thermoplastic resins can be used as the polymer phase.

The most important thermoplastic composites are made from flexible thermoplastics, or semicrystalline materials with a glass transition temperature below room temperature. Introduction of fillers brings stiffness at the cost of decrease in flexibility. Wood, clay or glass can be added to high-density polyethylene and polypropylene, decreasing the overall costs.

High-performance composites result in case of good adhesion between relatively polar polymers and polar fillers.

While thermoplastics are simply melted, curing in required for thermoset resins. Such chemical reactions result in cross-linking; therefore thermosets as opposed to thermoplastics, cannot be reshaped or recycled after curing. The most common thermoset composites are automobile tires, containing a number of polymers and fillers, and the most important of these are styrene–butadiene copolymer and a carbon-black filler, respectively.

Thermoplastics can be crystalline (polyamides, polyoxymethylene, polypropylene, polysulfones and polyesters) or amorphous. The former, often not transparent resins have such properties as superior resistance to organic solvents, a low melt viscosity, considerable mold shrinkage and warpage. As high-dimensional stability is only achieved below the glass transition temperature, introduction of reinforcing fillers above this temperature is required to get the dimensional stability and strength. The key features of the most important thermoplastics are provided in Table 3.1. Often thermoplastics are plastified prior to processing in, for example, continuous kneaders.

Table 3.1: Properties of thermoplastics [2].

Plastic name	Products		Properties
Polyamide (nylon)		Bearings, gear wheels, casings for power tools, hinges for small cupboards, curtain rail fittings and clothing	Creamy color, tough, fairly hard, resists wear, self-lubricating, good resistance to chemicals and machines
Polymethyl methacrylate (acrylic)		Signs, covers of storage boxes, aircraft canopies and windows, covers for car lights, wash basins and baths	Stiff, hard but scratches easily, durable, brittle in small sections, good electrical insulator, machines and polishes well

Table 3.1 (continued)

Plastic name	Products		Properties
Polypropylene		Medical equipment, laboratory equipment, containers with built-in hinges, "plastic" seats, string, rope, kitchen equipment	Light, hard but scratches easily, tough, good resistance to chemicals, resists work fatigue
Polystyrene		Toys, especially model kits, packaging, "plastic" boxes and containers	Light, hard, stiff, transparent, brittle, with good water resistance
Low-density polythene (LDPE)		Packaging, especially bottles, toys, packaging film and bags	Tough, good resistance to chemicals, flexible, fairly soft, good electrical insulator
High-density polythene (HDPE)		Plastic bottles, tubing, household equipment	Hard, stiff, able to be sterilized

Polymer blends are applied when the strengths of both polymers can be maintained while avoiding disadvantages. Engineering polymer blends have dimensional stability over a wide range of maximum temperatures, determined either by the component with the highest glass transition temperature or by combination of high crystallinity and a high melting point of the component with the highest melting point.

Figure 3.2 maps different engineering blends in terms of continuous use temperatures and impact strengths. As two polymers might have different miscibilities (complete, partial or immiscible), blending can result in properties that are not simple combinations of the two individual property profiles.

Phase separation in polymer blends can be avoided when different chemical groups in two polymers can interact with each other through dipole – dipole interactions, hydrogen bonding, etc. Alternatively, a decrease of the interaction energy within each component can be achieved by using random copolymers made from monomers forming themselves into immiscible homopolymers. The morphology of incompletely miscible systems depends on chemical interactions, melt viscosity ratio of polymers and shear rate dependence of melt viscosity. Conditions of compounding during melt blending are important for the final blend property profile. Blends are produced in batch or continuous mode using kneaders or compounding extruders.

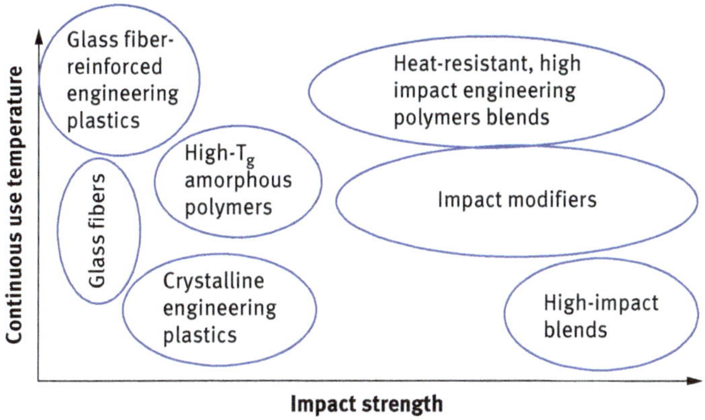

Figure 3.2: Second-generation polymer blends of engineering plastics (after [2]).

The requirements of the engineering plastics blends are related to heat and chemical resistance, high- and low-temperature impact strength, as well as stiffness and can be set by the end user or the standards of the specific industry segment.

Some selected requirements are discussed below. In automotive applications, mechanical properties of blends are mainly important for structural body parts even if high dimensional stability is also needed for nonstructural body parts. More radiation from the sun onto the horizontal exterior body means that a higher temperature resistance is required than for vertical exterior body parts. Bumper systems should have a crash-absorbing function, while the dashboard material in the passenger compartment, besides decorative functions, should have certain heat distortion temperature requirements, and reduce outside and engine noise. Moreover, good scratch resistance is highly appreciated by customers. Polymer blends used for cables and wiring should possess electrical, chemical and high-temperature resistance. Semicrystalline polymers find applications in the motor compartment being able to withstand harsh conditions in terms of temperature and chemical aggressiveness.

In electronic and electrical applications, high heat and chemical resistance as well as high dimensional stability in addition to low conductivity are required. Electrical equipment must also fulfill stringent flammability requirements.

In copolymers of ethylene with such polar monomers as acrylates, rheological properties can be adjusted by blending. Polyethylene blends used in the packaging industry display barrier properties, in particular, toward oxygen and water vapor, and chemical resistance toward aqueous systems, edible fats and oils.

Blends of polyvinyl chloride (PVC) that are otherwise rather brittle and have limited thermal stability, are used to improve their flowability by reduction of the molecular mass, stabilization and thermal control during molding to prevent overheating. PVC blends with polyurethane being UV-stable have also better thermal stability than PVC alone and are used for instance in boots, oil-resistant articles and calendared coatings.

Acrylonitrile–butadiene–styrene polymers (ABS) are blends of rubber-impact-modified styrene–acrylonitrile (SAN) copolymer with SAN or other SAN-miscible copolymers. Melt blending can be done during compounding. Alternatively blending takes place in a suspension or emulsion by coagulation. ABS polymers being cost-competitive have some other advantages including good to very high impact strength, acceptable mechanical properties below 90 °C, good chemical resistance and melt flow as well as low mold shrinkage and an excellent surface finish. Poor resistance toward prolonged exposure to light and heat, related to presence of unsaturated bonds in the butadiene rubber impact modifier systems, can be improved by adding acrylic rubbers (acrylonitrile styrene acrylate, ASA). Incorporation of α-methylstyrene into ABS enhances the actual use temperature. Blends of ABS (excellent surface features with high gloss but readiness to burning) and PVC (good impact behavior) combine advantages of individual polymers and are used in low-cost electrical panels.

Blending of polyamides is done to increase their heat distortion temperature, reduce their moisture sensitivity enhancing dimensional stability and to improve dry impact strength at low temperatures. Blends of polyamides with ABS combine chemical and abrasion resistance with high impact resistance and are applied, for example, as under-the-hood automotive components, power tools and appliances.

Another way of improving the properties of plastics, besides blending, is to use various additives, which can be subdivided into (a) stabilizers against degradation and aging during processing or use (antioxidants, light and heat stabilizers), (b) processing additives (lubricants, mold-release and blowing agents), and (c) performance additives bringing novel features, such as flame resistance, transparency or color, mechanical and electrical properties, dimensional stability or degradability. Impact modifiers, plasticizers, flame retardants, various fillers, dyes, pigments, antistatic and nucleating agents, as well as optical brighteners can be mentioned among such additives.

To illustrate the point of complexity it should be mentioned that, for example, in processing PVC stabilizers, lubricants, plasticizers, modifiers, fillers, pigments and special additives are used. The methods for processing of PVC blends and the major application areas are listed in Figure 3.3.

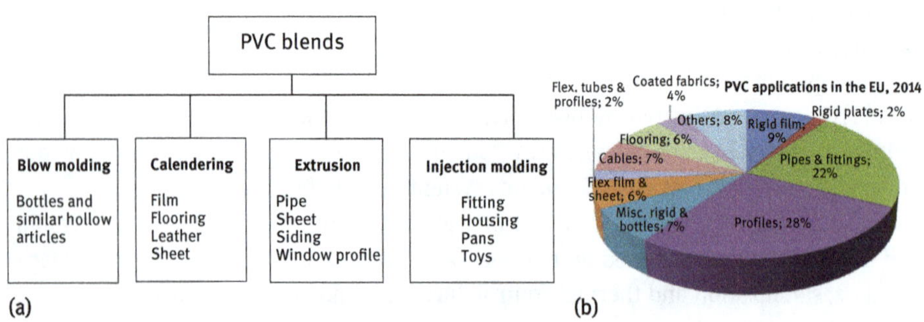

Figure 3.3: Processing PVC: (**a**) methods and (**b**) application areas [3].

Antioxidants that prevent or retard polymer autoxidation should be nontoxic, thermally stable, nonvolatile at processing conditions, soluble in polymers, resist leaching or extraction by food simulants, odorless, tasteless and not lead to corrosive or color-extensive products upon degradation. Sterically hindered phenols and amines, secondary aromatic amines, as well as phosphites, phosphonites and thioethers can be used as antioxidants.

Free-radical scavengers or primary antioxidants (hindered phenols and secondary aromatic amines) react with chain-propagating radicals such as peroxy, alkoxy and hydroxy radicals in chain-terminating reactions as illustrated below for scavenging oxy radicals:

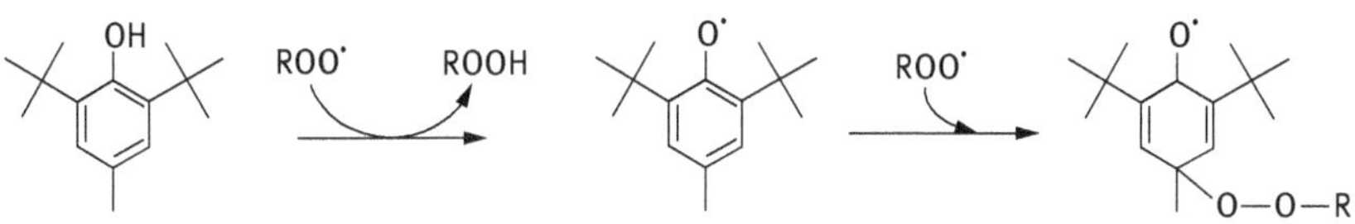

While secondary aromatic amines are more effective than hindered phenols as primary antioxidants, they result in substantial discoloration.

Hydroperoxides (ROOH) can be decomposed into nonreactive products instead of decomposition to alkoxy and hydroxy radicals by applying peroxide scavengers or secondary antioxidants such as phosphites,

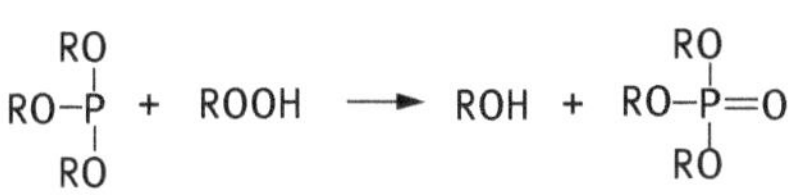

thioethers or organic sulfides.

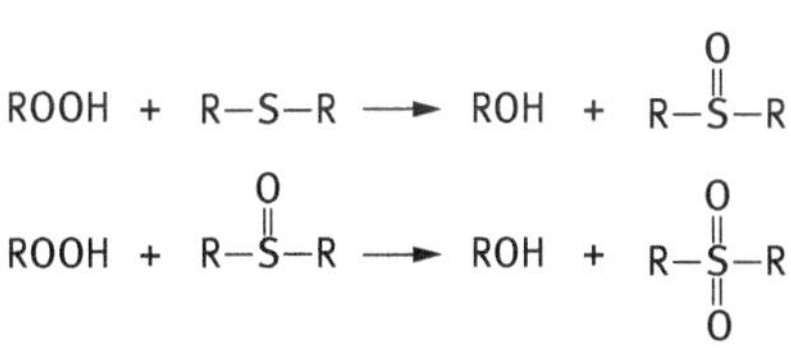

Secondary antioxidants are often used in combination with primary ones resulting in their synergistic effects.

Heat-stabilizing additives added in 0.3–5% amounts help a polymer to withstand processing temperatures without thermal degradation. In the case of thermally unstable poly(vinyl chloride) such decomposition results in release of HCl. Prevention of formation of free radicals through neutralizing labile chlorine atoms is done by adding stabilizers. Examples for rigid PVC include organotin stabilizers (tin mercaptides)

$$R{-}Sn(SR')_3 + HCl \rightarrow R{-}Sn(SR')_2Cl + HSR'$$

$$R{-}Sn(SR')_3 + HCl \rightarrow R{-}Sn(SR')Cl_2 + HSR', etc$$

tin carboxylates or Ca–Zn co-stabilizer systems. Ba–Cd and Ba–Zn liquid systems are applied for flexible PVC. The application of cadmium or lead stabilizers has some limitations if they come in contact with food as part of beverage bottles or packaging films.

Light stabilizers are used to protect plastics against photodegradation, which otherwise leads to yellowing, surface cracking, reduction of gloss and ultimate disintegration. Light stabilizers can be incorporated in the bulk plastic or alternatively the plastics can be coated with materials stable to UV radiation. Such stabilizers have certain requirements in terms of solubility in the polymer, color, stability, compatibility with other additives and toxicity if used in food related applications.

The stabilizers generally used in rather small concentrations (0.05–0.5%, exceptionally up to 3%) increase the service life of plastics substantially (10–20-fold) allowing also outdoor use (garden furniture, window frames, films for protective coverage, bumpers, sports equipment, etc.). Use of light stabilizers is especially important for polyolefins (polyethylene, polypropylene) prone to photodegradation and for acrylic and polyurethane coatings.

UV stabilizers reduce absorption of UV rays by the polymer matrix and the rate of weathering. Examples include oxanilides for polyamides, benzophenones for PVC, benzotriazoles and hydroxyphenyltriazines for polycarbonate.

Extremely efficient stabilizers against light-induced degradation of products especially from polypropylene and polyethylene are called hindered amine light stabilizers (HALS), being derivatives of 2,2,6,6-tetramethyl piperidine:

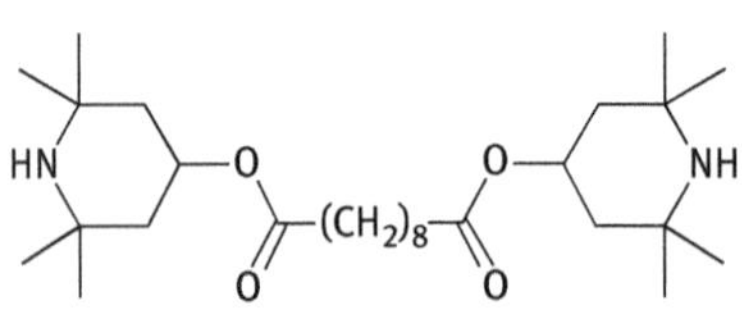

Contrary to UV stabilizers, HALS acts by inhibiting polymer degradation rather than by absorbing UV radiation allowing significant stabilization at relatively low concentrations. Stabilization results from trapping of radicals generated from the polymer upon exposure to UV. In particular the nitroxyl radical (R-O•) reacts with free radicals in polymers: R-O• + R'• → R-O-R'. HALS are regenerated during stabilization resulting in their involvement in cyclic reactions until HALS are degraded by themselves. HALS are extremely effective in polyolefins, but do not display sufficient effectiveness for PVC.

Lubricants control the frictional and adhesive properties of plastics during processing and use. Lubricants whose friction-lowering function is related to presence of long hydrocarbon chains (>C_{12}), can be subdivided into internal and external lubricants. The former, having high affinity for polymers, diminish friction between polymer particles and molecules during plastics melting and melt transport, thereby lowering melt viscosity and improving flow properties. As a result, plastication can be done at less harsh conditions with lower energy consumption. External lubricants possessing a low affinity for polymers decrease adhesion of polymer melts onto hot metal surfaces of the processing equipment. This improves melt flow, the gloss and smoothness. Some adhesion to the metal in any case is needed in extrusion and calendaring; thus special care should be taken to avoid over-lubrication.

From the chemical viewpoint many lubricants are amphoteric compounds containing a long hydrocarbon chain and polar groups (e.g., hydroxyl, ester or acid groups). Unlike plasticizers (completely soluble in the polymer), lubricants have only limited solubility because of their amphoteric nature. Another consequence of the dual character of lubricants is that their action depends on the polymer polarity. For instance, in the case of polar polymers (e.g., PVC) they act as internal lubricants being at the same time external lubricants in nonpolar polymers (e.g., polyolefins). Typically fatty (C_{12}–C_{22}) or montanic (octacosanoic) acids and their derivatives, paraffins (C_{20}–C_{70}), and polyolefin waxes (MM 2,000–10,000) can be used as lubricants, whose main applications are in PVC. In the case of rigid PVC the content of lubricants can be 1–4%. Lubricants are also used for styrene polymers, polyolefins, engineering thermoplastics and thermoplastic elastomers.

Fatty acid esters, calcium stearate and fatty alcohols are typical internal lubricants. Montan waxes, amide waxes, paraffins, oxidized polyethylene waxes and fatty acids act as external lubricants. Note that external lubricants often cause haze. Selection of lubricants also depends on the heat stabilizer type. In the case of rigid PVC pipes calcium stearate as an internal lubricant is applied along with an external lubricant (paraffin wax) and oxidized polyethylene wax as a mold-release agent. In flexible PVC requiring less lubricant (ca. 0.5%) glycerol monooleate or paraffin oils are used as lubricants, while amide waxes act as antiblocking agents. For engineering thermoplastics (polycarbonates, polyamides, polyethylene terephthalate) thermally stable pentaerythritol fatty acid esters are applied. As most lubricants are derived from natural sources, they are often approved for food packaging products.

Closely related to lubricants are mold-release, slip and antiblocking agents. Mold-release agents prevent adhesion of molded plastics to metal cooling rolls and molds by forming a film between the metal surface and the cooled plastic. Slip agents migrate to the plastic surface, improving their frictional properties and preventing adhesion between stacked films. Antiblocking agents create surface roughness allowing air to be trapped between stacked layers.

About 10% of all plastics (mainly PVC, ABS, polystyrene, unsaturated polyesters, polypropylene, polyethylene and polyurethanes) covering a very broad application range such as building construction, transportation, electrical industry (e.g., TV set components and household appliances) contain flame retardants. The latter can be either of additive type (ca. 90%) being incorporated before, during or often after polymerization during processing of polymers into final products, or alternatively reactive ones. Reactive flame retardants are added to thermosetting resins before cross-linking.

Flame retardants may be subdivided chemically into several classes. Inorganic substances such as aluminum or magnesium hydroxides mainly act as additive flame retardants and do not become chemically attached to the surrounding system. These mineral retardants decompose endothermally when exposed to high temperatures releasing water, which at the end cools and dilutes the flame zone. Aluminum trihydroxide is the most widely used flame retardant on a tonnage basis being inexpensive on one hand, while requiring very higher loadings in polymers (up to 60%). It decomposes at ca. 200 °C; thus another mineral flame retardant – magnesium hydroxide with decomposition temperature of ca. 300 °C is used in polymers requiring higher processing temperatures. Maximum polymer processing temperature could be thus determined by decomposition temperature of mineral flame retardants. Other inorganic fillers such as various carbonates and hydrates remove heat from the substrate while decomposing and as a result cool the polymer.

Organohalogen compounds include various chlorine- and bromine-containing substances and are applied together with an additive (e.g., antimony trioxide). Gas-phase radical quenching is the main mechanism of their action. Chlorinated and brominated materials upon thermal degradation release HCl and HBr or antimony halides if antimony trioxide is used as a synergist. When these halogen-containing compounds react with highly reactive $H^\bullet$ and $OH^\bullet$ radicals present in the flame-inactive molecules and less reactive $Cl^\bullet$ or $Br^\bullet$ radicals are formed. This in turn results in much lower propagation of radical oxidation reactions during combustion.

Brominated fire retardants are often used to prevent fires in electronics and electrical equipment, which accounts for more than 50% of their applications. There are several dozen commercial flame retardants of this type with different properties and toxicological behavior, adhering, however, to the same reaction mechanism, explained above. Some examples of the most important brominated flame retardants are presented in Figure 3.4.

Decabromodiphenylether (Figure 3.4a) is a commercial product from the polybrominated diphenylethers family, which is applied for styrenic polymers, polyolefins polyesters, nylons and textiles. Hexabromocyclododecane (Figure 3.4b) of a cycloaliphatic nature is commonly used at very low loadings in foamed polystyrene for building insulation. Higher loadings are required for high impact, compact polystyrene. Tight bonding of tetrabromobisphenol-A (Figure 3.4c) to the polymer is ensured when it reacts with epoxy resins. This flame retardant is, for example, applied for printed wiring boards, as an additive flame retardant in ABS plastics and in engineering plastics for electrical and electronic devices. Brominated polystyrene (Figure 3.4d) is itself a polymer being therefore immobile in the matrices of polyester or polyamides. Tetrabromophthalic anhydride (Figure 3.4e) is often applied as a reactive flame retardant

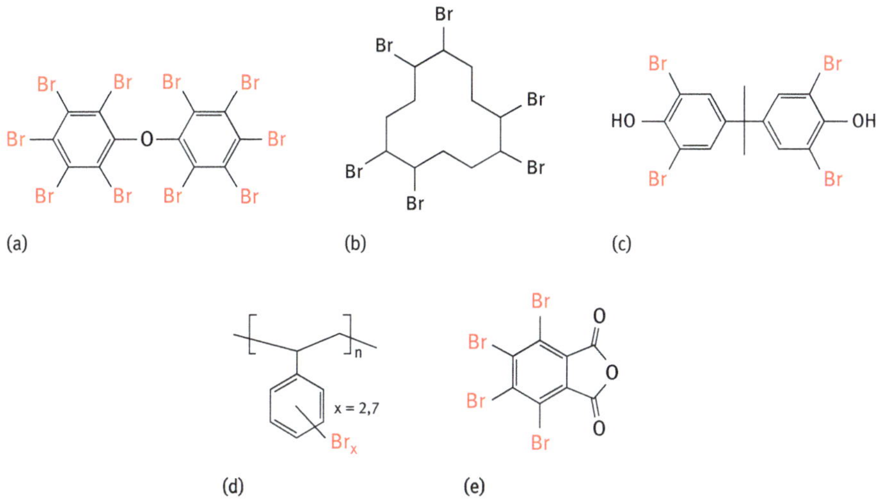

Figure 3.4: Some examples of brominated fire retardants: (**a**) polybrominated diphenylether (PBDE), (**b**) hexabromocyclododecane (HBCD), (**c**) tetrabromobisphenol-A (TBBPA), (**d**) brominated (poly)styrene and (**e**) tetrabromophthalic anhydride.

in unsaturated polyesters used for manufacturing of circuit boards and mobile phones. One of the concerns related to brominated flame retardants is the toxicity; therefore safety for human health should be always assessed. The same is valid for organophosphorus compounds as the public at large is very much concerned about flame retardant emissions.

Phosphorus-containing flame retardants include both reactive and additive compounds and have a broad application range. Some classes of these flame retardants are given in Figure 3.5. As particular examples, triphenyl phosphate and bisdiphenylphosphate (phosphates), or dimethyl methylphosphonate (phosphonate) and aluminum diethylphosphinate (phosphinate) can be mentioned.

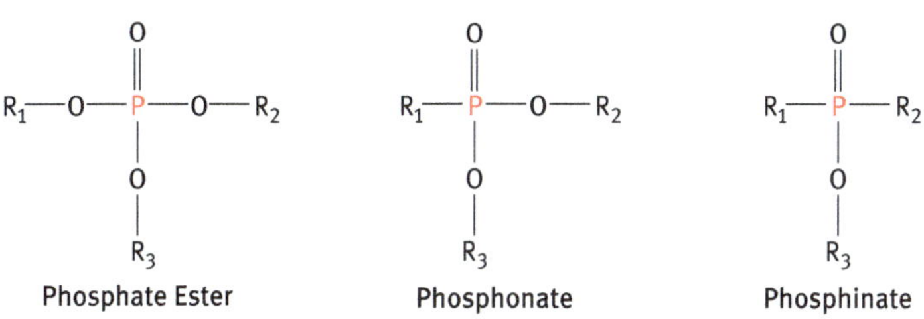

Figure 3.5: Examples of organophosphorus flame retardants.

Flame retardant compounds contain both phosphorus and a halogen. Some of these compounds, for example tris(2,3-dibromopropyl) phosphates, applied widely in the past as flame retardants in plastics and textiles, are mutagenic, and were banned to be used in fabrics worn by children.

Among nitrogen flame retardants, melamine and melamine derivatives and homologues can be mentioned. Several mechanisms are operative for these flame retardants. For example, melamine

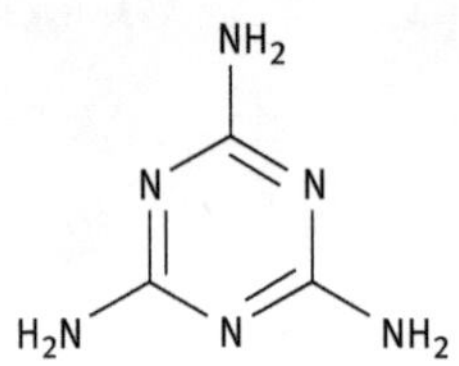

can be is transformed into cross-linked structures promoting char formation. Another mode of action is related to the release of molecular nitrogen diluting volatile polymer decomposition products.

Almost all thermosetting and thermoplastic resins contain fillers (20–50% wt% and even higher). Extenders such as calcium carbonate merely decrease the overall costs, while reinforcements improve polymer properties (strength, elongation, rigidity, impact strength, heat distortion temperature, dimensional stability, shrinkage, stability, chemical resistance, surface quality, etc.)

By adding special fillers, such features as flammability mentioned above, electrical conductivity, resistance to radiation and biodegradability can also be influenced. Some examples of fillers are provided in Table 3.2.

Table 3.2: Fillers and their properties.

Filler	Modification of resin properties
Alumina	Abrasion resistance, dimensional stability, electrical resistivity, thermal conductivity, toughness
Aluminum	Impact resistance, machinability, mechanical properties, thermal conductivity
Aluminum silicate	Chemical resistance, dimensional stability, extender, pigmentation
Beryllium oxide	Thermal conductivity
Calcium carbonate	Dimensional stability, extender, machinability, mechanical properties, pigmentation
Calcium sulfate	Dimensional stability, extender
Carbon black	Electrical and thermal conductivity, pigmentation, reinforcement, thermal resistance
Copper	Electrical and thermal conductivity, thermal resistance

Table 3.2 (continued)

Filler	Modification of resin properties
Glass microballoons	Density reduction
Mica	Chemical resistance, dielectric properties, electrical conductivity, lubricity, moisture resistance, toughness
Sand	Abrasion, thermal conductivity
Silver	Electrical conductivity, thermal conductivity
Titanium dioxide	Dielectric properties, extender, pigmentation
Zirconium silicate	Arc resistance

The modern plastic industry uses a range of dyes and pigments for making materials of various colors (Figure 3.6). Pigments represent particulate organic and inorganic solids virtually insoluble in the medium and therefore should be dispersed in this medium. Dyes being organic compounds are usually dissolved in the substrate. While inorganic pigments have superior heat resistance and weathering properties compared to dyes and organic pigments, they are considered to be dangerous to human health. As a result, there is a shift in industry from inorganic to organic pigments. Dyes are applied when powerful tinting (soaking the material in a dye) is needed. Some limitations of dyes are related to nonoptimal performance in outdoor applications.

Figure 3.6: Examples of plastic chairs of different colors [4].

Whatever the colorants used in plastics, they should be inert under processing conditions, not decomposing or reacting with the polymer, comparable with other additives, stable throughout the product service time, nontoxic and disposable.

Colorants can be introduced into plastics by blending dry colorants with the polymer before processing or more efficiently by using color concentrates.

Among white pigments, titanium dioxide needs to be mentioned. Possessing high hiding power and UV resistance, titanium can however have a negative effect on glass-filled resin systems. In the latter case, barium sulfide, which does not break down the glass, is preferred even if it should be used in higher concentration. Zinc oxide and antimony trioxide are other white pigments.

Iron oxides can deliver yellow, orange and red colors. Organic yellow pigments allow better transparency than inorganics, being, however, prone to fading when applied outdoors. The red color can be achieved with organic red pigments, which are obviously more expensive than red iron oxide.

Ferric ammonium ferrocyanide (iron blue) is used to achieve inorganic blue color, while adding chrome yellow to iron blue gives inorganic green color. Organic copper phthalocyanine pigments being more costly than their inorganic counterparts, also give blue and green colors. These pigments have overall good exterior durability, chemical resistivity and heat stability. Carbon blacks are typical black pigments adsorbing UV efficiently. The intensity of blackness depends on the particle size, the most prominent being for the smallest particle of 5–15 nm size. Other black pigments can be used for the purpose of increasing electrical conductivity (acetylene black) or acquiring a lighter shade of black (iron oxide blacks). Metallic appearance is achieved using aluminum flake pigments with smaller particles exhibiting darker appearance. On the contrary, larger particle sizes reflecting more light provide a more metallic appearance.

Antistatic agents are added to plastics to compensate electrostatic charging of surfaces because of friction, which eventually gives dirty surfaces due to dust, electric shock from walking on plastic floor coverings and adhesion deficiencies, to name a few problems.

External antistatic agents applied to the surface are removed by washing and require reapplication. Alternatively, internal chemical antistatic agents or migratory additives can be used. The mechanism of action of these antistatic agents is presented in Figure 3.7. Antistatic agents achieve an equilibrium level on the substrate surface; therefore antistatic agents beneath the surface act as reservoirs replacing antistatic agents removed from the surface by washing. After repeated washing the antistatic agent will be eventually removed. Resin crystallinity influences the ability of antistatic agents to migrate through the resin matrix with higher crystallinity bringing more difficulties.

Antistatic agents can be anionic, cationic (or simply “ionic”), and nonionic. The antistatic agents work by forming a conductive surface layer allowing fast charge dissipation. Cationic antistatic agents are mainly used for rigid PVC and in polystyrene,

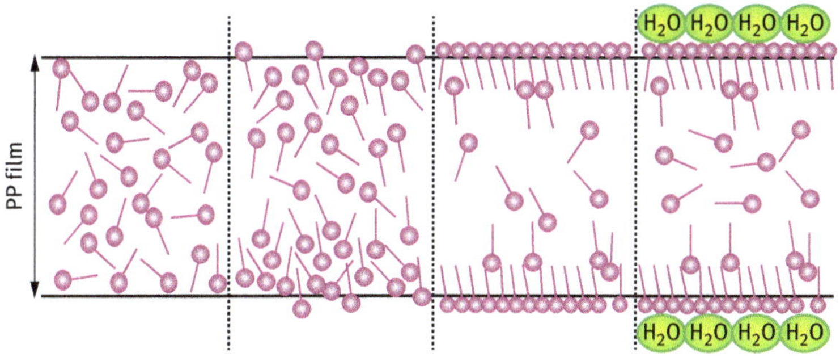

Figure 3.7: Mechanism of action of internal antistatic agents [5].

while not being recommended for polyethylene due to inherently low heat stability. Examples of cationic antistatic agents include quaternary ammonium salts as well as sulfonium or phosphonium salts (e.g., nitrates and chlorides). This type of antistatic agents is relatively expensive and is not used for food-related applications. Anionic antistatic agents (sodium alkyl sulfonates, alkyl phosphonates and alkyl dithiocarbamates) are also applied in rigid PVC. Nonionic antistatic agents are mainly used for polyolefins and include ethoxylated fatty alcohols, fatty amines or fatty acid amides, as well as polyethylene glycol esters of fatty acids and alkylphenols, glyceryl esters of fatty acids and sorbitol esters. These organic compounds have both hydrophilic and hydrophobic parts. Glycerol monostearate, being less efficient than ethoxylated fatty acid amines and diethanolamides in polyethylene and polypropylene, is recommended for short term performance of 1–2 months. On the other hand ethoxylated tertiary fatty acid amines, even if preferred for HDPE, have lower heat stability than glycerol monostearate and can cause some skin irritation. Diethanolamides, being similar in composition to ethoxylated fatty amines, are recommended for applications when low humidity and compatibility with polycarbonate components is required.

Phase 1	Phase 2	Phase 3	Phase 4
Homogeneous distribution of antistatic agent during extrusion	Antistatic agent begins to migrate to surface after extrusion	Hours or days later antistatic agent covers surface	Moisture pick up from the surrounding air

Impact modifiers (elastomeric copolymers with low glass transition temperatures dispersed as discrete soft phases in thermoplastics) are often used to improve impact strength, especially at low temperatures. They are mainly used in the range of concentrations between 5% and 15% and are frequently incorporated into the polymer during compounding. Examples include methylmethacrylate– butadiene–styrene copolymers; polyacrylate esters based on poly(butyl acrylate) and poly(2-ethylhexyl acrylate); graft copolymers of acrylonitrile and styrene on polybutadiene as well as chlorinated polyethylene for PVC and ethylene–propylene–diene copolymers for polypropylene (automotive parts, namely bumpers and instrument panels).

3.1.2 Plasticizers

Modification of a rigid polymer with plasticizers increases flexibility either chemically (internal plastication) or without a chemical reaction. Plasticizers then form a homogeneous physical unit with the polymer (Figure 3.8).

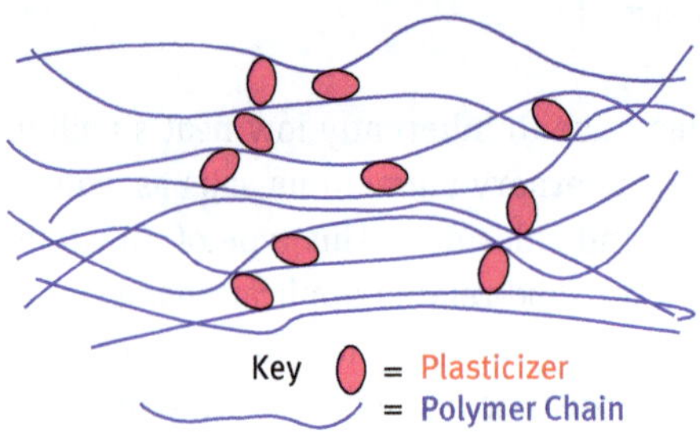

Figure 3.8: Interactions of plasticizer with a polymer chain [6].

Chemically speaking, plasticizers are inert, organic substances with low vapor pressures, predominantly esters, which after incorporation in polymers influence properties of the latter. Primary plasticizers diminish melt viscosity, the second-order transition temperature or the elastic modulus, while secondary plasticizers are not effective per se and enhance performance of the primary plasticizer. Hydrocarbons, chlorinated to varying amounts (30–70%) are the major secondary plasticizers, whose viscosity increases with an increase of chlorine content for a given hydrocarbon. Chlorinated hydrocarbons improve fire retardancy being also cheaper than phthalates, the primary plasticizers. The most important polymer requiring application of plasticizers is PVC, consuming around 80% of their production. Other plasticized polymers are polyvinyl butyral and acetate, acrylics and polyamides. PVC is different from other polymers in being able to accept and retain large concentrations of plasticizers. The reason is the morphology of PVC that consists of highly amorphous, semicrystalline and highly crystalline regions. Without the use of additives PVC is of little use. In some very hard PVC formulations, secondary plasticizers may be used alone. Special care should be taken regarding compatibility between chlorinated paraffin and phthalates, which worsens upon increase of the plasticizer content in PVC and phthalate molecular mass.

Several mechanisms were advanced to explain the behavior of plasticizers. The lubricity theory assumes that plasticizers introduced upon heating act at room temperature as lubricants for the polymer chains of the resin, whose rigidity results from intermolecular friction. The gel theory assumes that plasticizers break resin-resin interactions at so-called centers of attachment, masking the centers from each other. According to the free volume theory there is internal volume available in the polymer for the chains to move, which increases sharply at the glass transition temperature, T_g. Plasticizers decrease the glass transition temperature, thereby imparting increased flexibility to polymer at room temperature.

Plasticizers should satisfy several requirements. Besides being efficient in reducing T_g of the polymer, they should be compatible with polymers, nontoxic, thermally stable and nonflammable. Moreover, they should not possess any taste or odor. A serious issue is the volatility and extractability (leaching) of a plasticizer because such losses lead to an increase of the glass transition temperature gradually enhancing brittleness. While decreased volatility and hence increased permanence can be achieved by increasing the molecular weight of the plasticizer, this might come at the expense of decreased low-temperature flexibility.

As mentioned above, frequently a single plasticizer cannot satisfy the performance requirements; therefore several plasticizers are blended together. For example, dioctylphthalate (DOP), with the chemical structure presented in Table 3.3, can give flexibility for plasticized PVC at low temperatures, at the same time negatively affecting product hardness and performance at ambient temperatures. Aliphatic diesters are added to phthalate esters to improve low-temperature performance.

Several hundreds of plasticizers are currently manufactured with a worldwide production capacity ca. 5 million tons per year. Diesters of phthalic acid, synthesized from phthalic anhydride and linear or branched aliphatic alcohols with chain length ca. C_4–C_{11}, correspond to almost 85% of all plasticizers. Recently there were concerns raised about toxicity and carcinogenic behavior of DOP, the industry standard general purpose primary plasticizer often referred also as diethylhexylphthalate (DEHP). This plasticizer has good gelation characteristics, softening and adequate viscosity properties in emulsion PVC pastes, performing much better than di-*n*-octyl phthalate, a linear C_8 phthalate diester analogue of DEHP. In fact, despite somewhat higher volatilities of the branched structure and stronger degradation at high temperature compared to linear counterparts, the plasticizers with the branched chains have better low-temperature performance. Nevertheless, as a result of safety concerns, DOP is not allowed to be used in some countries in PVC toys and has been replaced by diisononyl phthalate or diisodecyl phthalate (DIDP). They have higher molecular weight, therefore being less volatile with slower migration rates. At the same time, a slightly higher molecular mass gives lower plasticizer efficiency, which has a consequence of requirement of a larger amount of the plasticizer to achieve the same softness as with DEHP. Ditridecyl phthalate, the highest molecular mass phthalate ester available in commercial quantities, has low plasticizing efficiency and some compatibility problems at higher plasticizer loadings required for efficient performance. However, low volatility gives superior high-temperature performance. Plasticizers with a higher molecular weight can be used in applications with strict electrical cable insulation specifications.

Besides phthalate esters, adipates, terephthalates and phosphate esters belonging to monomeric plasticizers are also used. Polymeric or permanent plasticizers made from dibasic, for example, adipic (hexanedioic) or sebacic (decanedioic) acids and a polyol are less volatile. Epoxy plasticizers are derived from vegetable oils improving heat and light stability of PVC products, giving, however, relatively poor low-temperature properties. Examples of various plasticizers are presented in Table 3.3.

As can be concluded from Table 3.3, plasticizers are liquids with high boiling points with an average molecular weight between 300 and 600. The relatively low molecular size of plasticizers allows them to occupy intermolecular spaces between polymer chains, thus reducing secondary forces and hydrogen bonding between the chains.

The plasticity of polymers depends on the plasticizer chemical structure, molecular weight and presence of functional groups. The low-temperature performance depends on the acid constituent of the plasticizer ester with linear aliphatic adipic, sebacic and azeleic (nonanedioic acid) acids giving much better low-temperature flexibility than corresponding phthalates and trimellitates. Linearity of the plasticizer also improves the low-temperature flexibility. Stability of plasticizers to thermal degradation can be improved by adding small amounts of an antioxidant.

Gelation properties, defined through either processing temperature to which the plasticizer and polymer must be heated or as a solution temperature at which a polymer grain is dissolved in plasticizer excess, are related to plasticizer polarity and molecular size. Higher polarity translates into a larger attraction for the PVC polymer chain and thus lower temperature at which elongation properties can be achieved. Plasticizer polarity depends on the acid type and alcohol chain length with, for example, more polar aromatic acids being more prone to gelation than aliphatic acid esters. Smaller plasticizers can easily penetrate the PVC matrix requiring less thermal energy for desired interactions with the polymer. Branching is preferential for gelation.

Plasticizers can be extracted from PVC by solvents, including water, at different rates depending on their molecular size and compatibility with the plasticizer and PVC. As can be expected water is not efficient, while low-molecular-mass solvents are the most aggressive. Extraction resistance depends on the molecular size of the plasticizer and branching – larger size and more branching make migration and extraction of plasticizers more difficult.

Plasticizers not only modify physical properties of polymers but also improve processing characteristics. Plastic products are prepared by shaping molten plastic followed by cooling. Hot compounding (e.g., calendaring, extrusion, injection and compression molding described in Chapter 2) can be significantly influenced by the plasticizer type and concentration. Lower viscosity, easier filler incorporation and dispersion give lower power demand and processing temperatures. Plasticizers can reduce the second-order transition temperature and the elasticity modulus improving cold flexibility.

The extensive use of polymers in such applications as short-term packaging, food and pharmaceutical industries, ignited an interest in biodegradable plasticizers and

Table 3.3: Types of plasticizers and their chemical structures.

Plasticizer type	Chemical structure	Example
Phthalate esters (dialkylphthalate)	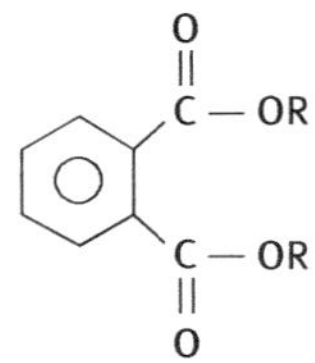	Di-2-ethylhexyl phthalate or dioctylphthalate (DOP) 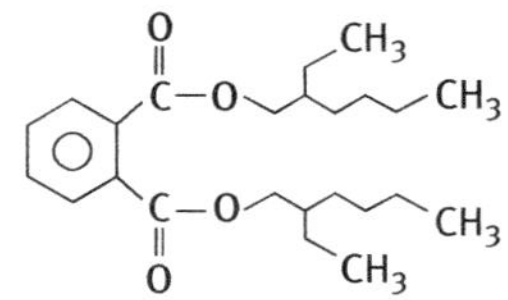
Phosphate esters (trialkyl-phosphate)	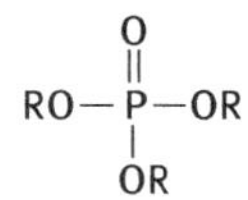	Tricresyl phosphate (TCP) 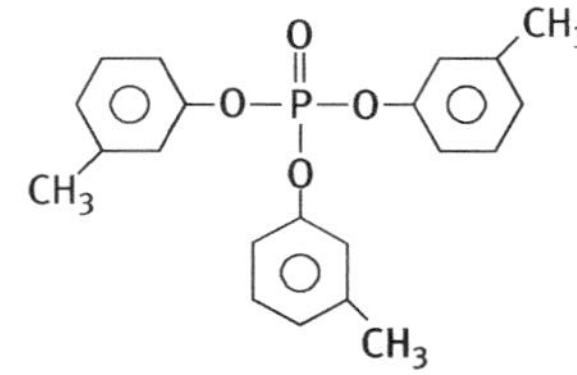
Adepates, azelates, oleates, sebacates (aliphatic diester)	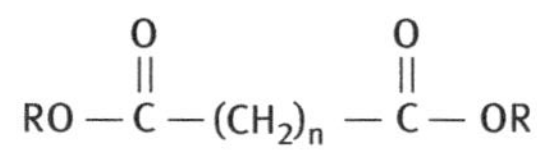	Di-2-ethylhexyl adepate (DOA) 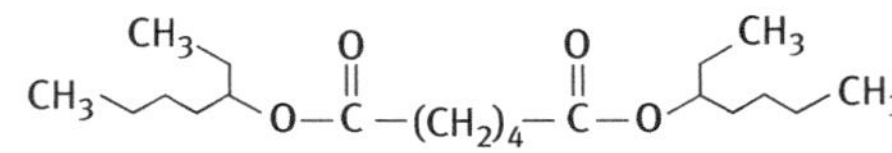

(continued)

Table 3.3 (continued)

Plasticizer type	Chemical structure	Example
Glycol derivatives	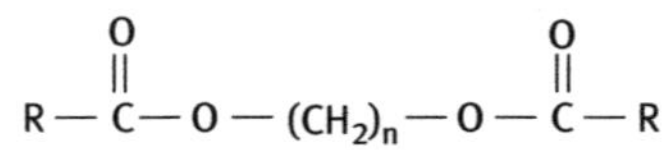	Dipropyleneglycol benzoate 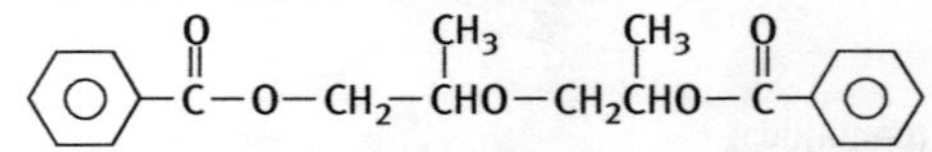
Trimellitates (trialkyltrimellitate)	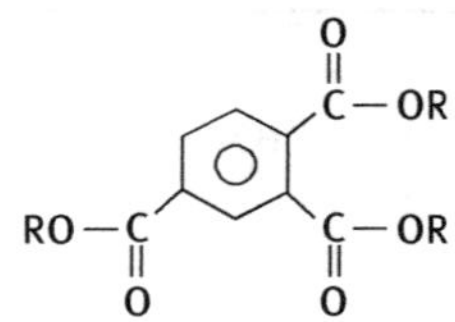	Trisethylhexy trimellitate (TOTM) 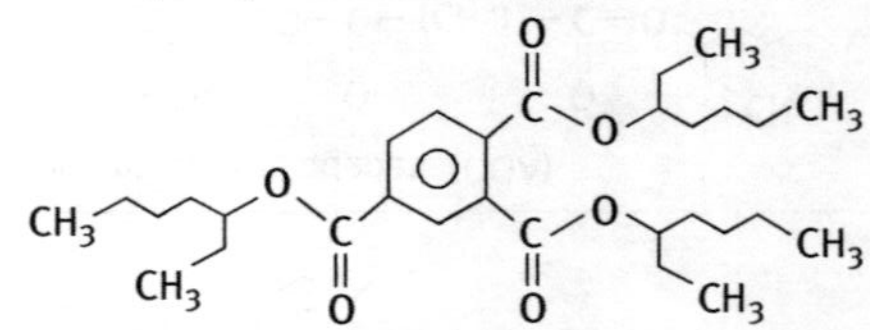

in plasticizers for natural polymers. Plasticizers for biopolymers should preferably also be biodegradable.

One of the most well-known biodegradable polymers is polylactic acid (PLA), which is brittle and has relatively poor impact strength and low thermal degradation temperature. In addition to these deficiencies a lack of reactive functional groups and high costs limit the applications of poly(lactic acid). Low-molecular-weight compounds such as oligomeric lactic acid, glycerol, triacetin and low-molecular-mass citrates can be used as plasticizers of PLA.

Another interesting natural polymer is starch, which during thermoplastic processing, is transformed from semicrystalline to homogeneous material with hydrogen bond cleavage between starch molecules disrupting long-range crystalline structure. Plasticizers penetrating starch and destroying the hydrogen bonds replace starch–starch interactions by starch–plasticizer interactions. The plasticized moldable thermoplastic starch, used for injection, extrusion or blow molding similar to other synthetic thermoplastic polymers, can be processed with different polyols (glycols of different carbon length, glycerol, xylitol, sorbitol), sugars (fructose, mannose, sucrose) or fatty acids as plasticizers.

Cellulose being an abundant, renewable, and biodegradable natural polymer constituting the skeletal part of plants, has been for a long time very attractive as a source of industrial materials because of its abundance, costs, environmental benefits and biocompatibility. Cellulose (a crystalline polymer of D-glucose) is poorly soluble in common solvents and cannot be processed in a melt decomposing prior melting. Therefore, derivatives of cellulose have been developed, such as cellulose acetate, which is produced by esterification of cellulose and acetic anhydride in the presence of sulfuric acid as a catalyst. Cellulose acetate is characterized by high glass transition and melting temperatures. A variety of plasticizers were reported for cellulose acetate plastics including such common additives as diethyl and diethyl phthalates, triphenyl phosphate and ethylhexyl adipate.

3.1.3 Foamed plastics

Foamed plastics are materials that have cells distributed throughout their entire mass. They are used in a variety of applications as insulators and adsorbents. Different polymers can be used for foaming, with the major share occupied by polyurethane (ca. 50%, Figure 3.9) followed by polystyrene (ca. 25%, Figure 3.10).

According to the size of the foam cells, polymer foams can be classified as: macrocellular (>100 μm), microcellular (1–100 μm), ultramicrocellular (0.1–1 μm) and nanocellular (0.1–100 nm) foam. In closed cell foams, the cells are isolated from each other resulting in lower permeability, better insulation and higher rigidity and strength, compared to open cell foams where cells are connected with each other. Open cell foams having softer and spongier appearance are effective for sound insulation and

Figure 3.9: Polyurethane polymer foams [7].

Figure 3.10: Polystyrene foams [8].

have good absorptive capability. Polymer foams are divided into thermoplastics and thermosets. The former are usually broken down and recycled, while recycling of thermosets is much more difficult as they are usually heavily cross-linked. Foams can be either rigid or flexible.

Rigid foams can be used in building insulation, packaging and food and drink containers. They can be applied in transportation and furniture, similar to flexible foams, which besides these application areas can be also used in bedding, textile, sports applications and shock and sound attenuation.

The three key starting materials for making foam are the polymer per se, a blowing agent and additives. The advantages of foams are related to their low density, low heat or sound transfer, flexibility and softness. Inferior mechanical strength, low thermal and dimensional stability and a need to apply blowing agents, which might be environmentally unfriendly, can be mentioned as disadvantages. For instance, chlorofluorocarbons used as blowing agents were found to cause ozone depletion in the upper atmosphere and were banned according to the Montreal Protocol. Other physical blowing agents include compressed gases (air, nitrogen) or low-boiling hydrocarbons (butane or pentane). For example, incorporation of pentane into polystyrene beads makes them heat-expandable. Contrary to physical blowing agents, chemical blowing agents are solids decomposing at elevated temperatures forming gases. Uniform distribution of chemical blowing agents throughout the substrate is needed;

thus they should have a small particle size and a narrow particle-size distribution. Another requirement is decomposition temperature, which should be harmonized with the polymer processing temperature and is typically in the range 150–250 °C. Moreover, decomposition should not be spontaneous and must occur within a relatively narrow temperature window (5–15 °C), thus being only slightly exo- or endothermic. Azodicarbonamide, often used as a chemical blowing agent,

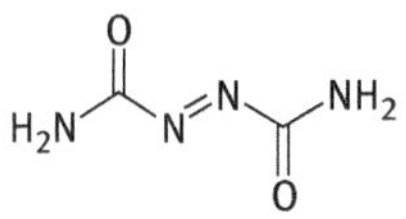

has decomposition temperature of ca. 200 °C, which can be lowered by different additives such as zinc oxide. The thermal decomposition of azodicarbonamide results in formation of nitrogen, carbon mono- and dioxide and ammonia, which are then trapped in the polymer as bubbles. In general, solid decomposition products should not interfere with processing, leading, for example, to discoloration.

During manufacturing of foams, some other additives are introduced. Nucleating agents (finely dispersed, often silicate-like solids) are important additives especially for physically blown, thermoplastic foams. For thermosetting plastics (when macromolecules are cross-linked) surfactants are required for emulsification of the liquid blowing agent, reducing interfacial tension and facilitating nucleation. Other additives can include dyes, inert or reinforcing fillers, fireproofing agents, antioxidants and UV stabilizers.

As an example, production of extruded polystyrene foams will be considered below. Such foams are suitable for thermal insulation when high compressive strength, good appearance in exposed locations and very low water adsorption are required. The starting point for expandable polystyrene (EPS) foams is expandable polystyrene, which is synthesized by suspension styrene polymerization with addition of blowing agents, particularly pentane. EPS beads contain 4–7 wt% of these substances. Other production methods for EPS include impregnation of standard polystyrene with blowing agents under pressure and extrusion of expandable or standard polystyrene with added blowing agents. Besides utilization of styrene as a starting component, foams can be also made from polystyrene copolymers.

Processing into foam is done in several steps: pre-expansion, drying, maturation and final foaming (or molding), as shown in Figure 3.11. During prefoaming, the EPS beads are fed to a vessel containing an agitator and controlled steam and air supplies. The blowing agent evaporates and as a result the bead volume expands substantially. After the pre-expander, the prepuff is either fed through a dryer to remove moisture or can be routed directly to a maturation silo. Maturation is followed by fully automated final foaming. In this final step, perforated molds are completely filled with pre-foamed beads and exposed to steam. Some of the water condensed in the foam evaporates and because of internal cooling the pressure of the foamed plastic decreases more rapidly allowing quick removal of the parts from the mold. When pentane is used as a blowing agent, temperature during prefoaming and final foaming is 85–120 °C.

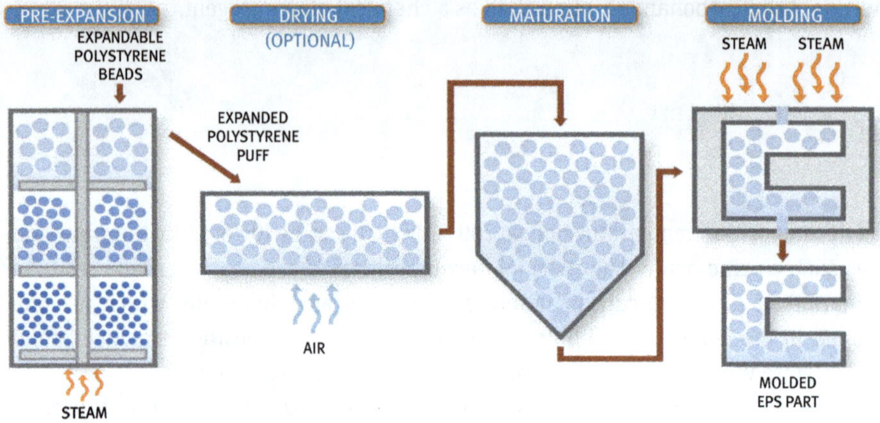

Figure 3.11: Processing of expandable polystyrene (EPS) foam into molded parts [9].

Boards are manufactured batchwise using block molds and board molding machines or continuously with moving-belt machines. Stabilization for several hours to complete cooling and air uptake is followed by trimming and then cutting blocks into boards on cutters resulting in sheets of the desired length and 1–6 mm thickness. Storage of plastic prior to use helps remove internal moisture (steam condensate and residual cooling water) and diminish foam, post shrinkage.

Another example of plastic foam manufacturing process is Dunlop Flexible Foam continuous process for making polyurethane foams when the polyurethane-forming machine is vertical to save space, but could be also horizontal. In this process (Figure 3.12) a mixture of toluene-2, 4-diisocyanate (major component) and toluene-2, 6-diisocyanate reacts with a polyalcohol (molecular mass can be as large as 20,000) in a highly exothermic reaction forming polyurethane:

A rather low melting point of toluene di-isocyanate (17 °C) means that the storage tanks should be kept at 25 °C to avoid solidification.

Upon introduction of the blowing agents, there is a rapid expansion of the reacting mixture, which is pumped to the bottom of the settling chamber as a liquid. The reaction mixture gets the chamber shape and the foam gradually solidifies traveling upward the settling chamber with the aid of a paper conveyor. The residence time is

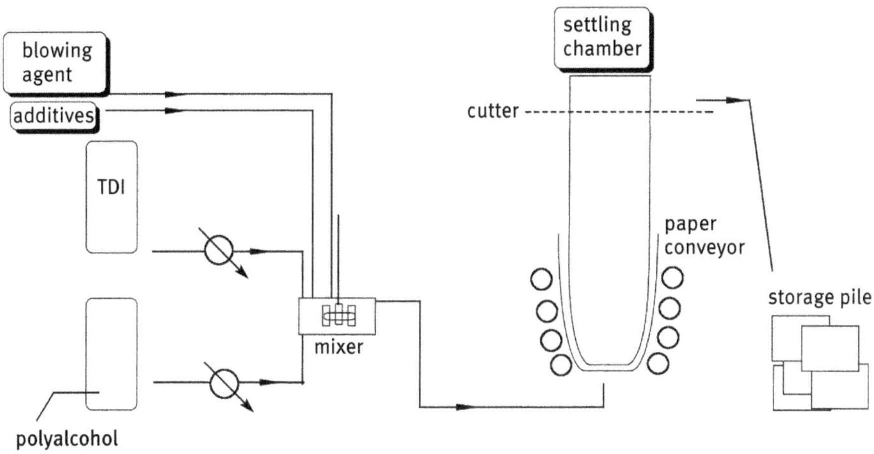

Figure 3.12: Schematic diagram of the Dunlop polyurethane process [10].

ca. 9 min allowing full hardening of the foam, which is thereafter cut into blocks by an electric cutter (Figure 3.12). When the blocks are transported to storage, they are still rather hot and it might take at least 18 h for curing. After that, fully cured foam blocks are cut in smaller sizes or delivered to customers. Vacuum packing during transportation is used to save space and decrease costs. The foams, decreased in size after exposure to vacuum, are reverted to the original size at atmospheric pressure.

3.1.4 Adhesives

Adhesives are used to hold surfaces together. After wetting the surface an adhesive must adhere to the surfaces and strongly bind them remaining stable.

From the fundamental viewpoint adhesion or an ability to bind to the surface is a specific interfacial phenomenon. Adhesion depends on surface energy, roughness, contamination and surface wettability and is a combination of adsorption, electrostatic attraction and diffusion. Reliable adhesion needs not only liquid adhesive but also reliable surface conditions. Surface preparation or, in particular surface wetting, is of primary importance before applying the adhesive. Figure 3.13 illustrates this point showing a case of limited contact (left) when a liquid stays off the surface. For optimum contact the liquid should be smooth on the surface (Figure 3.13, right). This can be done through surface preparation, for example, cleaning and pretreatment.

The market of adhesives is highly specialized and fragmented in ca. 50 segments, including packaging, construction, automobile, footwear and laminate industry to name a few. An illustration of application of adhesives for labeling food products in presented in Figure 3.14.

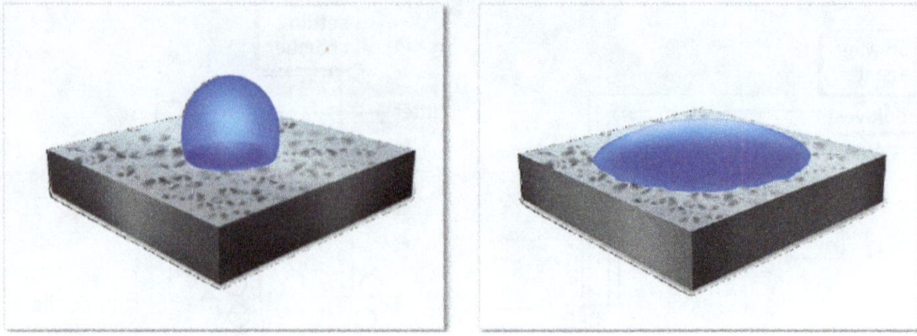

Figure 3.13: Contact of adhesive with the surface [11].

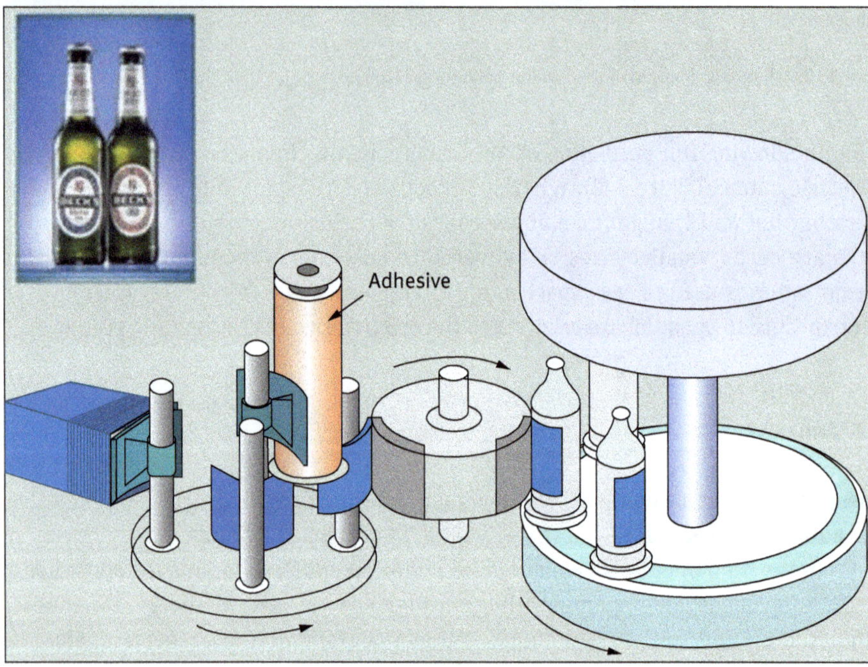

Figure 3.14: Application of adhesive in product labeling [12].

Adhesives can join dissimilar substances of different geometrics, sizes and composition and do not produce any deformation in the materials by eliminating metal grinding. Application of adhesives, which can be easily automated, provides flexibility in the design of products and allows improvements in their appearance. Other advantages of using adhesives are reduction of the product weight, even stress distribution and filling of large gaps. Among disadvantages of adhesives, the need for waiting time for solidification, sensitivity to surface cleanness, nonresistance to temperature and action of chemical agents or UV can be mentioned.

Adhesives are made mainly from natural or synthetic polymeric materials. Examples of the former polymers are starch, dextrin, gelatin, asphalt, bitumen or natural rubber. Among synthetic polymers are various vinyls (polyvinyl- acetate, -alcohol, -butyral, -chloride, -ether), acrylics, rubber (polychloroprene, styrene–butadiene, acrylonitrile, polyurethane, etc.), epoxy-, aldehyde condensation (e.g., resorcinol or melamine-based), amine (e.g., polyamide) or polyester resins. An important product segment is related to acrylic adhesives accounting for more than one-third of the total market. Aqueous polymer dispersions or alternativeely dispersions in organic solvents (often toxic and volatile, but taking shorter time to dry) contain ca. 25–50% of the solid.

Besides the base material fillers, pigments, stabilizers, plasticizers and other additives are also required during compounding of adhesives.

Low- to medium-performance products are based on natural substances including starch, dextrin, and natural rubber or synthetic polymers such as polyvinyl acetate:

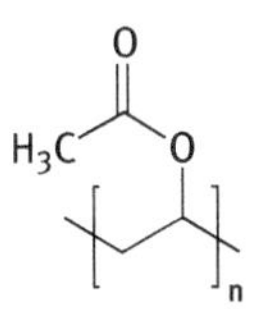

Aqueous dispersion (latex) of polyvinyl acetate is nontoxic, easy to use and sets fairly quickly, being, however, difficult to remove once cured. Other examples of this kind include polyvinyl alcohol, polyesters, acrylics, neoprene, butyl rubber and phenolics.

High-performance products are based on polymers such as polyepoxides, where the resin is made by contacting epichlorohydrin with bispenol A (Figure 3.15). These products require polyamines, phenols and aminoamides as hardeners, work on most surfaces, allow strong adhesion and exhibit fast curing and long-term durability along with heat and chemical resistance including water. These advantages simultaneously imply short handling time and inability to be reapplied. Mixing ratio is also important for good adhesion.

Alternatively, polysulfide, polyurethane, cyanoacrylate and silicone can be used as the base of high-performance adhesive, which give improved bond strength, elongation capacity, durability and environmental resistance.

Adhesives can be also classified based on the application areas (metals, textiles, wood, cement, glass, etc.). Starch suspensions are used as adhesives for such packaging materials as corrugated board, while for the lamination of paper and board adhesives based on starch, dextrin, glutin and polyvinyl alcohol, as well as emulsion adhesives, are used.

Polyacrylate emulsions and hot-melt adhesives discussed below are growing in importance for adhesive coating of paper and board being solvent-free. The adhesives for laminated films are one- or two-component, moisture-cross-linking polyurethane adhesives. For labeling glass, dextrin, starch and casein glues are applied, which can-

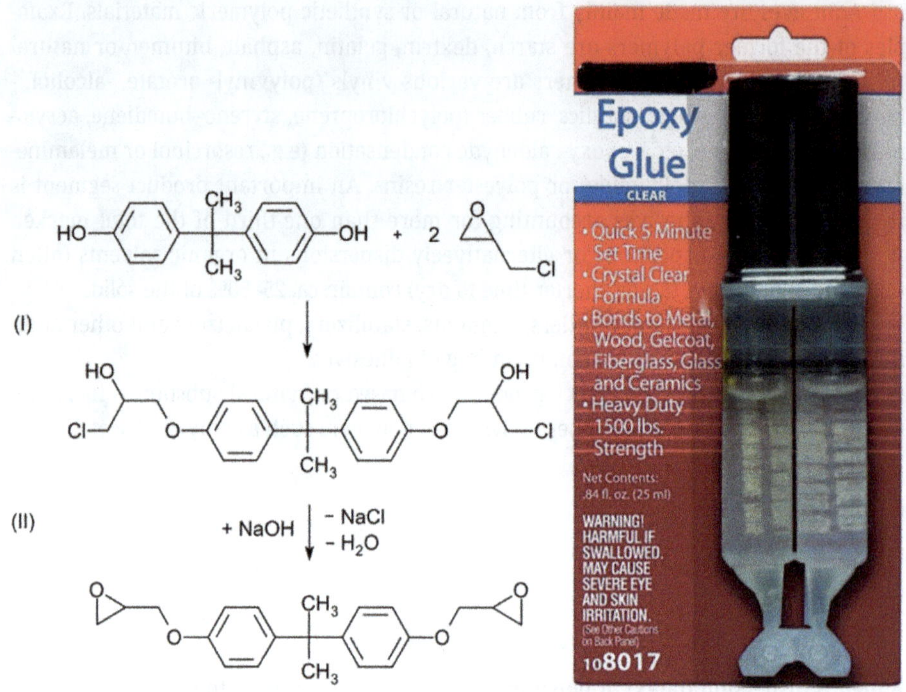

Figure 3.15: Synthesis of polyepoxide resin by reacting bisphenol A with epichlorohydrin [13].

not be applied for plastic containers not adhering properly to, for instance, plastic bottles. Modified copolymer emulsion-based and hot-melt adhesives are used for labeling with polystyrene, polyethylene terephthalate and PVC. For polyethylene and especially polypropylene, adhesives with stronger adhesive properties should be used.

Different types of adhesives are used in households for gluing, assembling or repairing. Natural and/or synthetic polymers are used as adhesive components in glue sticks for gluing paper, cardboard, photos and labels. Adhesive rollers in cassette form are also applied for the same purpose. The solvent-free multipurpose adhesives contain transparent to opaque dispersions or solutions based on polyurethane or acrylate. Adhesive tapes and pads are primarily used in households for packaging, masking renovation work, small-scale assembly, fixing posters and other lightweight objects. Adhesives based on methyl cyanoacrylic acid ester are used in households for repair work being suitable for small parts.

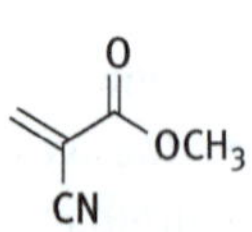

Alternatively, ethyl- or butyl cyanoacrylic acid esters can be used for this purpose.

Two-component adhesives with, for example, epoxy resin and a hardener (amine or mercaptan) or methacrylate with a powder hardener (dibenzoyl peroxide in gypsum) are used for household applications for difficult materials. The former combination is suited for glass, china, ceramics and metals, while the latter one can be applied also for stone and plastics in addition to ceramics and metal. Plastic adhesives able to bond various plastics, for example, rigid PVC, ABS or Plexiglas contain solvents such as butyl acetate and solvate the surface to be bonded. Wood glues are typically water-based with poly(vinyl acetate) as a base.

Adhesives are also classified based on the way they react with the surfaces to be joined being either nonreactive or adhesives setting by chemical reaction (Table 3.4).

Table 3.4: Classification of adhesives based on reactivity.

Nonreactive adhesives	Reactive adhesives
Drying adhesives	UV light-curing adhesives
Hot melt adhesives	Heat-curing adhesives
Pressure-sensitive adhesives	Moisture-curing adhesives

Nonreactive adhesives include drying, hot melt and pressure sensitive adhesives. Drying adhesives (either solvent or water based) are cured by drying. An apparent drawback of solvent-based adhesives is significant levels of volatile organic compounds (VOCs), which are often subject to regulations. As a result of environmental concerns, there is increased interest toward water-based adhesives.

Solvent-based adhesives (Figure 3.16) are a mixture of ingredients (e.g., vinyl polymers, natural rubber, synthetic rubbers of low polarity) dissolved in organic solvents.

In wet bonding adhesives, the substrates are joined when adhesive is wet followed by evaporation. In contact adhesives, the solvent is allowed to evaporate before formation of a bond and such bond is formed by bringing together two coated substrates. A special case is adhesives for plastics, when the plastic surfaces swell in a solvent allowing them to join by migration processes. The role of the polymer in such adhesives is to establish required flow properties and retain the solvent during dissolution.

Solvent-containing adhesives are used for a wide range of substrates and contain as binders high-molecular-mass compounds such as nitrocellulose and poly(vinyl acetate) with affinity to various types of surfaces. Mixtures of esters, ketones and, sometimes alcohols are used as solvents, providing the required adhesive fluidity and setting time.

In these adhesives also acting as wet bonding and contact ones water is either allowed to evaporate or is absorbed by the substrate. An apparent advantage of waterborne adhesives is the absence of VOCs. From the chemical viewpoint water-based

Figure 3.16: Solvent-based adhesives [14].

adhesives, accounting for ca. 50% of the total market, are based on vinyl acetate and vinyl propionate polymers and copolymers of vinyl acetate with ethylene

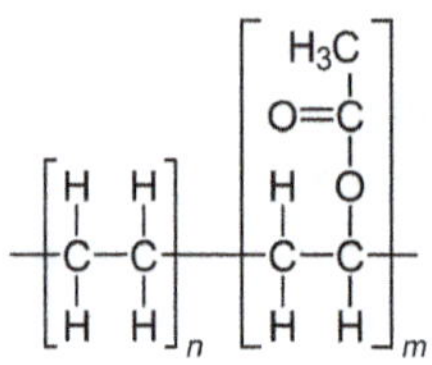

or maleic esters; styrene copolymers (styrene–butadiene rubber); polyacrylic esters; polyurethane; natural rubber latex and synthetic elastomers; starch; dextrins; casein and cellulose ethers.

Hot-melt adhesives are solvent-free, solid compounds with negligible or no volatile organics. They melt and flow when heat is applied and thereafter solidify upon cooling, giving strong adhesion. Hot-melt adhesives are commonly supplied in solid cylindrical sticks of various diameters and are melted in an electric hot glue gun (Figure 3.17). For hot-melt adhesives, a minimum pressure level should be used until solidification of the hot-melt. It then develops sufficient tack holding the substrates. The tack of an adhesive refers to the holt-melt stickiness while changing from a liquid to a solid state.

Hot-melt adhesives are suited for repair and small-scale assembly work, allowing bonded parts to be exposed to stress or further processing in a few minutes after application. An apparent drawback is relatively high processing temperature of ca. 200 °C, limiting application of these adhesives to temperature-sensitive materials (e.g., plasticized PVC). Low-melt cartridges can operate, however, already at ca. 110 °C. Hot-melt adhesives are usually based on thermoplastic polymers, such as ethylene–vinyl acetate copolymers, polypropylene, polyamides and polyesters. Due to thermoplasticity, melting and resolidification processes are repeatable.

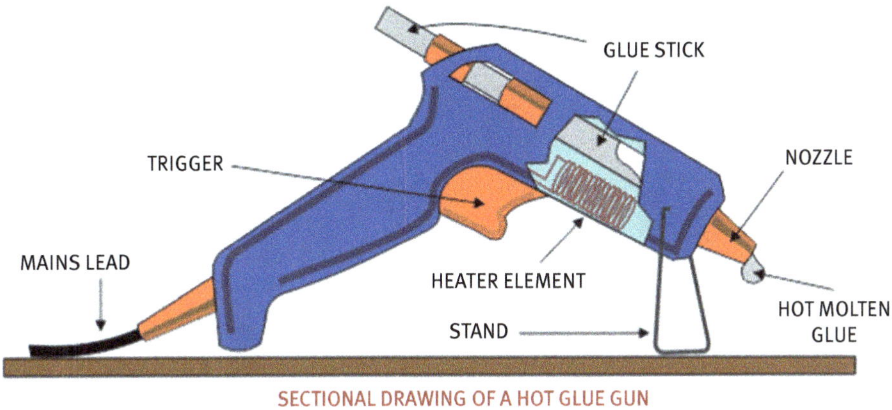

Figure 3.17: Hot glue gun [15].

In hot-melt adhesives, viscosity that controls the extent of surface wetting, decreases with temperature. Temperature of the melt and the application equipment should be, therefore, carefully controlled and maintained as constant as possible. High strength of a holt-melt adhesive giving high melt viscosity apparently contradicts the need for low melt viscosity from the application viewpoint. This along with a limited adhesion of pure thermoplastics requires application of diluents (waxes, plasticizers, stabilizers, extenders, and pigments), which are added to polymers. These diluents decrease viscosity and enhance wettability, adhesive strength, rigidity or flexibility.

Hot-melt adhesives are applied at temperatures (so-called running temperatures) that provide sufficient viscosity for proper surface wetting. Running temperatures exceeding adhesive degradation temperatures deteriorate the overall properties; thus introduction of stabilizers helps improve stability of the material.

In the manufacturing of hot-melt adhesives, the components are melted in heated stirring vessels and then solidified on a cooled steel belt and subsequently cut. Another option is to use strip granulation, when the melt passes through a strip-casting device and is laid down in strip form onto the cooling belt. The basic principles of strip casting are presented in Figure 3.18, while an alternative way of making drops is shown in Figure 3.19 illustrating that the hot melt is deposited onto a continuously running belt made of steel in the form of defined droplets.

At the end of the belt system, the strips are chopped into rectangular granules. A general overview of the manufacturing process offered by Sandvik is presented in Figure 3.20.

A hot-melt adhesive production system consists of remelting equipment, mixing reactor, strip former and cutter or alternatively Rotoform drop depositor, steel belt cooler and downstream bagging and weighing equipment. Two reactors ensure continuous pastillation.

When heat is transferred from the product to cooling water, the droplets are solidified. Cross-contamination is avoided as the cooling water and product do not come

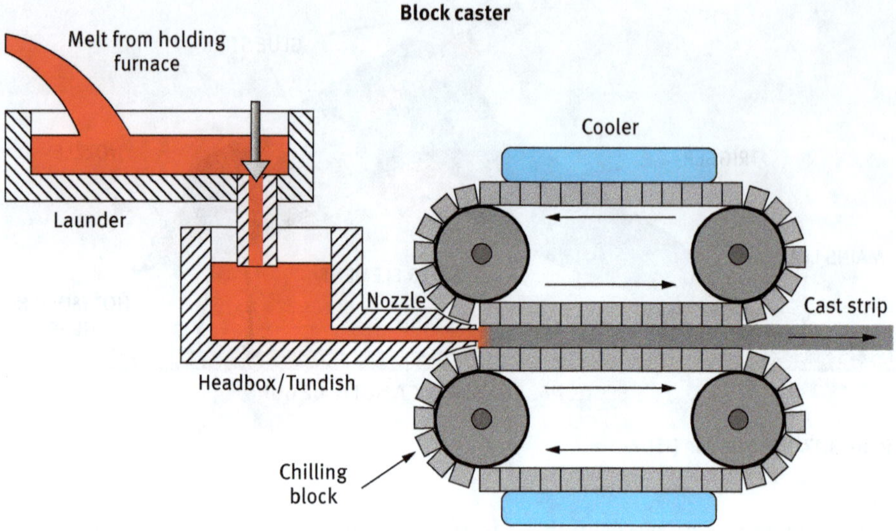

Figure 3.18: Basic principles of strip casting [16].

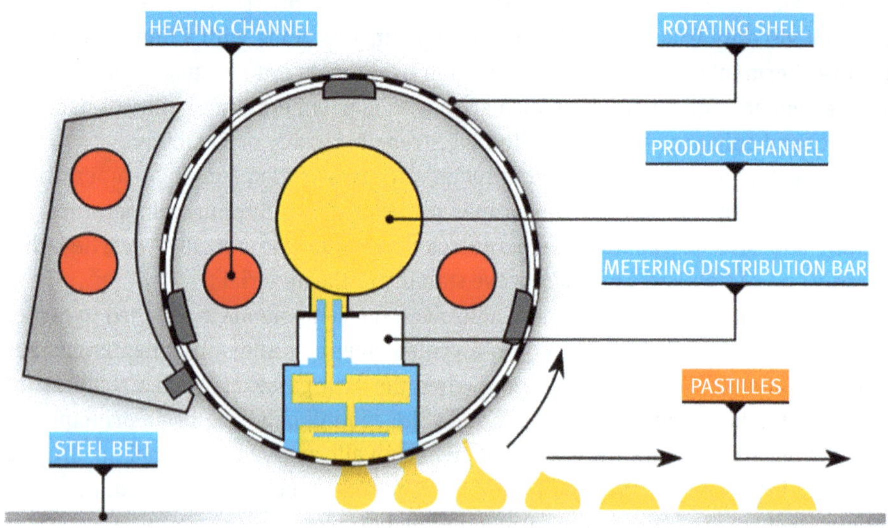

Figure 3.19: Rotoform drop depositor [17].

in contact. At the end of the cooling belt, air is used for additional cooling. The solidified product is collected in a buffer vessel, which is followed by weighing and bagging. This technology allows accurate control of pastille dimensions and shape.

Pressure-sensitive adhesives, needed in many applications ranging from automotive to medical ones, form a bond with the substrate upon exposure to light pressure.

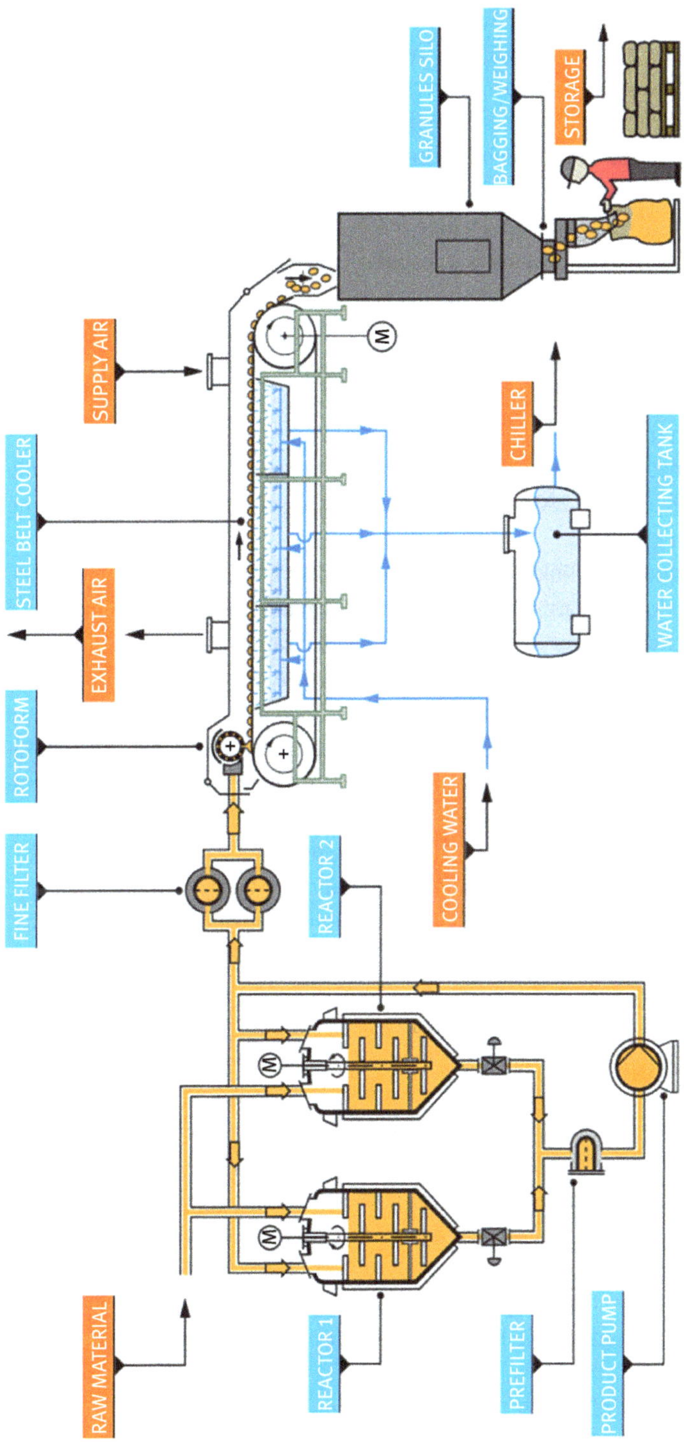

Figure 3.20: Production of hot-melt adhesives [17].

They can be water-based, solvent-based or hot melt and provide nonspecific adhesion to virtually any surface, in particular, for the permanently tacky coating of tapes, films, and labels. Natural and synthetic rubbers in conjunction with modified natural resins and/or synthetic resins have the properties of pressure-sensitive adhesives.

Reactive adhesives (Table 3.4), which can be cured either by UV light, heat or moisture, are predominantly low-molecular-mass monomers and/or oligomers. During curing, they are transformed into high molecular mass, often three-dimensionally cross-linked polymers as a result of different polymerization, polyaddition and polycondensation reactions. Reactive adhesives can be one-pack, which are easier to handle, or two-pack requiring complete mixing in a correct ratio prior to use. Moreover, a third option of no-mix adhesives exists to avoid mixing errors with two-pack adhesives. In this case, the bond is formed instantly when the two surfaces, one pretreated with the adhesive resin and another with the adhesive primer, are brought together.

Warm-setting adhesives generally require temperatures of ca. 80–100 °C, while higher temperature levels (100–250 °C) are needed for hot-setting adhesives affording adhesive joints of high quality.

When formation and cross-linking of the adhesive film is done by polymerization, two-pack adhesives include unsaturated polyesters with monomeric vinyl compounds, such as styrene or methyl methacrylate. One-pack polymerization adhesives comprising cyanoacrylate adhesives setting under influence of moisture or alkalinity can be used for various substrates (metals, plastics, rubber or wood). Cyanoacrylate adhesives are somewhat exceptional being able to cure on several substrates within seconds or minutes, while typically cold-setting reactive adhesives require hours to days at room temperature or slightly elevated temperature for hardening.

Epoxy resin adhesives in combination with amines, polyamidoamines, or dicarboxylic acid anhydrides follows polyaddition mechanism and are used for metals, silicate-containing materials and plastics. Polycondensation (elimination of low-molecular-mass cleavage products during setting) is operative, for example, in the case of phenol (or urea, melamine, resorcinol)–formaldehyde resins and when adhesives are applied for wood or metals. Adhesion can be also obtained by cross-linking halogenated polymers with cross-linking agents under elevated temperature and pressure making rubber-to-metal bonding agents.

Ultraviolet curing requires less time compared to heat-cured adhesives reducing the overall production costs. These types of adhesives based on special acrylic acid esters are applied in various areas, including dental fillings, construction, interiors, wood and furniture to name a few. Finally, moisture-curing adhesives, as the name says, require moisture to trigger the curing reaction, which can be used in the case of viscous adhesives typically consisting of nonvolatile urethane polymers. One of the current trends is development of bio-based alternatives for moisture-based urethane adhesives.

An approximate guide of which type of glues should be used for different purposes is presented in Table 3.5.

Table 3.5: A guide for adhesives [18].

		Two-component epoxy resin glues			Polyurethane glues		Cyanacrylate glues		UV glues		Silicone glues	Dispersion and contact glues		
		CG 500-35	**Uhu Plus endfest 300**	**Aradite 2011**	**Aradite 2026**	**Scotch Weld DP 610**	**UHU instant Adhesive**	**Cyberbond 2,999**	**Photobond GB 368**	**Photobond GB 345**	**Elastosil N2199**	**Konstruvit**	**Chrisanne**	**Bostik 1475**
Inorganic materials	Crystal	✓			✓	✓			✓	✓	✓			
	Glass	✓			✓	✓			✓	✓	✓			
	Ceramics	✓	✓	✓							✓			
	Stone	✓	✓	✓	✓						✓			
Metals	Aluminum	✓	✓	✓							✓			
	Brass	✓	✓	✓							✓			
	Silver	✓	✓	✓							✓			
	Steel	✓	✓	✓							✓			
Gluable synthetic materials	PC	✓	✓	✓	✓	✓	✓	✓						
	PS	✓			✓		✓							
	PVC/ABS	✓	✓	✓	✓	✓	✓	✓			✓			
	Rubber	✓			✓		✓	✓						
Organic materials	Paper	✓					✓	✓			✓	✓	✓	✓
	Cork	✓									✓	✓	✓	✓
	Wood	✓	✓	✓							✓	✓	✓	✓
	Textiles					✓						✓	✓	✓

Finally, it is worth considering a few issues related to joint design. Failures in application of adhesives can be related to inappropriate joint design (Figure 3.21) and could be due to failure in adhesion, cohesion or the substrate.

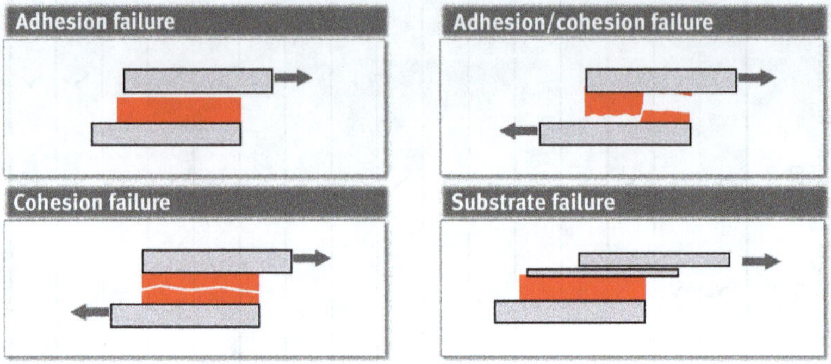

Figure 3.21: Failures in application of adhesives [11].

Some alternatives to shear joint design (Figure 3.22) can mitigate problems with adhesives per se.

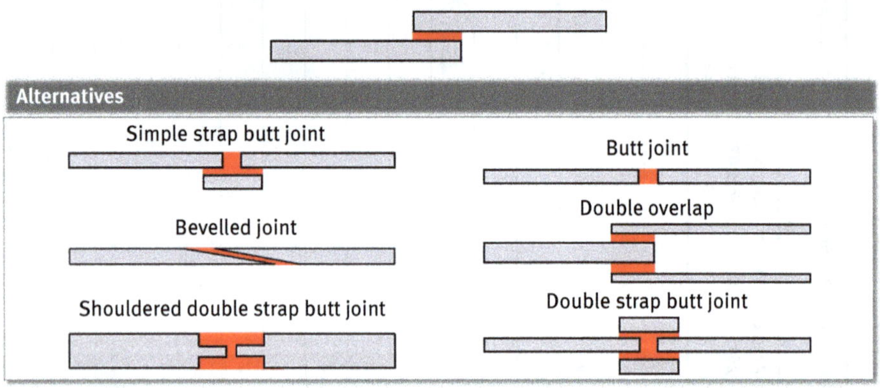

Figure 3.22: Alternative design of shear joint [11].

3.2 Paints, coatings and dyes

3.2.1 Properties and composition of paints and coatings

Paints and coatings are liquid, paste or powder products, which when applied to surfaces give layers of certain thickness for the purposes of decoration (automotive coatings, household appliances, furniture); protect against corrosion, weathering, and mechanical damage; or provide information (traffic and information signs).

Pigments should not be confused with dyes also applied for decoration purposes. Contrary to dyes, pigments are practically insoluble in the medium for application being dispersed in it as solid particles. The dyes are used in a dissolved form and are fixed on the surface by adsorption or chemical reaction.

In fact, coatings can also have other properties as illustrated in Table 3.6.

Table 3.6: Properties of functional coatings.

Electrical/magnetic properties	– Conductive – Insulation – Dielectric – Shielding – Antistatic – Electromagnetic
Mechanical properties	– Wear – Abrasion – Hardness – Lubrication
Physicochemical properties	– Corrosion protection – Antifouling – Antimicrobial – Fire-retardant – Hydrophilic and hydrophobic – Self-sealing – Thermal barrier
Optical properties	– Antireflection – Photocatalytic – Photochromatic – Photoluminescence

Sometimes paints are defined as drying oils and/or natural resins, while coatings are associated with synthetic resins or inorganic silicate polymers and are thought to have superior performance in terms of, for instance, adhesion, toughness and resistance to chemicals.

The range of applications is thus very broad. The industrial segments include for example automotive industry, railroad transportation, aviation, marine coating and road marking, etc. and in addition to private customers there are numerous commercial organizations (hotels, shopping malls, restaurants, manufacturing facilities, offices, hospitals, warehouses, etc.) where decisions are made on which type of paints and coating to use.

When coatings are applied, adherent films are formed on the surfaces either physically or chemically; thus knowledge of the specific interactions between the paint components and substrates (e.g., wood, plastic, metal and glass) is important. Moreover, for many applications, several layers should be used to ensure that the required characteristics of the coating are reached. In addition to the primer, two or three coats needed to provide strong adhesion to the substrate; further coating layers are applied bringing durability, mechanical properties, corrosion and chemical resistance as well as the visual appearance (opacity, color, gloss, smoothness). Subsequently purposeful use of paints and coatings should be based on fundamental understanding of particle–surface adhesion, interactions between their components and properties of coatings.

Physical film formation can be arranged by drying or melting for liquid and powder coatings, respectively, and is only possible with solid and nontacky coatings. Drying is done by either evaporation of the organic solvents from solvent-containing paints or by water evaporation from waterborne paints. Physical drying is done for paints with high-molecular-mass polymer binders (cellulose nitrate and cellulose esters, chlorinated rubber, vinyl resins, polyacrylates, styrene copolymers, thermoplastic polyesters, and polyamide and polyolefin copolymers). For liquid, tacky, or pasty coating components, film formation is a result of a chemical reaction between the components – low-molecular-mass products react with other low or medium molecular mass binder components by polymerization or cross-linking forming macromolecules. The reactive components can be a part of the coating or added upon application. An example of polymerization is a reaction of unsaturated polyesters with styrene or acrylate monomers when one component serves as a reactive solvent.

Polyaddition cross-linked macromolecules are formed when low-molecular-mass reactive polymers (alkyd resins, saturated polyesters, or polyacrylates) react with polyisocyanates or epoxy resins at room temperature. The binder components are mixed shortly before application. In case of polycondensation addition of catalysts or application of higher temperatures are required. Oxidation of coating components is also a possibility giving a chemical film.

There is also an option to combine physical and chemical film formation and in practice not just one method is used as physical and chemical drying can take place simultaneously.

Application of paints can be done by several methods such as dipping, flow coating and spraying. Other methods include brushing (Figure 3.23), rolling, printing and silk screening.

A simple coating method not requiring any sophisticated equipment is dip coating giving 95 to 100% theoretical transfer efficiency (TE). In this method, a product is immersed into a tank of coating material and then withdrawn. This is followed by draining in the solvent-rich area above the coating surface and subsequent drying. Viscosity, flow, the solid loading by volume and the withdrawal rate define the film thickness.

The wedge effect implies a differential of the film thickness from top to bottom, along with fatty edges on lower parts of products; runs and sags can be considered as

Figure 3.23: Painting by brushing (photo by the author).

disadvantages of dip-coating. Care should be taken that the light products are not floating off the hanger and hooks and falling into the dip tank.

Dip-coating can be also arranged in a continuous mode as illustrated in Figure 3.24 for coating of a fabric.

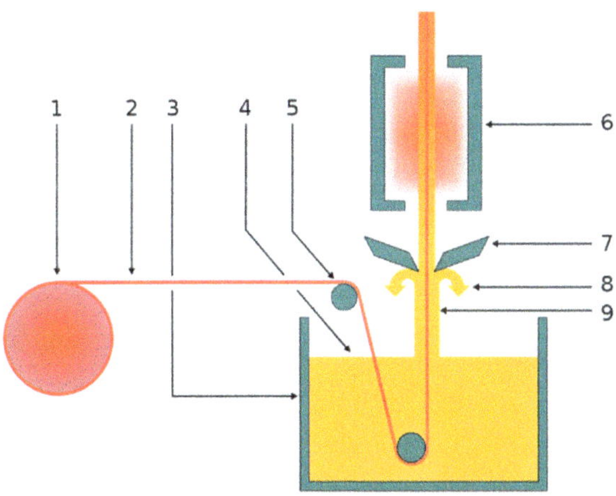

Figure 3.24: A schematic of the continuous dip-coating process [19]: 1, roll of coarse cloth; 2, cloth; 3, bath; 4, liquid material; 5, rollers; 6, oven; 7, scrapers; 8, excess liquid falls back; 9, coating remains on the fabric cloth.

Electrocoating with organic coatings deposited from an electrolytic bath has high theoretical TE (95–100%) and can be done from an alkaline bath when the coating is deposited on the anode or from an acidic bath leading to coatings on the cathode (Figure 3.25). This method is less labor-intensive and generates lower solvent emissions giving uniform coatings in the axial direction as well as inside and outside on products of complex shapes. On the other hand, electrocoating being more expensive with higher capital and material costs requires high operation skills.

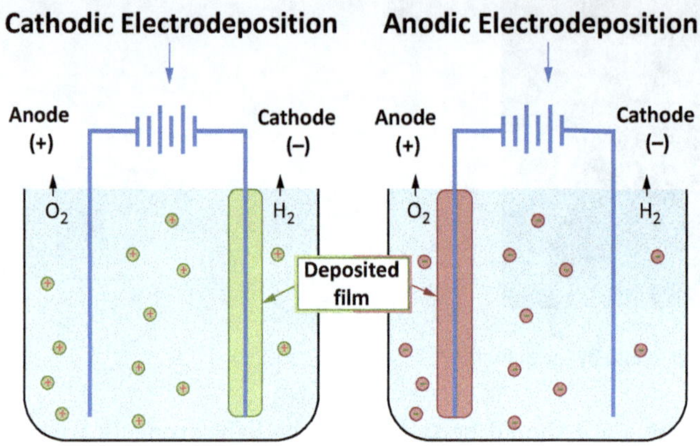

Figure 3.25: Electrocoating [20].

Spray coating (Figure 3.26) with a much broader range of theoretical TE between 30% and 90%, is the most common and versatile method, which can be easily automated, giving excellent surface quality and uniform thickness.

Figure 3.26: Application of paints by spraying [21].

Among several versions of this method, air-atomized spray coating with 30–40% TE can be mentioned when (Figure 3.27) compressed air is mixed with a stream of a liquid coating material in adjustable ratios and flows giving small droplets, which are coated on the material. Apparent disadvantages are low TE and a need to spray only low-viscosity coatings, leading to emissions of VOCs.

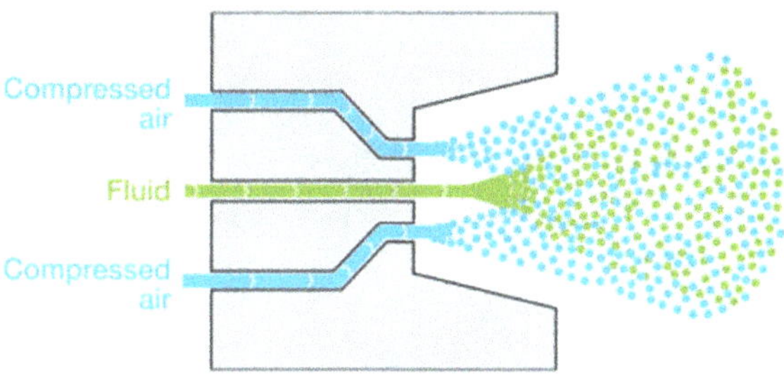

Figure 3.27: Air-atomized spray coating [22].

In-mold painting (Figure 3.28) used for automotive applications eliminates paint facility expenses and provides better design flexibility. In essence, in mold coating plastic components are coated in a special tool via so-called negative molds during production. In an open process option, coating and application of the reinforcing material takes place in a nonclosed tool, while in an alternative close variant, the tool is closed after applying the coating layer before the reinforcing material is injected (Figure 3.28).

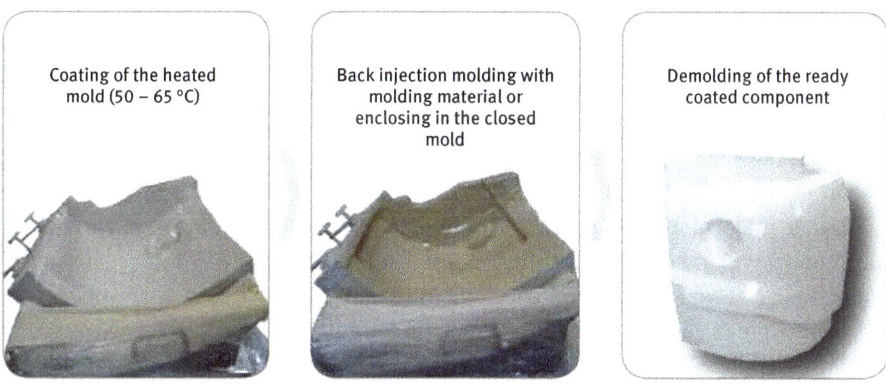

Figure 3.28: Application of paints by in-mold painting [23].

The properties of a paint are determined not only by its composition, which will be discussed below, but also surface conditions, in particular its cleanliness.

Paint is essentially composed of nonvolatile binders (resin), pigments, volatile solvents and various additives (Figure 3.29). Solvents and pigments are not always present in a coating formulation. Thus, solvent-free paints and pigment-free varnishes are commercially available.

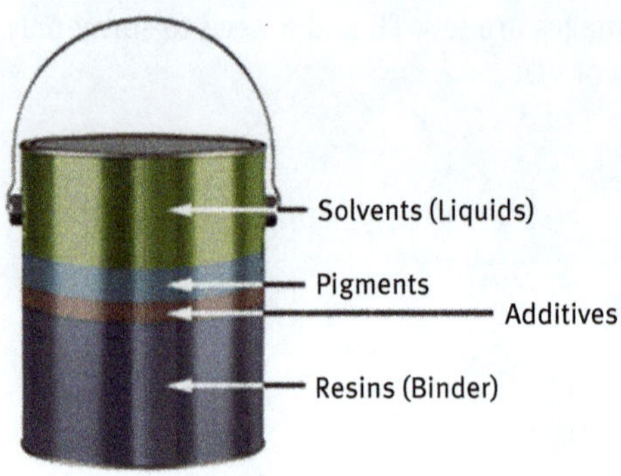

Figure 3.29: Composition of paint [24].

For emulsion paints the solvent is water, while for resin-based paints a variety of organic compounds, such as aliphatic, aromatic and chlorinated hydrocarbons, alcohols, ketones, esters or glycol ethers can be used. Among nonvolatile components besides binders and pigments, also plasticizers, various paint additives, dyes and extenders can be mentioned.

Binders, being the most important component of a paint formulation, determine how paints are applied. This holds for drying and hardening behavior, as well as adhesion to the substrate. Binders also define performance properties (mechanical properties, chemical, UV and moisture resistance, etc.). There is a broad range of binders available of different origin (Figure 3.30), which are macromolecular products with a molecular mass between 500 and ca. 30,000.

Acrylic, polyester, alkyd and polyurethane resins comprise ca. 80% of the global paints and coating market with the largest share (ca. 45%) belonging to acrylic resins. The performance of common coating resins is presented in Table 3.7.

Higher molecular mass products (cellulose nitrate, polyacrylate and vinyl chloride copolymers) are suitable for physical film formation, while low-molecular-mass products (e.g., alkyd or epoxy resins) should be chemically hardened after application to the substrate to generate cross-linked macromolecules with high molecular mass.

Higher molecular mass of the binder in the polymer film improves elasticity, hardness and impact deformation at the expense of higher solution viscosity. Low viscosity combined with low solvent content is desirable from the environmental viewpoint and ease of application. This can be achieved by applying low-molecular-mass binders giving low solution viscosity and low-emission paints with high solids content. Moreover, solvent-free paints can be obtained. As a result, there should be a compromise in terms of the molecular mass of the binder, which can, in fact, consist of a mixture of several reactive components.

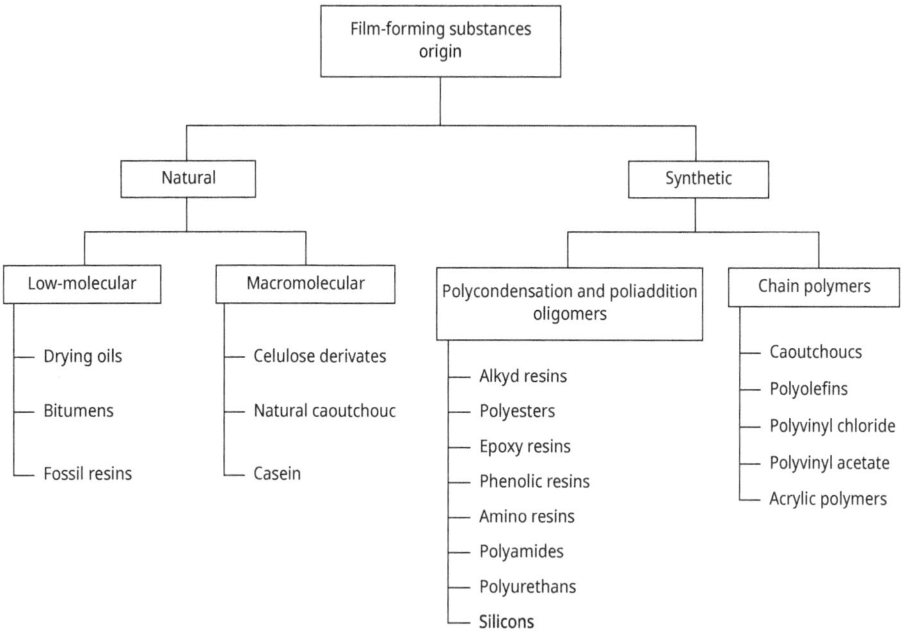

Figure 3.30: Binders applied for paints and coatings.

Table 3.7: Performance properties of common coating resins [25].

Resin type	Humidity resistance	Corrosion resistance	Exterior durability	Chemical resistance	Resistance to light abrasion
Acrylic	E*	E	E	G	E
Alkyd	F	F	P	G	G
Polyester	E	G	G	G	G
Polyurethane	E	G	E	G	E

*E, excellent; G, good; F, fair; P, poor.

Alkyd resins are the most common resins in solvent-based paints being used for both air-drying and heat-cured paints. These resins are extremely important for corrosion protection coatings. Glycerol and phthalic anhydride are often used as feedstock for making alkyd resins resulting in complex glycerylphthalate resins (Figure 3.31).

The quality of films made just by an alcohol and an acid (or anhydride of a dibasic acid such as phthalic anhydride) is far from being sufficient. By incorporating oils in the reaction mixture and introducing long-chain carboxylic acids films with good durability, excellent color retention and superior gloss can be obtained. Some other additives to alkyd resins are given in Table 3.8.

Alkyd resins containing oil (fatty acid), dicarboxylic acid (mainly phthalic acid or phthalic anhydride), and polyhydric alcohol (e.g., glycerol) are classified based on the

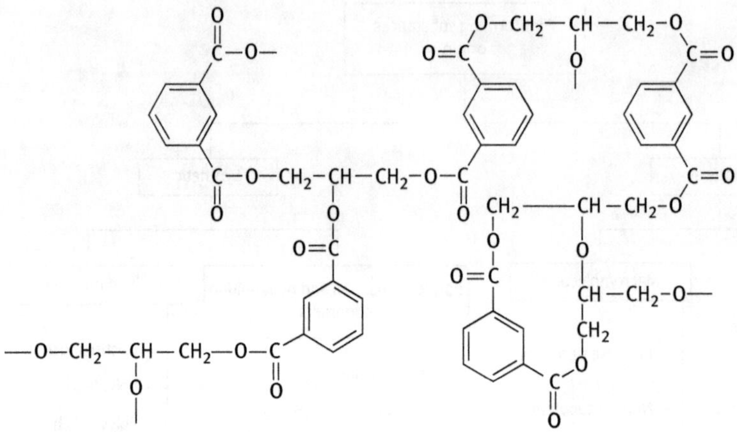

Figure 3.31: A fragment of glycerylphthalate resin.

triglyceride (fatty acid) content relative to the solvent-free resin as short oil (≤40%), medium oil (41–60%), and long oil (61–70% oil content) resins. The latter alkyd resins

Table 3.8: Additives for alkyd resins.

Additive	Function
Styrene 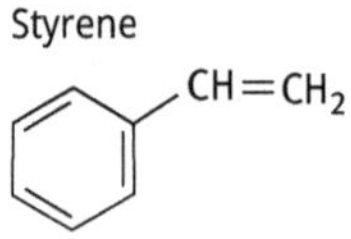	- Reduce drying time - Improve durability of paint
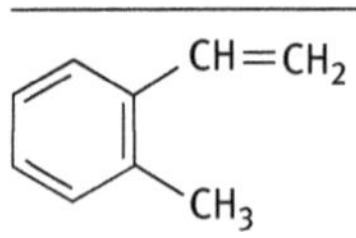	- Reduce drying time - Improve durability of paint
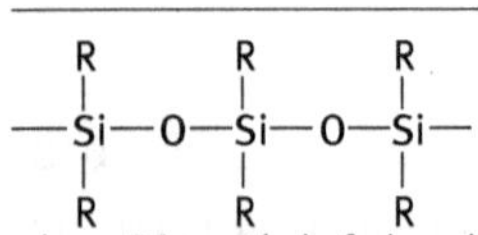 where R is methyl of phenyl	- Co-polymerization for more durable glossy paints - Effective with dark paints
Abietic acid 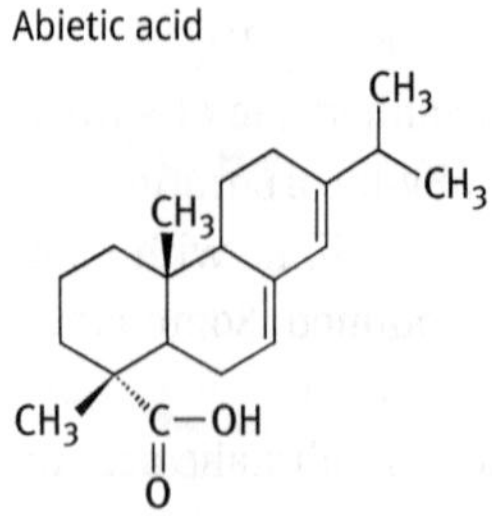	- Dries quickly - Produce glossy paints - Soluble in aliphatic solvents

Table 3.8 (continued)

Additive	Function
Phenolic resins (e.g., bisphenol A) 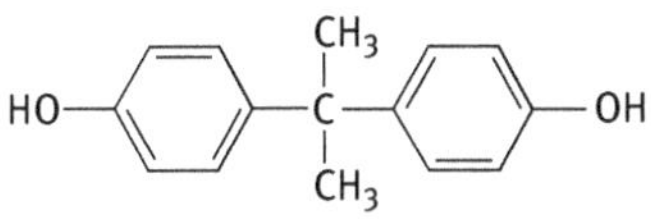	– Resistance to water, alkali, grease and oil – Hard surface – Glossy finish

are used as binders for architectural and do-it-yourself paints. They are mainly modified with soybean oil and do not yellow in the absence of direct light. Medium oil alkyd resins based on drying oils or fatty acid mixtures are applied for air-drying and forced-drying machinery coatings, industrial coatings and some other applications (car repair finishes). Alkyd resins can be combined with other binders, for instance, short oil resins can be used together with urea resins giving acid-curing paints for wood. Mixing short oil resins with nitrocellulose gives cost-effective and easily applied furniture lacquers and automotive refinishing systems. Alkyd resins can be also mixed with chlorinated rubber or amino resins, etc.

Alkyd paints are widely applied for protection and decoration in many sectors ranging from industrial to home applications.

Among other resins used in paints, are epoxy resins, nitrocellulose, chlorinated rubber and polyurethanes. Paints based on polyurethane give tough, durable and easy to clean films retaining their gloss. They can be used, for example, in painting aircrafts, where durability and ease of cleaning are important. Reinforcement of the polyurethanes properties can be done by combining with other resins, for example, acrylics giving excellent outdoor weathering properties.

Paints containing acrylic resins as binders are now one of the largest products in the paint and coatings sector. In these paints, polyacrylates (copolymers of acrylate and methacrylate esters) are applied as binders. The resins can be manufactured as solids; solutions in organic solvents or water; emulsions, or dispersions.

The advantages of acrylate resins compared to other paint binders are related to their chemical structure, namely, absence of unstable double bonds. This results in high chemical resistance, stability to UV radiation and hydrolysis, colorless, transparent appearance with excellent gloss and gloss retention. Cross-linked polymers are even more resistant to chemicals than linear polymers, which is extremely important for high-grade coatings.

Epoxy resins are applied in the appliance/electric insulation and automotive industries due their heat-resistant properties. At the same time, ultraviolet light resistance is poor eventually leading to discoloration when used outdoors. Subsequently two-component epoxy coatings should be used for such purposes and in aggressive environments.

Nitrocellulose lacquers are a mixture of binders (nitrocellulose and resins) dissolved in organic solvents and include besides binders also plasticizers and pigments. Applied at low solids using expensive solvents, nitrocellulose lacquers do not meet the environmental standards regarding VOC emissions and thus are modified with other resins such as alkyds.

Chlorinated polymers with the mean molecular masses of 50,000–350,000 form coating films by physical drying and require plasticizers (ca. 35%) because brittle films are formed otherwise.

Pigments (substances insoluble in the application medium), besides the decorative function giving the paint its color and finish, also protect the surface from corrosion and weathering. In addition pigments help to hold the paint together. Inorganic (Table 3.9) and more expensive organic substances are used as pigments. Organic pigments give clearer colors being also more transparent and stable.

Table 3.9: Some common inorganic pigments [26].

Material	Production	Advantages	Disadvantages	Color
Carbon black	Decomposition of carbonaceous matter	High strength, good color, light and weather resistance	Thickens paint	Black
Azurite $Na_7A_{l9}Si_4O_{24}S_2$	Kaolin, sodium carbonate, sulfur and carbon heated above 800 °C	Rich colors	Fades on contact with acid	Blue
Cadmium sulfides	Reacting cadmium salts with sodium sulfide	Heat and light resistant, clear pigment, high opacity	Expensive, poor weather resistance	Greenish yellow to red to Bordeaux
Chromium oxide		Light, weather, alkali and acid resistant, thermally stable	No clear colors	Green, blue
Iron oxides	Mined or synthesized	Nonreactive, light and weather resistant	Cannot produce clean shades	Red, yellow, brown, black
Zinc chromate	Synthetic (1)	Corrosion by reaction (2) is inhibited by reaction (3)	Thickens paint	yellow
Titanium dioxide	Synthetic	High strength and opacity, good UV resistance, low costs	Forms radicals degrading the binder	White

Table 3.9 (continued)

Material	Production	Advantages	Disadvantages	Color
Lithopone ZnS mixed with $BaSO_4$	Synthesized (4)	Pure tints when mixed with organics, unreactive, easily wetted	Poor weather resistance	White

(1) $4\ ZnO + 2CrO_3 + K_2Cr_2O_7 + H_2O \rightarrow K_2CrO_4 + 3ZnCrO_4\ Zn(OH)_2$
(2) $4\ Fe + 3H_2O + 3O_2 \rightarrow 2\ F_2O_3\ H_2O$ (spongy mass)
(3) $2Cr^{6+} + 2Fe + 6OH^- \rightarrow 2Cr^{3+} + F_2O_3$ (hard, protecting coating) $+ 3H_2O$
(4) $ZnSO_4 + BaS \rightarrow BaSO_4(s) + ZnS(s)$

Lower amounts required to get an equally strong color and better gloss development can be mentioned as another advantage of organic pigments. Moreover, some organic pigments can absorb UV light, thus preventing the binder from being damaged if it is sensitive to UV radiation. Widespread use of inorganic pigments is justified not only by costs but also resistance to bleeding as well as stability to heat and light. Some special anticorrosion pigments and special color (black and white) pigments have inorganic nature, as there are no organic counterparts giving the same properties.

The most important white pigment for surface coatings is titanium dioxide, being the strongest known pigment in terms of both opacity and tinting power. Properties of other inorganic pigments are mentioned in Table 3.9. Some of the pigments that have, for instance, high opacity, hiding power and heat stability such as cadmium pigments or anticorrosive properties as red lead, Pb_3O_4, have to be taken out from the market because of toxicity and environmental concerns. Organic pigments have gained importance not only because of the legal restrictions but also due to some interesting features that they can offer. Organic pigments have a high light absorption and a low scattering power. In addition, they possess lower density, higher surface area, better color purity and tinting strength than inorganic pigments, and can dissolve at high temperature. Azo pigments (Table 3.10) belong to the most important class of organic pigments. Among other pigments metal-complex ones (e.g., copper phthalocyanine) and higher polycyclic compounds (e.g., anthraquinone, quinacridone, etc.) can be applied. Organic pigments are incorporated in paints as dry powders consisting of agglomerates of pigment particles.

Some pigments can be partially replaced by cheaper secondary pigments or extenders (Table 3.11) or mineral compounds, which are used for economic reasons and to adjust coating properties (e.g., layer thickness and structure, volume, optical features, sticking/adhesion and corrosion resistance).

Besides inorganic extenders also synthetic organic extenders, which include polymer fibers, are used for reinforcement.

Particle size distribution of extenders is similar to pigments and they are incorporated into coatings in the same way. In general, the hiding power and tinting strength of paint depend on the pigment particle size, which is typically in the range 0.1–2.0 μm.

Table 3.10: Examples of azo dyes [26].

Group	Azo dyes	Colors						
		1	2	3	4	5	6	7
Monoazo	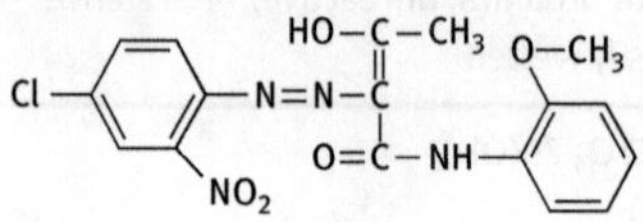 Arylamide yellow (PY 73)	+	+	+	+	+		
Diazo	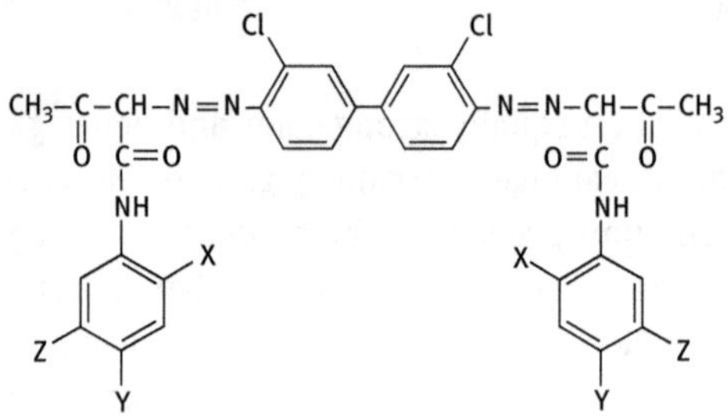 Diarylide yellows	+	+	+	+		+	
Azocondensation	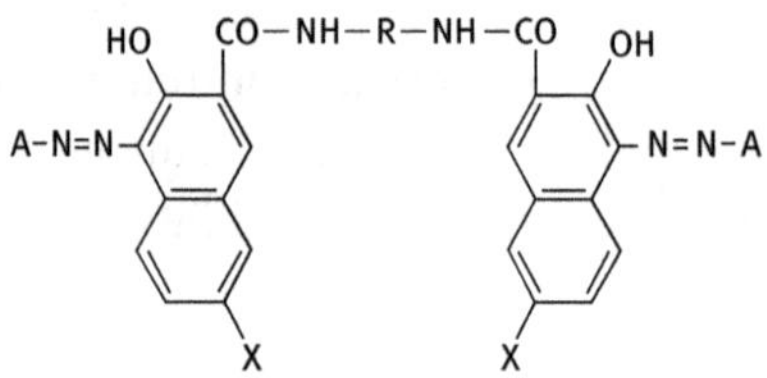	+	+	+	+			
Azo salt	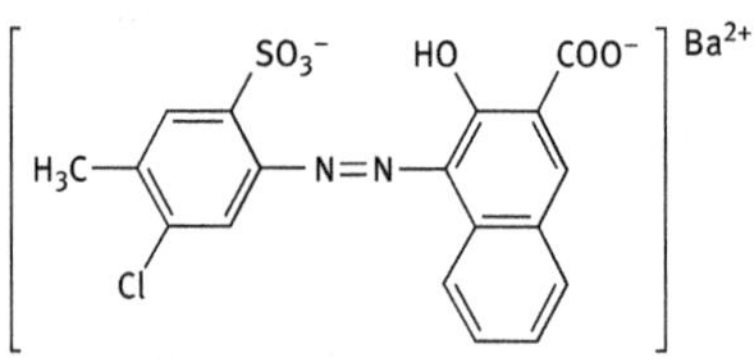 Barium red 2B toner (PR 48.1)	+	+	+	+			

1, yellow; 2, orange; 3, red; 4, brown; 5, violet; 6, blue; 7, green.

Table 3.11: Common extenders [26].

Common name	Formula	Uses
Whiting	Calcium carbonate	Undercoats and flat paints
Kaolin	Hydrated aluminum silicate	Assists TiO_2 dispersion, decreases viscosity
Talc	Hydrated magnesium silicate	Assists TiO_2 dispersion, improves sanding

Table 3.11 (continued)

Common name	Formula	Uses
Silica	SiO_2	Flatting agent, traffic paint (wear resistant)
Mica	Hydrous aluminum potassium silicate	Chemically and solar resistant, improves water resistance
Barytes	Barium sulfate	Traffic paints (wear resistant), pigment extender

As a consequence, pigments have high surface area, which should be effectively wetted by ensuring close contact between the pigment and binder. Such contact is achieved by using high shear forces. This is done by grinding pigment agglomerates into primary particles with simultaneous incorporation into the binder solution. Such deagglomeration influences the color and the depth of color; however, an excessively long dispersion process might be required to reach the maximum depth of color especially with very fine pigments. An alternative to longer dispersion could be an increase in the pigment concentration. Higher pigment dispersion increases the color depth and changes the color tone as already apparent from the text above. In addition, it enhances the gloss and improves viscosity. The latter was discussed in Section 2.6.

Efficient wetting of the pigment surface depends on the critical pigment volume concentration, which in turn is lower at higher pigment dispersion. Above a certain pigment concentration, termed the critical pigment volume concentration, the pigment particles in the coating film are no longer fully wetted by the binder. Viscosity approaches a limiting value asymptotically and the coating film properties deteriorate substantially.

In fact, application of different relative concentration of solvents, binders and pigments results in different types of pigmented paints (Figure 32).

Besides the particle size per se, the particle size distribution (PSD) is an important parameter influencing the pigment powders. Narrower PSD allows easier dispersion as otherwise the small particles are linking the larger ones. This makes dispersion more difficult as there is less available space in the agglomerates worsening solvent penetration.

In addition to resins, solvents and pigments, various additives in quantities between 0.01% and 1% are added to paints for a number of reasons. Table 3.12 provides an overview of some common additives. A few of them are discussed in more detail below.

Film-forming promoters, being closely related to flow agents, give pore-free and uniform surface decreasing the film-forming temperature. High-boiling glycol ethers and glycol ether esters together with hydrocarbons are applied.

Surfactants with a typical polar–nonpolar structure are used as wetting agents helping wetting of pigments by binders and preventing pigment flocculation. Wetting

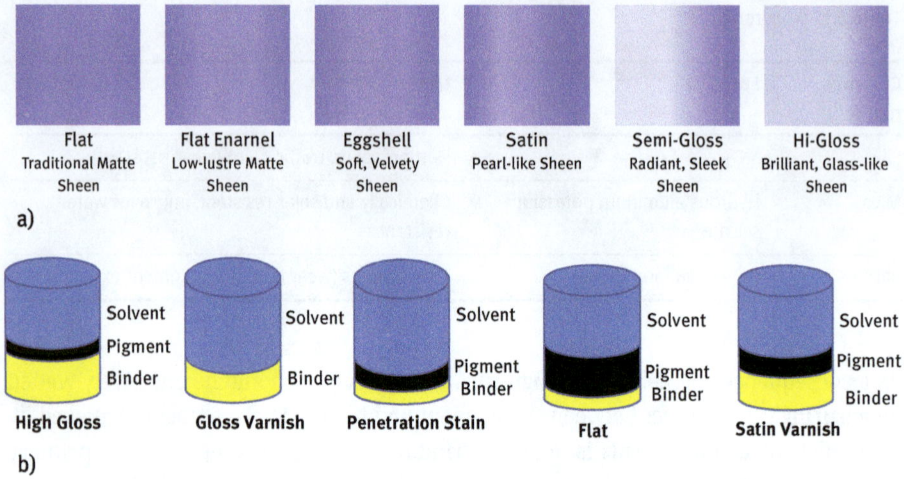

Figure 3.32: Different types of pigmented paints: (**a**) visual appearance [27] and (**b**) relative concentration of solvent, binder and pigment.

Table 3.12: Categories of coatings additives [28].

Manufacturing process aid	Application aid	Performance enhancement	Paint stability	Other
Antifoam	Anti-float	Adhesion promoter	Antifreezing	Matting agent
dispersant	Anti-sag	Abrasion resistance	Anti-settle	Optical brighteners
	Defoamer	Anti-block	Bactericide	Slip aid
	Driers	Anti-Mar	Dispersant	
	Flow and levelling	Corrosion inhibitor		
	Wetting agent	Flame retardancy		
		Mildewcide		
		Rheology modifier		
		UV resistance		
		Water repellent		

additives diminish the binder solution surface tension, making the contact angle close to zero, which is required for efficient wetting.

Dispersing additives are adsorbed onto the pigment surface leading to repulsion between individual pigment particles and stabilization of paint systems.

Antifoaming agents, used to prevent foaming during manufacturing and application, are often very system-specific. For waterborne paints, mineral oils are often used along with finely dispersed hydrophobic particles such as silicone defoamers. The latter do not give a decrease of gloss observed with mineral oils. In solvent-borne coatings, silicones are also the main defoamer components. Acrylates and acrylic copolymers can be used as alternatives to silicone defoamers.

Drying agents with air-drying binders accelerate the decomposition of peroxides and hydroperoxides formed during drying, promoting radical polymerization of the binders. Metallic soaps of monocarboxylic acids with 8–11 carbon atoms, mainly naphthenates and octanoates of cobalt or manganese are used. Since driers can cause skin formation during paint storage, antiskinning agents (antioxidants such as oximes or alkyphenols) are added to air-drying paints to prevent skinning. These agents negatively affecting the driers should be correctly dosed; otherwise they would prolong the paint drying time. Antiskinning agents evaporate together with solvents during drying.

Catalysts are used to accelerate cross-linking of binders at room temperature. For polyester–melamine resins free acids or their ammonium salts can be used, while in the case of polyester–isocyanate resins, tertiary amines or dibutyltin dilaurate can be applied.

Antifloating and antiflooding agents prevent horizontal and vertical segregation of pigments and thus differences in the color and luster of the film surface.

Matting agents include natural minerals (talc or diatomites) and synthetic materials (e.g., pyrogenic silica). They are applied to get coatings with a matte, semimatte, or silk finish. Illustration of matting is presented in Figure 3.33.
Neutralizing agents (ammonia and various alkylated aminoalcohols) are used in waterborne paints for the purpose of neutralizing binders and stabilizing the product.

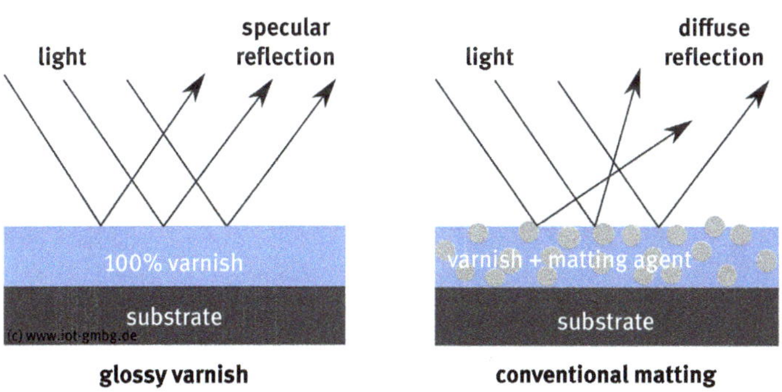

Figure 3.33: Influence of matting agent on light reflection [29].

Waterborne coatings systems are prone to deterioration because of microorganism attack, and therefore biocides and fungicides are applied as preservatives. While solvent-borne coatings per se are not influenced by microorganisms, this is not the case with dry paint films and therefore biocides are applied for coating protection. Preservatives should easily evaporate and also have low solubility in water.

Corrosion inhibitors are used when waterborne paints are applied to metallic substrates. They include oxidizing salts (e.g., chromates or nitrates), organic amines or organic salts (benzoates, naphthenates, octoates).

From the paint rheology viewpoint, when paint and a coating are applied with a spray gun, brush, or a roller, transfer of the paint from the bulk container to the applicator, transfer of the paint to the surface to form a thin and even film, and finally the flow-out of the film surface should be considered. Shear rates during different stages are obviously different as exemplified in Figure 3.34.

Low viscosity of the paint is needed for facile use in an applicator. The paint should be able to penetrate between the bristles of a brush or the porous surface of the roller and be held by capillary/surface tension forces during the transfer to the surface. Too high loading of a brush implies that the capillary forces will not be able hold the weight of the paint leading to paint drip or runoff the brush. Alternatively, a thin paint film is formed for too low brush load. The latter case can also result in non-uniformity of the film, with areas of thicker films.

An increase of the brush-loading is achieved at higher bulk paint viscosity or preferably by adding rheology modifiers as the latter give a shear-thinning system, preventing dripping or running of the paint, and moreover high residual viscosity mitigates potential sedimentation of particles in the paint.

Some attempts have been made to calculate the range of shear rates involved in paint brush applications. For a linear velocity (V) of 0.5 m/s and the layer thickness (d) of 0.2 mm (2×10^{-4} m), the shear rate $V/d = 2{,}500/\text{s}$. Obviously higher linear velocity and lower thickness result in even higher shear rates.

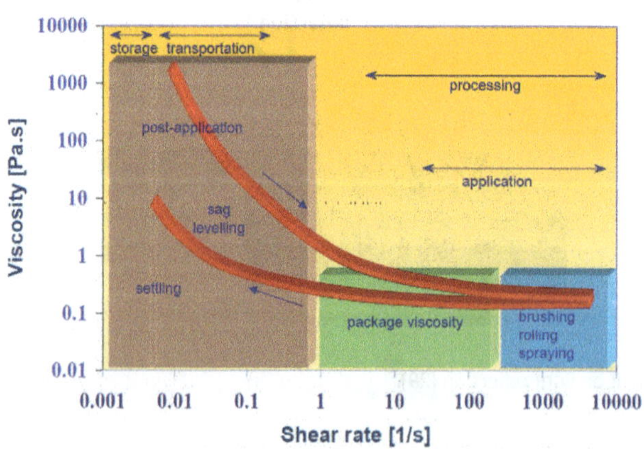

Figure 3.34: Paint properties and shear rate [30].

Not only in brushing, but also in spraying the paint rheology should be carefully controlled. Spraying requires viscosity low enough that the paint can be thinned right before the application. Too small size of the droplet should be avoided, resulting otherwise in drifting on application; thus rheological modifiers are applied to achieve the optimum droplet size distribution. In general, the droplet size of the spray influences the appearance of the finished surface significantly. Glossy or matte surfaces are influenced not only by the presence of a matting agent, but also depend on whether a fine or a coarse spray, respectively, is applied. The size of droplets depends on the surface tension, viscosity and density of the paint per se, as well as the spraying conditions (the flow rate and relative velocities of the liquid and air).

It is apparent that for roller-coating thicker paints are used than in spraying. At the same time, if viscosity is too high, the paint would not be able to flow under gravity or low pumping energy and cannot spread to an even layer when a roller is used.

Solvent evaporation during application has a major effect on color uniformity, hiding power, rheological properties as well as flow faults such as sagging and slumping.

In summary, a paint formulation should have a sufficiently low viscosity allowing a transfer to the applicator and simultaneously sufficiently high residual viscosity preventing particle sedimentation and high modulus to prevent syneresis. Such requirements can be fulfilled with the aid of appropriate rheology modifiers. The same strategy is applied not only for paints but for other formulation products as well, for example, concentrates of agrochemicals.

For painting applications, an optimum viscosity is needed, as low viscosity will result in flowing of paints from vertical surfaces, while too high viscosity will result in lumps of thickened paint and irregular films. Subsequently the initial low viscosity essential for ease of spreading and coating on the substrate must be followed by a sharp increase in viscosity upon solvent evaporation. The latter process should not be too slow, as otherwise it may result in dripping and sagging; or too fast leading to a film with the brush marks of the applicator. All such brush marks should disappear during drying.

From the above discussion on the rheological requirements of the paints it is apparent that the flow properties of paints can be modified with rheology additives (Table 3.13) providing good levelling or, in other words, formation of smooth uniform surfaces upon paint drying. Efficiency of the flow agents, which decreases paint viscosity during drying, depends on the binder and drying temperature.

In emulsion paints mainly cellulose derivatives or polyacrylates are used as thickeners. Polyurethane thickeners (associative thickeners) have more favorable levelling properties.

Light stabilizers were discussed in Section 3.1.1. For industrial coatings, such UV absorbers such as hydroxyphenylbenzotriazoles are the most important, while sterically hindered amines are used as radical scavengers.

Table 3.13: Types of rheology modifiers.

Rheology modifiers	Waterborne	Cellulosics Acrylates Associate thickeners Clays
	Solvent-borne	Organoclays Hydrogenated castor oils Polyamides Overbased sulfonates

3.2.2 Paint systems

3.2.2.1 Solvent-borne paints

These paints when physically dried are solutions of thermoplastic polymers with high molecular masses (>20,000) and a high solvent content (>60%). Examples include thermoplastic and optoelectronic coatings as well as metallic basecoats used in the production of automobiles. The reasons for a high solvent content are related to thin layer thicknesses when spin coating is applied; low solubility of polyimides needed for electrical insulation coatings or optimal rheological behavior for metallic basecoats.

Solvent-borne paints are flammable and have strong odor because of high content of VOCs. They have harder finish than waterborne paint, better tolerance to poor weather conditions, can be wiped easier, provide a higher gloss and exhibit excellent resistance when two painted surfaces are put into contact. These paints, however, become brittle as the paint ages.

Environmental regulations regarding application of solvents led to development of paints with a low solvent content (medium solid and high solid (>85%) coatings) or even powder coatings, which are solvent-free. Solvent-free and low-solvent paints allow less pollution improving overall safety. Moreover, application of these paints is more economical providing savings in materials, transportation and energy costs, as well as time because of higher layer thicknesses in an application cycle.

A decrease in the solvent content of paint can be achieved to a certain extent by introducing special additives, using solvents with solubility parameters corresponding to the binder or adding cosolvents, which decrease viscosity. Challenges with application of low-solvent paints are related to properties determined at least partially by the solvent (flowability, anti-sagging on vertical surfaces, anti-wrinkling or drying behavior). Even high-viscosity binders with high glass transition temperatures undergoing physical drying can be used with appropriate solvents. In the case of low solvent content, the properties mentioned above should be carefully considered because a poorly selected binder leads to sagging or wrinkling.

Difficulties in application of low-solvent paints are related not only to absence of physical drying, but also to longer cross-linking time. To overcome apparent shortcoming with curing time, reactive polymers should be used with a higher degree of unsaturation or a larger proportion of functional groups. Binders could be also applied together with nonvolatile reactive diluents (low-molecular-mass reactants acting as solvents with very low volatility). The latter take part in cross-linking because they have the same or similar functional groups as in the principal binder. Some examples of reactive diluents include polyfunctional methacrylate esters or low-viscosity metal compounds such as metal (e.g., aluminum, titanium or zirconium) alcoholates. In the latter case, a coordination compound with the binder is formed upon removal of an alcohol. Incorporation of reactive diluents in the polymer network during curing is needed to avoid segregation of components. Otherwise, hazy films with reduced gloss can be formed.

While these methods can be successful in lowering the solvent amounts in the paint, the high solid content (above 70%) requires resins with a very low viscosity, which should be achieved by adjusting the polymer structure of the binders, namely by decreasing the mean number-average molecular mass or making a polymer with narrower molecular mass distribution. Oligomers with a molecular mass of ca. 1,000–3,000 are required for high-solids paints giving a highly polymeric system formed after cross-linking and curing. Changes in the chemical structure of the polymer can be also done, for instance, by incorporation of functional groups.

3.2.2.2 Waterborne paints

A way to reduce solvent emissions from paints without lowering the molecular mass of the binder is to use nonaqueous dispersions, which are commonly referred to as emulsion paints and have become the largest product group in the paint and coating industry. Acrylates mainly used as binders for waterborne paints in addition to a low viscosity have other attractive features, which make them competitive in comparison with high-solids paints and powder coatings.

Waterborne paints result in lower emissions of VOCs than solvent-borne counterparts and therefore lower odor. Not only these paints themselves are thinned and cleaned up with water, but also washing up in water is easily done for paint tools. Besides advantages with their application, which include also faster drying, there are certain performance benefits, such as better durability and thermoplastic properties.

Waterborne coatings containing polymer dispersions can be applied without stringent safety measures and are typically dried physically just by evaporation of water. Due to the presence of water, emulsion paints are noncombustible in the liquid form allowing easy storage and transportation. Moreover, equipment cleaning can be done with water.

Use of waterborne paints requires a few precautions regarding corrosion-resistant and wear-resistant materials for the equipment, which are used not only for application but also for production of these paints. Moreover, since water can be easily frozen giving ice, waterborne paints require protection against frost during storage.

Typically, in emulsion paints fine (particle diameter 0.1–0.3 μm) and medium (particle diameter 0.3–2 μm) dispersions are used as binders.

As discussed above, emulsion paints besides binders contain also pigments, extenders, and various other additives. The polymers for waterborne polymer dispersions are synthesized by emulsion polymerization when the monomer droplets are polymerized in water-containing surfactants (anionic or nonionic) and protective colloids (e.g., poly(vinyl alcohols) and cellulose ethers). The stirring speed and nature of the protective colloids and emulsifiers influence the size and size distribution of the dispersed polymer particles.

In polymer dispersions, vinyl acetate and vinyl propionate copolymers (e.g., with acrylate esters); acrylate–methacrylate and acrylate–styrene copolymers are applied as binders. Some special types of binders with core-shell structure have been also developed. For example, aqueous metallic base paints for automotive coatings and finishes have acrylic core – shell polymers as binders where core and shell have a different monomer composition. The presence of anionic groups in the shell interacting with water and stabilizing the dispersion allows pseudo-plastic flow behavior. This in turn gives parallel orientation of the aluminum pigments in the wet paint film. Such orientation along with a low solids content results in the metallic gloss and high color flop (a change in color of a vehicle finish when viewed from different angles).

Waterborne coatings also contain some organic cosolvents (2–20%, typically 5%), such as low-molecular weight polar ketones, alcohols and esters in addition to water (typically ca. 50%) and solids (ca. 45%). Low solvent content results in more stringent requirements for surface cleanliness. Contrary to most solvent-borne coatings where solvents may dissolve some surface oils and contaminates, the cosolvent concentration in waterborne coatings in much less than in solvent-borne coatings.

The waterborne coatings also contain pigments. The same type of pigments as in conventional paints and coatings can be applied for waterborne paints, apart from some pigments with limited alkali resistance. Other components of waterborne coatings are dispersants or wetting agents to disperse the extenders and pigments; defoamers; preservatives; thickeners; and film-forming auxiliaries promoting film formation and reducing the minimum film-forming temperature. A coating film is formed when the minimum film-forming temperature is reached or exceeded during drying. Below this temperature, which depends on the binder structure and glass transition temperature, nature and content of film-forming auxiliaries and other additives, a noncoherent and mechanically unstable film is formed. The minimum-film forming temperature for house and wall paints is in the range of +5 to +10 °C.

Water and cosolvents are removed from the surface of the paint coating via evaporation, which is a physical process without chemical cross-linking. Coating thickness, amount of water and the evaporation rate determine the drying time. During air-dried curing of waterborne paints, first the organic cosolvent evaporates followed by evaporation of water and coalescence of the resin particles to form a continuous coating film. The curing time for waterborne coatings is much longer than for solvent-borne paints. An important issue with the waterborne paints is the need to maintain the surface temperature higher than the dew point to prevent condensation, which might be an issue in cold climates. Application of waterborne paints in places of high humidity can be challenging as the film might not cure because of difficulties with water vapor release during drying. To assist curing, moderate air flow and increased temperature are used. Recommended conditions for application of waterborne coatings are shown in Figure 3.35.

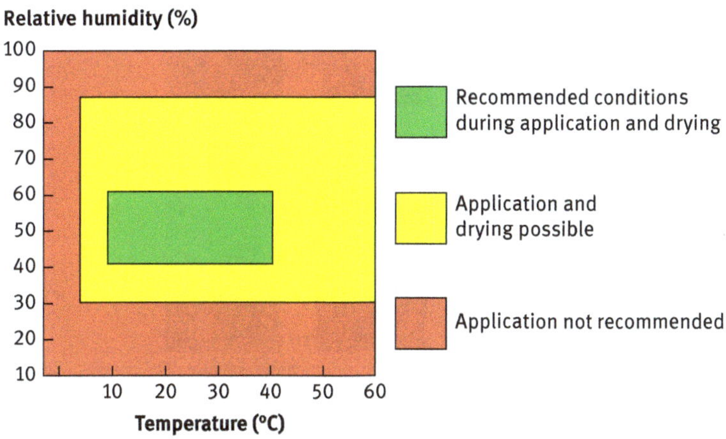

Figure 3.35: Recommended conditions for application of waterborne coatings [31].

Various spray methods including traditional air atomization and dip coating can be used with waterborne coatings, contrary to high-solid paints with high viscosity, which cannot be dip-coated. For optimal spraying operations relative humidity and temperature should be 40–60% and 10–25 °C, respectively. For drying, less stringent conditions are possible (relative humidity below 60%, temperature between 10 and 40 °C).

3.2.2.3 Coating defects and stability

An example of potential risks when coating can be done in a wrong way will be illustrated below using the Tu-154M RA-85763 aircraft flight on the Krasnodar-Novosibirsk route on 12.01.2000. During landing on a foggy night, the middle engine failed at the height of 700 meters. At an altitude of about 300 meters the right engine failed with only one engine still in operation. The latter failed at a height of 6 meters, but fortunately the plane was already above the runway and landed using emergency wheel braking, in fact wearing down all 12 wheels to the brake discs. The cause of the incident was the quality of fuel at the Krasnodar airport: a kerosene tank was painted with the wrong paint on the inside, which then dissolved in jet fuel. When the airplane's fuel tanks were opened, there was no liquid but rather a jelly substance. The spool pairs of the engine regulator pumps were so stuck that they could only be knocked out with a hammer.

Some of the potential coating defects (Figure 3.36) and their mitigation strategy will be discussed below, based substantially on [32]

Figure 3.36: Some coating defects [32].

Adhesion failure, which originates form surface contamination or condensation, incompatibility between coating systems, or too high overcoating time, can be prevented by using a clean and dry surface, free from contamination, and applying the correct coating specification as well as overcoating time.

Formation of wide crisscross cracks (crocodiling) is a result of the surface shrinking faster than the paint film body because of internal stresses. The latter are related to excessive film thickness and limited paint flexibility. To avoid crocodiling, a hard topcoat should not be used over a more flexible softer undercoat. Moreover the topcoat should be used after properly drying the undercoat.

Bleeding can occur when strong solvents are used in the topcoats resulting in a (partial) dissolution of the previous coat, while the local loss of adhesion gives blistering or formation of dome-shaped projections. Both types of defects can be prevented by using a suitable coating system.

Entrapped air/solvent within the coating not released before the surface dries leads to formation of temporary or permanent bubbles of air or solvent vapor. Bubbling can be mitigated by applying a correct mixing equipment and use of defoaming agents.

Formulation errors leading to generation of stresses result in formation of breaks in the surface of a paint (checking). This can be mitigated by correct formulations, as also the case of cissing when the surface is contaminated by different impurities, which are not removed when incorrect solvent blends have been used. Cratering is similar to cissing and happens for an insufficient time to low into a uniform film. Avoiding air entrainment during mixing, adding a thinning agent and improving spraying mitigates cratering.

Formation of minute crisscross cracks on the surface of a paint (crazing) is caused by too low application temperature, incompatibility with previous coating, aging, or high film thickness. Subsequently thinning the coat and adding a slower drying solvent is recommended for prevention of crazing.

Application of incorrect pigments, which are not stable under light, such as some organic pigments can result in fading, which apparently is prevented by using coating systems with light stable pigments. A similar advice of using recommended application techniques with suitably formulated products holds for such failures as the impact damage often occurring on glass-fiber reinforced plastics, sagging and for mud cracking. The latter case is associated with overapplication of heavily pigmented primers such as inorganic zinc silicates for highly pigmented waterborne paints, leading to a network of deep cracks. Sagging can originate from overapplication of paint, too high content of thinning agents and a lack of a curing agent.

Failure of the paint film to flow out giving orange peeling is caused by poor application techniques, incorrect solvent blend or too high thixotropy.

Similar to orange peeling, small holes (pinholes) also can be caused by incorrect spray application or a solvent blend.

Overapplication of a paint, too high content of the thinning agents, absence or a wrong curing agent can lead to runs. The latter can happen also when there is a lack of experience in depositing paint.

Oxidation of iron in the case of metallic contamination of a coated surface leads to formation of ferric hydroxide/oxide and thus rust spotting. The latter is prevented by protecting coating from contamination and ensuring an adequate thickness of a primer coat.

Formation of spiral and worm structures on the substrate surface, i.e., solvent lifting, is caused by incompatible paint systems, as topcoats with a strong solvent blend can react with previous and weaker solvent-blended coatings. Application of paint to a corroded substrate gives undercutting, while initial formation of a surface skin with solvent-based paints when overcoating is done before the previous coat has adequately hardened results in wrinkling. The latter is mitigated by using a correct coating procedure.

A specific problem can be thermal aging of coatings working at elevated temperatures, such as coatings for aircrafts, space ships, radio electronics, various heating elements, etc. High temperatures obviously have an impact on the polymer films, weakening of the intermolecular interaction in polymer and even resulting in a rupture of chemical bonds. Thermal stability can be enhanced by, for example, introduction of cyclic polymers into the chain or presence of pigments, having a reflective ability or performing the functions of heat stabilizers. On the contrary, presence of plasticizers and residual solvents in the film can often cause an increase in degradation.

Fire-resistant (noncombustible) coatings and coatings with reduced flammability (not capable of sustaining combustion) are obtained by using paints that do not contain organic components (silicate, lime, cement), applying halogen, phosphorus and silicon-containing film formers and especially by introduction of flame retardants into the composition of paints.

The most widespread as fire retardants are inorganic substances (aluminum hydroxide, borate, sulfate and ammonium chloride, etc.), which are easily decomposed when heated, as well as various halogen and phosphorus-containing organic compounds (chloroparaffins, chlorinated biphenyl, tetrabromoxylene, etc.)

Fire-retardant intumescent coatings swell, forming a porous heat-insulating layer during decomposition, and subsequently creating a thermal barrier. They are obtained using different film formers: polyvinylidene chloride, polyurethanes, epoxy oligomers, polyvinyl acetate latexes, polyacrylates and vinyl copolymers.

The resistance of coatings to UV aging depends on their composition. Film formers that do not have active functional groups are the least susceptible to destruction: fluoroplasts, polystyrene, polyacrylates, oil-free polyesters, polysiloxanes, chlorosulfonated polyethylene (in cured form). Melamine, urea and phenol-formaldehyde, alkyd, vinyl film formers are less efficient against UV aging.

Aging is accelerated by using pigments with increased photochemical activity (e.g., ZnO), while radiation shielding pigments (e.g., carbon black, bronzes) delay aging.

Various photostabilizers such as oxybenzophenone derivatives, benzotriazoles, arylsalicylates, etc., while being effective against photochemical aging, also delay drying.

A significant problem is the biological damage of coatings in warm (24–30 °C) sea waters, rich in plant and animal organisms (Figure 3.37).

Figure 3.37: Coating fouling because of algae and shellfish [33].

An initially slim bacterial film is formed on the surface, consisting of bacteria, protozoa and diatoms; while increasing in size it can reach a thickness exceeding 40 kg/m^2 per year. Besides accelerated degradation of coatings and premature metal corrosion, the operational characteristics of ships are worsened decreasing the speed by 50–100% and increasing the wave load on hydraulic structures.

To protect marine vessels and hydraulic structures from fouling, antifouling coatings are subsequently used, comprising a film-forming base (e.g., perchlorovinyl, chlorine rubber, copolymers of vinyl chloride with vinyl acetate, chlorosulfonated polyethylene, synthetic rubber) and toxins. The latter (e.g., compounds of carbamic, thio and dithio carbamic acids, etc.) act as poisons.

3.2.2.4 Production of paints and coatings

The market is very fragmented, ranging from small paint manufacturers making a few hundred tons to large ones producing several hundred thousand tons on an annual basis. These large companies offering sometimes up to 20,000 formulations still account for a small fraction (ca. 20%) of the global market.

Different paint formulations are produced in a limited number of production steps, requiring careful scale-up to achieve the quality specifications without additional correction stages; proper selection of ingredients (pigments, film formers, solvents, additives, etc.) and even mixing sequence. The equipment (mixers, tanks, filling, packaging, etc.) should be of a different scale reflecting the fact that customers order production batches of different sizes ranging from kilograms to dozens of tons still maintaining the required quality.

The most important processes in the production of coating materials are homogeneous mixing of the liquid components and uniform distribution of submicron size pigments in the liquid medium for pigmented systems. Unit operations in manufacturing consist of (a) mixing of raw materials (resin, pigments, additive agents and solvents), (b) milling in a disperser to make finely dispersed pigment particles, (c) blending of the mill base with the required components, (d) filtering and (e) packing. A general scheme of the manufacturing is presented in Figure 3.38.

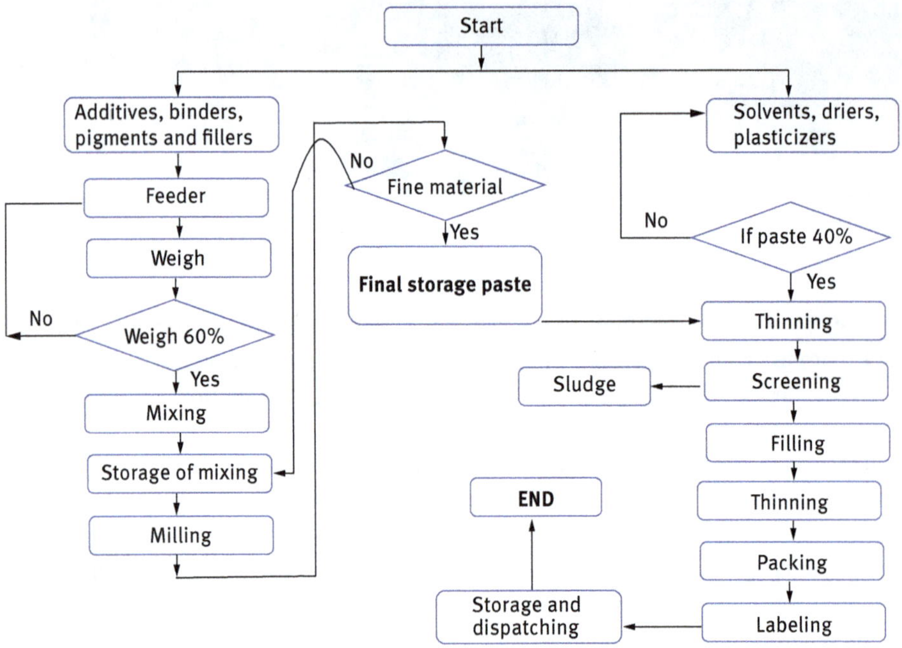

Figure 3.38: A generic scheme of paint manufacturing.

The pigment is first premixed with resin (binder) and some additives to form a paste. Thereafter grinding is done in, for example, sand mills dispersing the pigment particles throughout the mixture. The sand particles should be then filtered to remove them from the mixture.

An alternative to sand mills, which is used in the case of water-based latex paints, is to use high-speed dispersion tanks. In such dispersers, high-speed agitation is provided by a circular, toothed blade attached to a rotating shaft. In this way high shear forces are produced allowing high dispersing efficiency.

The step of making the paste (either in a sand mill or a dispersion tank) is followed by its thinning producing the final product by adding solvent, resins and additives.

Powder coatings, being solvent-less systems, are produced by melting and fusion of binders with other additives. These thermosetting or thermoplastic coatings can include various binders (epoxies, polyurethane, polyesters, PVC or acrylics), which should be efficiently mixed with pigments, catalysts, hardeners, and additives (Figure 3.39) in a pre-mixer. The mixture is routed through a metering device to an extruder. Heating to a temperature above the softening point of the binder results in melting and homogenization. The extruded paste of the size 2–3 mm passes through a cooling conveyor belt to a cutter (coarse granulator) and then to a fine grinder. After sieving to get the desired size the powder coating goes to packaging. Oversized particles are returned to grinding.

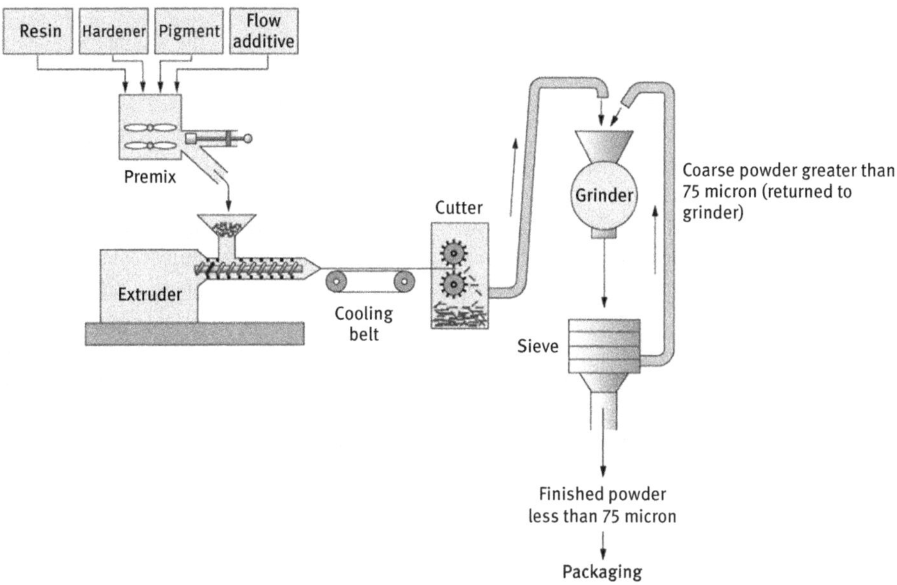

Figure 3.39: Production of powder coatings [34].

3.2.3 Dyes

The objective of dyeing is to produce uniform coloration of a substrate usually to match a preselected color. The color should be uniform throughout the substrate and be of a solid shade with no unlevelness or change in shade over the whole substrate. There are many factors that will influence the appearance of the final shade, including: texture of the substrate, construction of the substrate (both chemical and physical), pretreatments applied to the substrate prior to dyeing and post-treatments applied after the dyeing process.

Dying of textiles requires that the dyes should be fixed on the substrate to be dyed chemically or by forming insoluble salts and pigments. Azo or anthraquinone dyes (Figure 3.40) are able to dye cellulose directly from aqueous solutions. Transformation of such dyes into insoluble compounds by treatment with cationic substances or complex formation with copper is typically required.

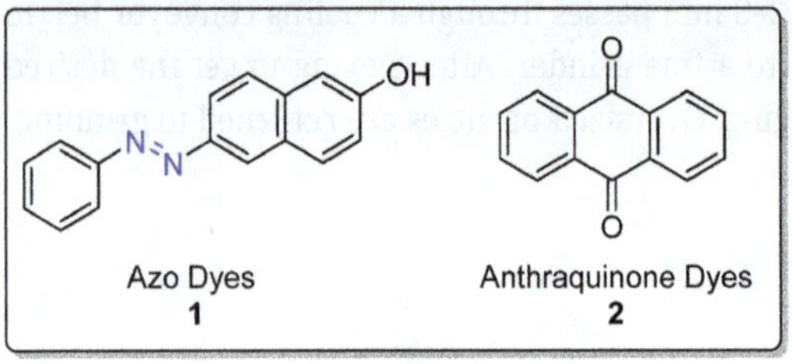

Figure 3.40: Monoazo and antraquinone dyes [35].

Sulfonic acid groups forming alts with basic groups are present in acid dyes, which are used for dyeing and printing of wool and polyamides. They can be also applied for silk or leather, and for food application as soluble sodium salts (Figure 3.41).

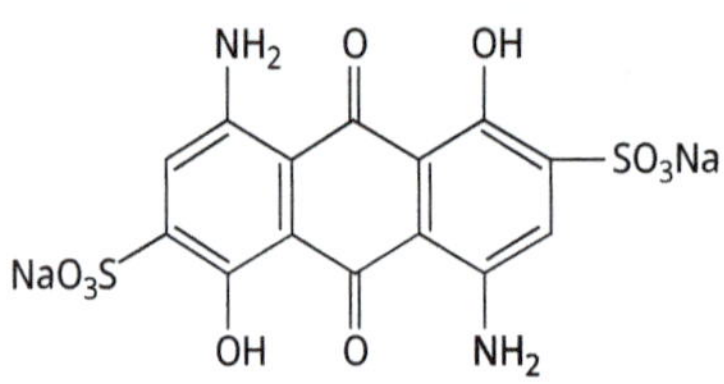

Figure 3.41: A sodium salt of an acid dye [36].

Dyeing of semisynthetic or synthetic fibers such as polyamide, polyester and polyacrylonitrile requires dispersion dyes (Figure 3.42) that are poorly soluble in water and are able to be adsorbed by the fibers because of preferential affinity to the hydrophobic fibers than for water.

Dyeing of polyacrylonitrile and acid-modified polyester fibers, along with paper and leather is done by cationic or basic dyes containing positively charged nitrogen atoms (Figure 3.43).

Some dyes, termed vat dyes, operate by adsorption in the reduced water-soluble form (leuco form) followed by oxidation into the insoluble pigment with atmospheric oxygen (Figure 3.44).

Naphthol AS azo dyes and phthalocyanine developing dyes (Figure 3.45) similarly to vat dyes are only transformed into their final colored form after adsorption.

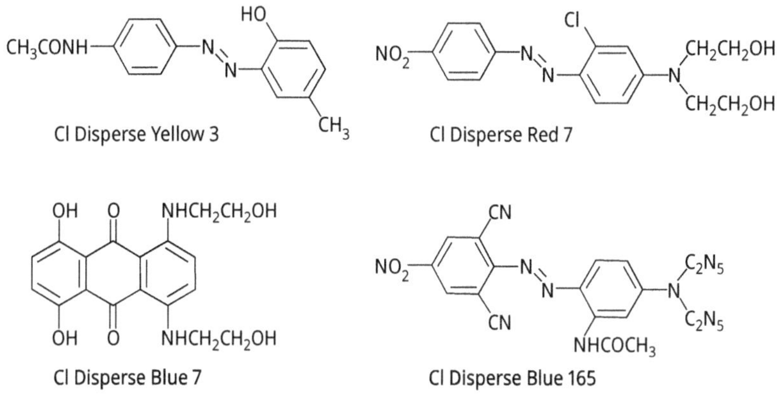

Figure 3.42: Dispersed dyes [37].

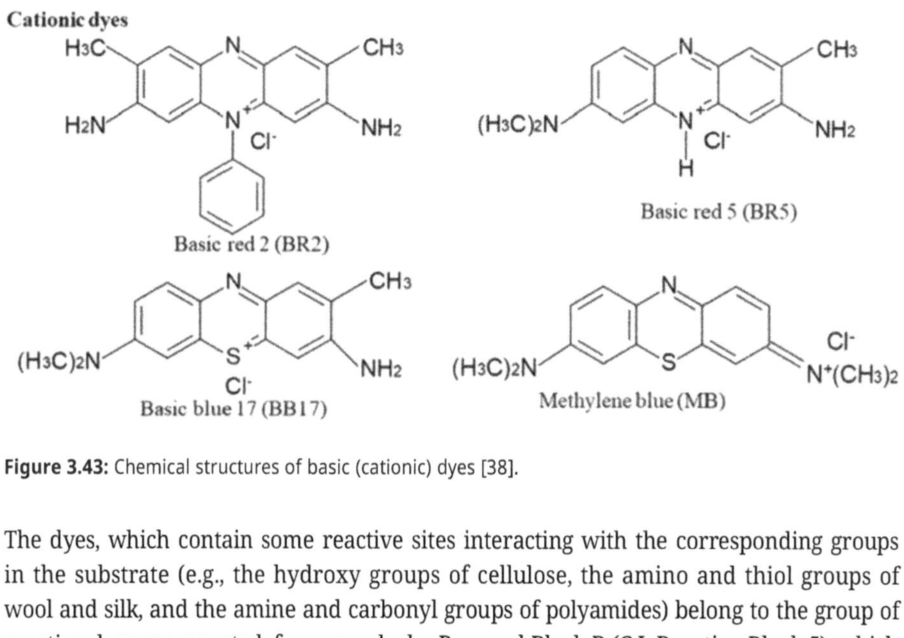

Figure 3.43: Chemical structures of basic (cationic) dyes [38].

The dyes, which contain some reactive sites interacting with the corresponding groups in the substrate (e.g., the hydroxy groups of cellulose, the amino and thiol groups of wool and silk, and the amine and carbonyl groups of polyamides) belong to the group of reactive dyes represented, for example, by Remazol Black B (C.I. Reactive Black 5), which gives navy blue and black shades on wool, cotton and viscose (Figure 3.46)

Commercial dyes are applied as powders, granules, aqueous fluid pastes or concentrated solutions, containing in addition to the dye per se also different agents, such as different neutral or alkaline inorganic salts (e.g., sodium chloride or carbonate), buffers (e.g., mono- and disodium phosphate), nonelectrolytes (e.g., dextrin or urea), anti-dusting agents (e.g., mineral oil), dispersants (e.g., lignosulfonates), wetting agents (e.g., long-chain alkyl sulfonates or alkyl naphthalene sulfonates), anti-gelling and antifoaming agents, and some other ingredients. Additional emulsifiers (e.g., polyethoxylated

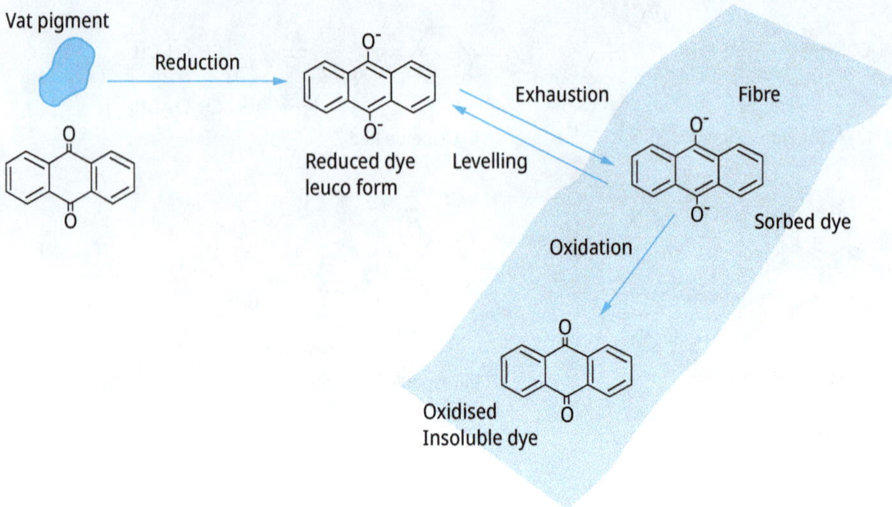

Figure 3.44: Basic steps in utilizing vat dyes [39].

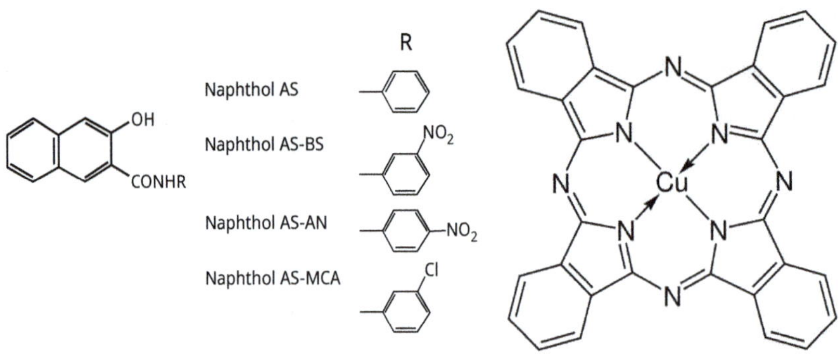

Figure 3.45: Examples of developing dyes: (a) naphthol based and (b) copper phthalocyanine [40].

products) and solubilizers can be added to, for example, convert the dyestuff into the most stable crystal modification, as the dissolution rate depends on the crystal size. Moreover, the particle size of the dispersion dyes should be in the right range, which is ensured by grinding. To slow or retard drying of aqueous suspensions glycols or glycerine can be added as humectants, while for dried dispersion dyes a substantial amount of a dispersant (2:1 ratio) is added preventing aggregation.

For the sake of lower exposure of humans to dye dust, concentrated solutions are more frequently applied with polyols, ether alcohols and other solvents used in addition to water.

For dyeing of a fabric, the dye liquor can be sprayed (Figure 3.47) or printed onto the substrate, or alternatively the fabric is immersed in a dyebath and then the

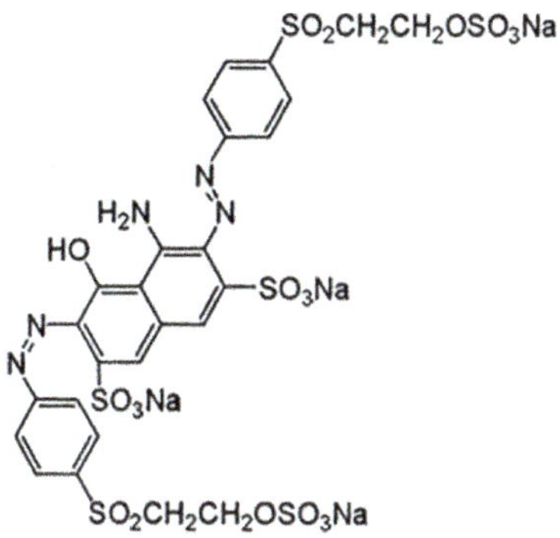

Figure 3.46: The chemical structure of C.I. Reactive Black 5 dye.

excess dye liquor is removed by padding (Figure 3.48), meaning passing the substrate through a pad.

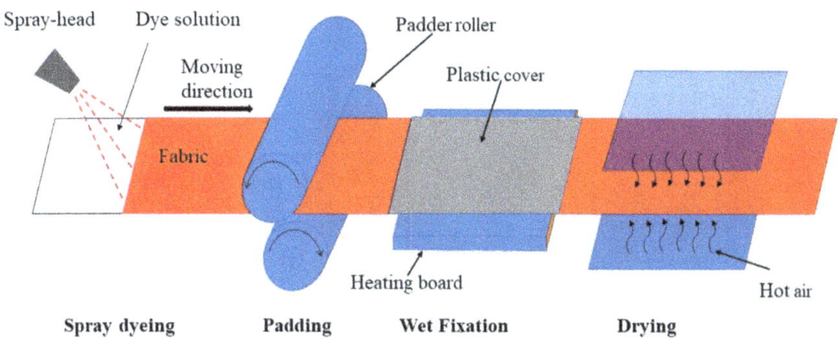

Figure 3.47: Dyeing of a fabric with spraying [41].

The dissolved dye should be first absorbed on the fabric penetrating then the interior of the fiber. This is followed by fixation, drying or chemical conversion, which stabilize the absorbed dye. Adsorption is from the thermodynamic viewpoint favored at lower temperatures, while higher temperatures are preferred for faster diffusion. The pressure of the squeeze rollers defines the amount of dye liquor retained by the substrate. Drying by hot air steam or infrared heat is done to ensure uniform fixation of dyes. Exposure to heat also helps in fixation of the dyes, which includes penetration in the fabric. Finally, washing removes unfixed dye and auxiliaries.

Printing of a dye can be done from a print paste containing thickeners and auxiliaries, which is applied to the fabric via a roller (Figure 3.49) or by screens. Concentration of thickeners (natural compounds, for example, starch or guar derivatives, alginates or polyacrylic acid as a synthetic thickener) is 2–10%.

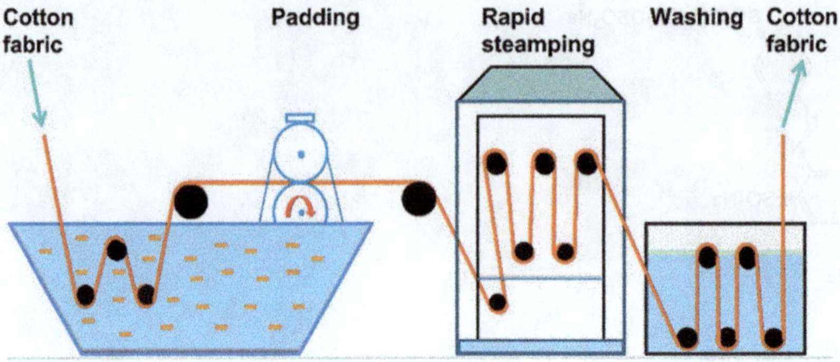

Figure 3.48: Dyeing from a bath [42]. Reproduced with permission. Copyright 2018 American Chemical Society.

Figure 3.49: Roller printing [43].

3.3 Laundry detergents

3.3.1 General

Laundry detergents contain surfactants, which by altering interfacial properties make possible dirt removal from solid surfaces. An important issue in detergency is hardness of water, which is related to presence of calcium and magnesium salts. Quality of water depends on the region, being relatively hard in most European regions, compared to, for instance North America, or Japan. Presence of Ca and Mg and in fact other minerals results in several cleaning related issues. In particular, when water is hard, it will not lather with soap (i.e., produce soap); glasses will have spots and residue. Moreover, clothes will look dingy, while hair might be sticky and look dull. At the same time, hard water is preferred for drinking. Soft water, containing only sodium, lathers with soap, gives cleaner glasses and healthier appearance of hair.

The reason for such strong influence of water hardness can be explained by possible ion exchange with the counterions of the soaps and surfactants of the detergent. This gives insoluble calcium or magnesium compounds, which contrary to soluble sodium compounds, bind the surfactant hindering its contributing to soil removal.

Surfactants are needed to influence the surface tension of water, because high water surface tension (72 mN/m) adversely influences the wetting behavior. Therefore, the surface tension of water should be decreased to the level of organic solvents (30 mN/m and below) justifying application of surfactants, which are key components in detergents helping remove soil or oil deposits (Figure 3.50).

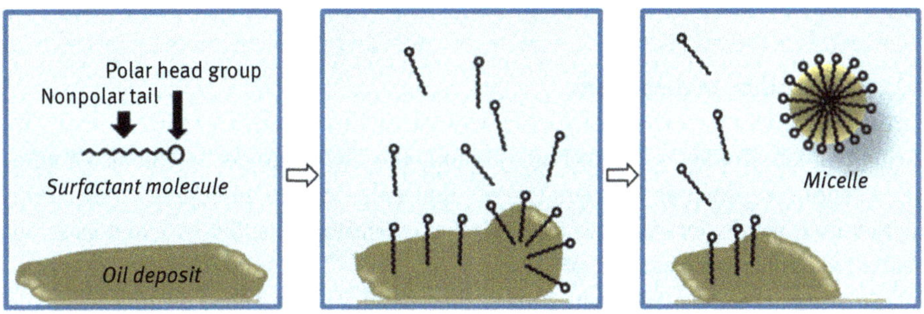

Figure 3.50: Illustration of the action of surfactants as detergents [44].

A range of substances constituting soil can be removed using surfactants, which are one of the main components of detergents. Other substances require cleavage by enzymes (e.g., proteins) or chemical reactions (e.g., wine, coffee and tea).

Wetting of a solid is quantified through a contact angle θ between the solid and a drop of a liquid on a solid surface (Figure 3.51).

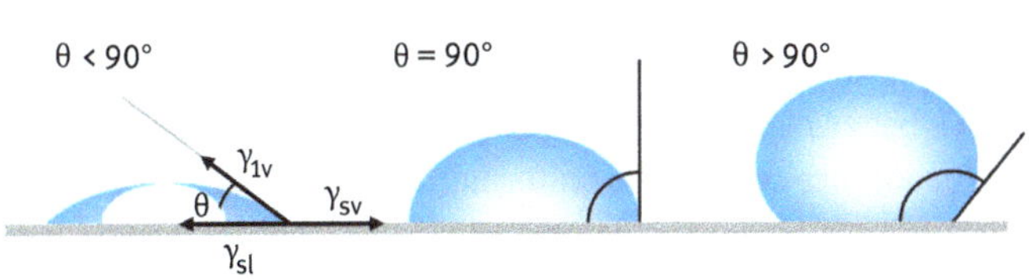

Figure 3.51: Illustration of contact angles formed by sessile liquid drops on a smooth homogeneous solid surface (reproduced with permission from [45]).

This angle depends on the surface structure and the surface tension of the liquid. Appropriate wetting is possible when a liquid has a surface tension equal to or lower than the critical surface tension of a certain solid. With addition of a surfactant, the amount of liquid staying on the surface passes through a maximum, as at high concentrations the

Gibbs–Marangoni effect retarding the liquid flow in the interface, is compensated by diffusion of the surfactants. As a consequence, a high enough surfactant concentration is needed. Anionic and nonionic surfactants are different in their adsorption kinetics and mode of action. For instance, contrary to ionic surfactants, nonionic surfactants have negligible effect on the surface charge and therefore adsorption determines the properties of the adsorption layer. Adsorption of nonionic surfactants is strong on hydrophobic, mildly polar interfaces.

Cationic surfactants reduce the magnitude of the negative surface charge, which is acquired by fibers and pigments in aqueous media. In case of electrical neutrality, redeposition of the previously removed soil can happen; therefore in detergents cationic surfactants are less suitable than anionic ones.

The extent of negative charge increases with increasing pH, thus introduction of alkali improves the wash performance.

3.3.2 Composition of detergents

A common laundry detergent includes not only the surfactant per se, but also a number of additives, such as builders (agents that enhance the detergent action), anti-redeposition agents, brighteners and bleaches, co-surfactants, biocides and sales appeal ingredients (e.g., fragrances).

To be an acceptable detergent, a surfactant, besides being a good wetting agent and economically competitive, should be able to displace soil into the washing fluid and act as an anti-redeposition agent. Other requirements include low sensitivity to water hardness, high solubility, neutral odor and color, low toxicity and desired foaming.

The detergent action of surfactants is practically not affected by micelles, the main function of which is to serve as a reservoir for replenishing surfactant and for solubilizing greases and oils.

Anionic surfactants are the most common ingredients in laundry and dishwashing detergents. Some examples of anionic surfactants include linear alkyl sulfonates, alkyl aryl sulfonates (dodecylbenzene sulfonate) and alcohol ether sulfates (e.g., sodium lauryl ether sulfate) having a negatively charged polar head group balanced by a cation (Na^+). Linear chains surfactants have better biodegradability than branched alkyl chains.

Linear alkylbenzenesulfonates are sensitive to water hardness and have high foaming ability, which can be controlled by foam regulators. Alkyl ether sulfates possess low sensitivity to hardness of water, high solubility and good storage stability at low temperature, at the same time form foams very intensively. This makes such high-foam surfactants suitable for vertical-axis washing machines, rather than for horizontal-axis drum-type counterparts. Application areas of alkyl ether sulfates include delicate washing, foam baths, hair shampoos, and agents for manual dishwashing. The least sensitive to water hardness are alcohol ether sulfates. They have a much lower critical micelle concentration than linear alkylbenzenesulfonates. Lower detergent concentrations can be used, therefore, which is preferred in some geographical markets.

Nonionic surfactants with a nonpolar head group are not deactivated by ions in hard water. Such nonionic surfactants as alcohol ethoxylates are becoming increasingly important and can be used together with anionic surfactants. Their favorable properties are related to low critical micelle concentration, meaning high detergency performance even at relatively low concentrations.

As anionic surfactants have poor cleaning efficiency for fabrics cationic surfactants are applied for this purpose. Cationic surfactants have a positively charged polar head group, such as long-chain quaternary ammonium.

Sometimes several surfactants are used in one detergent because of their synergy. Obviously opposite head-group charges of anionic and cationic surfactants make them incompatible.

The fiber type, in particular, hydrophobicity/hydrophilicity of the surface groups, has a significant impact on soil removal. Cotton acquires a negative surface charge in water behaving differently than synthetic fibers. Moreover, because of large calcium content on the surface of cotton, removal of soil is more difficult.

Detergency builders are used to enhance performance of the detergent by raising pH and complexing Ca^{2+} and Mg^{2+} (Figure 3.52) to avoid interference of these cations with surfactants.

This can be done with phosphates (sodium tripolyphosphate, hexametaphosphate, tetra-sodium- or tetra-potassium pyrophosphates), which, however, have some concerns because of sensitivity to skin and environmental impact. In fact, builders should satisfy a number of requirements including eye irritation, oral toxicity, compatibility with bleaching agents and alkalinity. The range of nonphosphates includes citrates, tartrates, succinates, ethylenediamine tetraacetic acid (EDTA) or dihydroxyethyl glycine, to name a few. Precipitating water-soluble builders such as sodium carbonate act by precipitating calcium and magnesium carbonates.

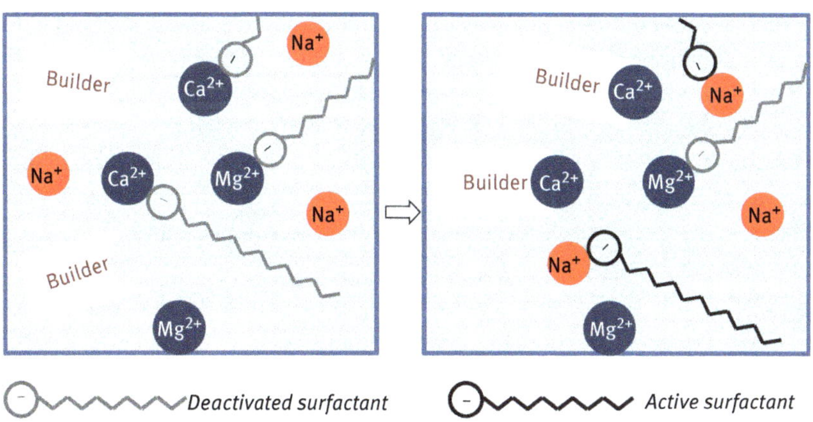

Figure 3.52: Builders binding to calcium and magnesium ions in water [46].

Ion exchange insoluble builders (zeolites or layered crystalline sodium silicate $Na_2Si_2O_5$) form insoluble complexes with Ca^{2+} and Mg^{2+}. Zeolites, such as the most often used small-pore zeolite A, besides ion exchange, provide adsorption of water-soluble substances (e.g., dyes), heterocoagulation of pigments and solid fats and even act as crystallization nuclei for sparingly soluble salts. Sodium silicate is included in modern detergents as a corrosion inhibitor. It protects metallic surfaces by deposition of a thin layer of colloidal silicate. This type of protection is used even if current washing machines are constructed using corrosion-resistant stainless steel.

Bleachers in laundry detergents based mainly on various peroxides generate hydrogen peroxide anion in alkaline media:

$$H_2O_2 + OH^- \rightarrow H_2O + HO_2^-$$

which oxidizes bleachable soils and stains. Inorganic peroxides and peroxohydrates (sodium peroxoborate tetrahydrate, $NaBO_3 \times 4\ H_2O$) added directly to a powder laundry product are applied as a source of hydrogen peroxide. The latter is formed when peroxodiborate is hydrolyzed in water. Sodium perborate monohydrate with better storage stability than tetrahydrate is preferred in geographical locations with higher ambient temperatures. In countries with a tradition of cold water-washing, sodium perborate is less effective and hypochlorite is used for bleaching forming hypochlorite anion under alkaline conditions:

$$HOCl + OH^- \rightarrow ClO^- + H_2O$$

Liquid sodium hypochlorite with overall limited storage stability must be added separately in either the wash or the rinse cycle independent on temperature. From the application viewpoint it is strongly dependent on the user experience. A potential danger is incorrect dosage, which can be detrimental to laundry and colors. Excellent reactivity of hypochlorite bleaches allows, on one hand their application even at very low temperature, while on the other hand, textile dyes and most fluorescent whitening agents exhibit poor stability in the presence of chlorine.

Bleach activators are used to improve performance of less reactive sodium perborate and sodium percarbonate at temperatures below 60 °C. Acylating agents used as bleach activators at pH 9–12 preferentially react with hydrogen peroxide forming organic peroxyacids. These acids possess higher reactivity than hydrogen peroxide and thereby improve low-temperature bleaching properties. Due to low concentration, organic peroxyacids are less aggressive on fabric dyes.

Laundry detergents contain also antimicrobial agents, which either kill or inhibit the growth of microorganisms (bacteria, fungi and viruses). Examples include quaternary ammonium chlorides and alcohols.

Fragrances are added in low quantities (<1%) to detergents providing a pleasant odor as well as masking certain unpleasant odors.

Removal of various type of stains from proteins, fat or carbohydrates (starch) can be done with the aid of enzymes (respectively proteases, lipases and amylases), which also provide color and fabric care. Enzymes, breaking down the large molecules mentioned above, should be able to operate effectively at low wash temperatures (e.g., 20–40 °C) and not be destroyed at wash temperatures up to 60 °C. Besides efficiency and stability in this temperature window, the enzymes in detergents should have optimum activity at alkaline pH, stability to other detergent ingredients such as surfactants and builders and low specificity to soils.

Optical brighteners enhance the light reflected from the fabric surface making fabrics that yellow naturally with time look brighter. Such brightening is done by converting UV into visible blue-violet light (Figure 3.53a), which by interacting with the yellow light emitted by the fabric produces white light. Optical brighteners for paper, textiles and detergents are predominantly di- and tetra-sulfonated triazole-stilbenes and di-sulfonated stilbene-biphenyl derivatives. Two examples of optical brighteners are shown in Figure 3.53b.

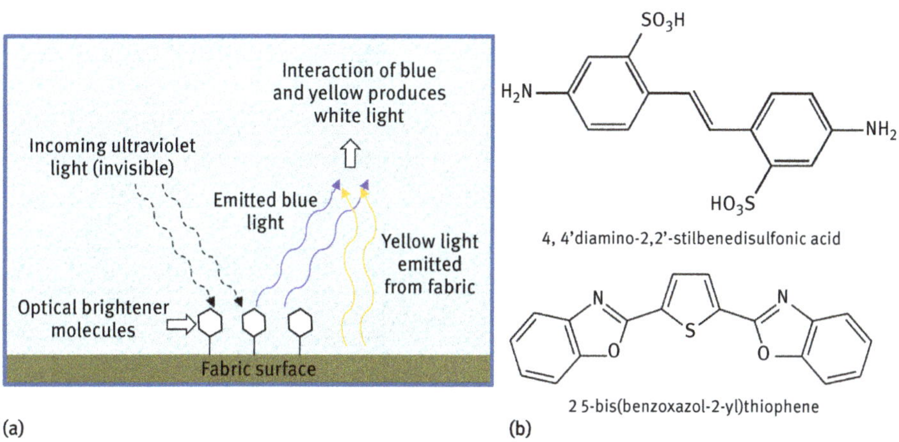

Figure 3.53: Optical brighteners: (**a**) mode of action [46] and (**b**) typical examples.

Other ingredients of laundry detergents include preservatives to prevent spoilage during storage (e.g., glutaraldehyde), hydrotropes to keep the pouring characteristics, various processing aids as well as foam regulators.

Hydrotropes, which prevent gel formation or separation into layers in a bottle, have a polar domain and a nonpolar domain, similar to surfactants. They act by interrupting formation of surfactant micelles, which can lead subsequently to viscosity gels. Examples of hydrotropes include short-chain aromatic sulfonates (e.g., xylenesulfonate) or glycol ether sulfates. Hydrotropes are typically smaller and less linear than surfactant molecules.

Among processing aids desiccants (e.g., sodium sulfate) to capture moisture and solvents can be mentioned. The latter help dissolve detergent ingredients or lower the freezing point of liquid laundry detergents.

Finally, foam regulators (soaps, siloxanes and paraffins) inhibit formation of foam, which is especially important in front-loading washing machines.

General formulations of powder specialty detergents are presented in Table 3.14, while Table 3.15 contains a comparison between two popular brands, showing that although some of the ingredients are common for both detergents, there is also a substantial difference.

Table 3.14: Formulations of powder specialty detergents in Europe.

Ingredients	Composition (%)			
	Detergents for delicate textiles	**Detergents for woolens**	**Detergents for curtains**	**Handwashing detergents**
Anionic surfactants (alkylbenzenesulfonates, fatty alcohol either sulfates)	5–20	10–20	10–16	12–25
Nonionic surfactants (alcohol ethoxylates)	2–10	2–20	2–7	1–4
Soaps	1–5	0–5	2–6	0–5
Cationic surfactants				0–5
Zeolite A	20–40	5–40	25–40	7–30*
Sodium perborate			0–15	
Sodium silicate	0–7	0–3	2–5	2–9
Sodium polycarboxylate	2–5	2–5	2–5	2–5
Antiredeposition agents	0–2–1.5	0–1.5	0.5–1.5	0.5–1.5
Enzymes	0–1			0–1
Optical brighteners	0–0.2		0.1–0.2	0–0.1
Dye transfer inhibitors	0–0.5			
Fragrances	+	+	+	+
Fillers and water	Balance	Balance	Balance	Balance

*Some handwashing detergents contain instead 15–30% sodium triphosphate.

Note: Reproduced with permission from [47].

Detergents can be divided into four performance tiers. The lowest level of ultra-economical detergents might contain basically an inexpensive anionic surfactant (e.g., linear alkylbenzene sulfonate) with some fragrances. At the next level a builder (e.g., sodium citrate), a viscosity enhancer and a second surfactant (alcohol ethoxylate of non-ionic type) can be added. For the third-tier products just below premium, optical brighteners, better surfactants, chelating agents, additional builders, and anti-redeposition polymers preventing soil redepositing on fabrics are included. The premium level contains besides the main surfactants at high loading, also additional ones (alcohol sulfates, alcohol ethoxysulfates, amine oxides, fatty acid soaps and cationic surfactants). Some companies include advanced soil-capturing polymers and a variety of enzymes (protease, amylase, mannanase, lipase, cellulase, glucanase) for such formulations.

Table 3.15: Detergents.

Purpose	Ingredients	Ingredients
Soil-capturing, soil-buffering agent	Monoethanolamine citrate	Tetrasodium ethylenediamine tetraacetic acid
Solvent	Diethylene glycol	
Soil-release or soil-capturing polymer	Polyethyleneimine propoxyethoxylate	
Emulsifying and dispersing agent	Sodium cumene sulfonate	
Optical brightener	Disodium diaminostilbene disulfonate	Disodium distyryl biphenyl disulfonate
Chelating agent	Diethylenetriaminepentaacetic acid	Trisodium ethylenediamine disuccinate
Builder and chelating agent		Sodium citrate
Breaking down polysaccharide-based soils and stains	Glucanase	
Breaking down fat-based soils and stains		Lipase
Removal of soil from cotton		Cellulase
Enzyme stabilizer	Calcium formate	
Foam regulator		Alcohol ethoxylate
Process aid	*Water*	*Water*
Defoaming agent	*Polydimethylsiloxane (dimethicone)*	*Polydimethylsiloxane (dimethicone)*
Traditional surfactant	*Sodium fatty acids*	*Sodium fatty acids*
Premium surfactant for hard water	*Alcohol ethoxysulfate*	*Alcohol ethoxysulfate*

Table 3.15 (continued)

Purpose	Ingredients	Ingredients
Low-cost general-purpose surfactant	*Alkylbenzene sulfonate*	*Alkylbenzene sulfonate*
Soil-release or soil-capturing polymer	*Polyethylenimine ethoxylate*	*Polyethylenimine ethoxylate*
Processing aid	*Ethanol*	*Ethanol*
Solvent and enzyme stabilizer	*Propylene glycol*	*Propylene glycol*
Soil-capturing agent, enzyme stabilizer	*Sodium borate*	*Sodium borate*
Breaking down protein-based soils and stains	*Protease*	*Protease*
Breaking down starch-based soils and stains	*Amylase*	*Amylase*
Breaking down guar gums	*Mannanase*	*Mannanase*
Balances electrolytes	*Sodium formate*	*Sodium formate*
Colorant	*Blue dye*	*Blue dye*

Note: Common ingredients are in italics. Modified from [48].

3.3.3 Production of powder detergents

Various options for making laundry detergents are shown in Figure 3.54.

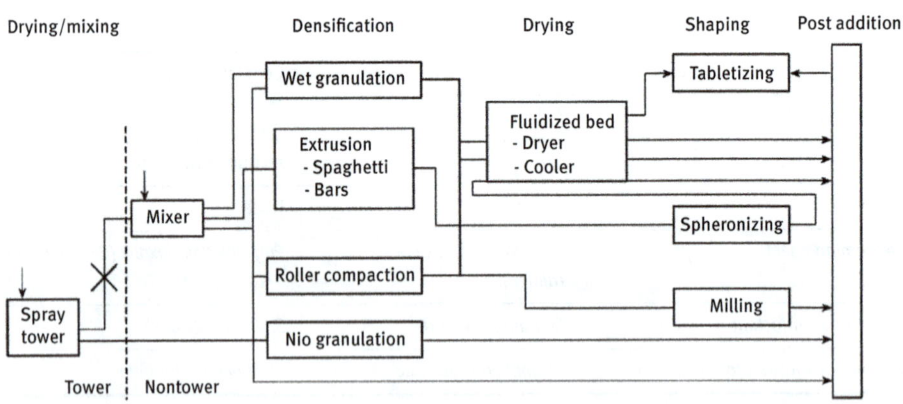

Figure 3.54: Chains of different processes (reproduced with permission from [47]. Copyright Wiley-VCH Verlag GmbH & Co. KGaA).

In, for example, a combination of tower/nontower "premix- process" in the drying step one part is spray dried, another part is granulated with water and/or surfactants and the third part is dry mixed. The densifying step can be integrated in the high granulation steps or the mixture can be extruded, tableted or roll-compacted.

For compact powder detergents, the traditional technology includes: spray drying of slurry containing most detergent components; subsequent addition of temperature-sensitive components and finally a densifying step (Figure 3.55).

First, dry and liquid ingredients are combined into a slurry. This is followed by spray drying of the slurry through nozzles under high pressure. When the small droplets created because of spraying fall, they are exposed to hot air, which results in formation of hollow granules. The granules collected at the spray tower bottom are screened. The out-of- specification granules are reprocessed. Producers of laundry detergents are apparently interested in high-density detergent powders, which can be packed in much smaller packages. The traditional spray drying gave relatively low-density detergent powders; however, in more advanced processes the amount of air inside the granules during spray drying has been reduced allowing higher densities.

The spray-drying process operates most common countercurrently. This implies that drying starts in an area of high humidity and lower temperature. The final product as a result has thicker walls, being heavier and coarser than in an alternative co-current spray drying.

Heat-sensitive ingredients and or chemically incompatible ingredients, which cannot be processed at spray drying temperatures (e.g., bleaching components and activators, enzymes, fragrances, soil repellents, foam inhibitor concentrate) are added to cooled granules. The size and distribution of granules/powders added after spray drying should match the size of the base powder to avoid segregation in the package.

An alternative to spray drying is nontower technology, which relies on agglomeration process carried out in special types of continuous mixers or combinations of high- and low-shear mixers. This technology utilizes extra amounts of zeolites to bind surplus water. In the first step the powder components (e.g., zeolites, applied to bind surplus water in larger quantities than in spray drying) are continuously added along with the separately dosed liquids (e.g., nonionic surfactants, solutions of polymers, fatty acids and linear alkylbenzenesulfonic acid). High-energy input results in formation of granules. Neutralization happens under the action of NaOH or Na_2CO_3. Due to limited loading of nonionic surfactants, the product might be sticky requiring an additional step of surface treatment with a zeolite. If not all water is removed by, for example, zeolites and soda ash, a fluidized-bed dryer is used for drying with hot air (140 °C). Cooling of the granules at the outlet is normally done using cold air (5–15 °C).

Drying apparently is not needed if water is efficiently removed. In such a case, the fluidized bed is used only for cooling. The post-addition step performed after sieving and milling, has some special features. The amounts of water, acid in the feed material, and the nonionic surfactants have some limitations. The temperatures of the granules should be kept under control; therefore heat of neutralization with acids

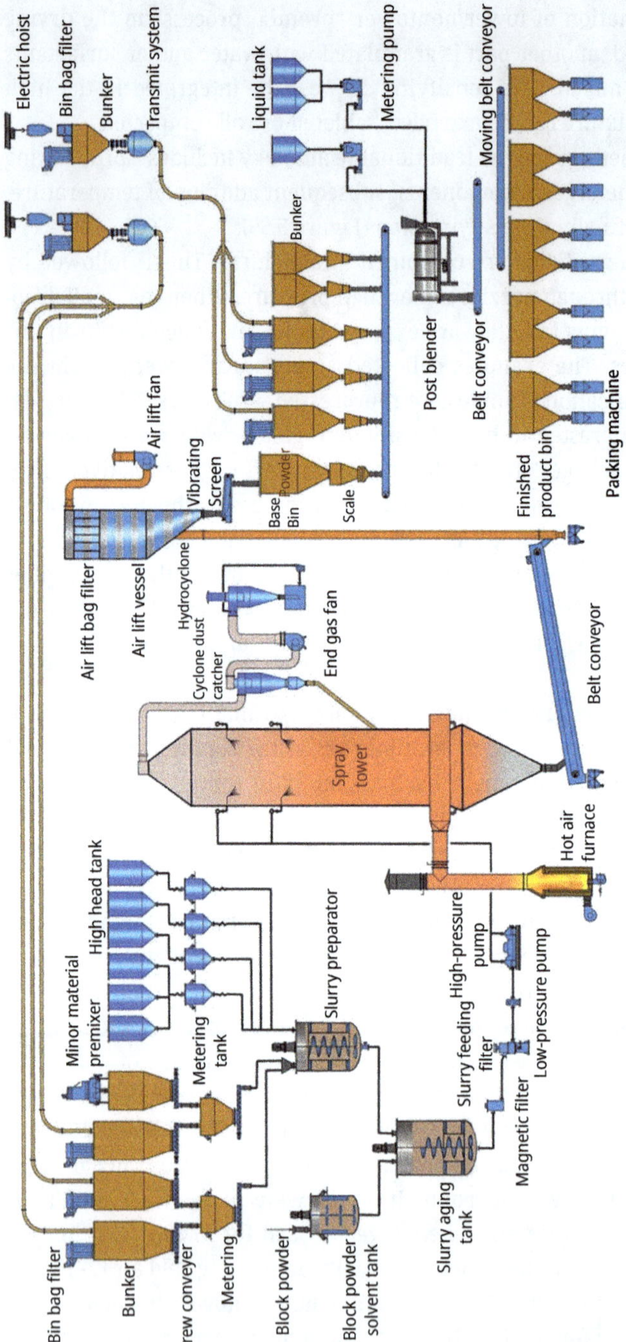

Figure 3.55: Compact laundry detergent production process flow scheme [49].

should not be excessive, limiting the acid amount as a result. A high content of nonionics makes the granules too soft and sticky.

A nontower process can be arranged also through compounding by blending various separately made components, for example, in the following sequence: anionic surfactant, nonionic surfactant, silicate, citrate, soil release, foam regulator, tetraacetylethylenediamine (TAED), percarbonate, perfume, enzyme and liquid fragrance.

Densification to increase the bulk density of the final product above 650 g/L is done by introducing densifying steps and/or application of denser raw materials. Typically, the size adjustment is done by sieving and milling, giving values similar to tower powder (0.2–0.8 mm). This size is needed, because oversized products dissolve slower. One of the common densification methods is wet granulation (Figure 3.56) when water, polymer solutions and anionic surfactant pastes are introduced at high shear rates. The process steps can include grinding in a high-speed mixer; granulation and conditioning in another high- or medium-shear-rate mixer; water evaporation in a fluidized bed dryer; sieving of coarse particles and milling. The oversized particles are recycled, while fine particles remain in the product. Such process details as mix-

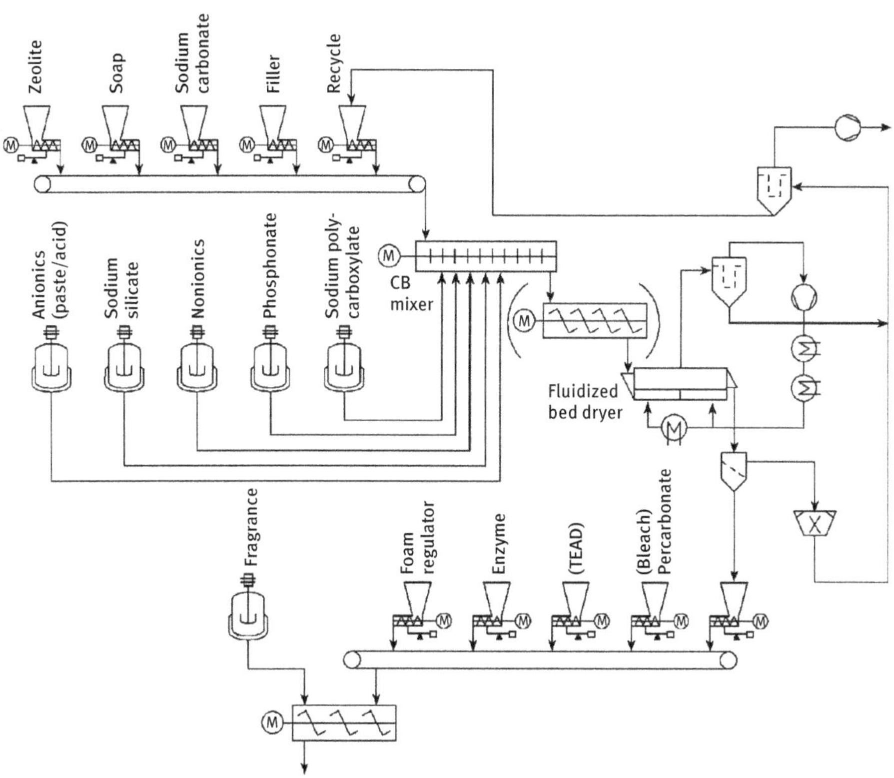

Figure 3.56: Flow sheet of a typical nontower granulation process (reproduced with permission from [47]. Copyright Wiley-VCH Verlag GmbH & Co. KGaA).

ing type, granulation temperature, dosage, etc. are typically not reported by detergent manufacturers and constitute their know-how.

Detergents can be used not only in the powder form but also as tablets consisting of one or more layers. In this way, ingredients that might interfere with each other in a negative way during storage (e.g., enzymes and bleach) can be introduced into different layers of the same tablet. The challenges in using tablets are related to low dissolution rate because of low surface-to-volume ratio and possible formation of gels in the presence of surfactants adversely effecting dissolution. The latter should be rapid from the start of a wash cycle. Manufacturing of tablets is done using rotary die presses. The tableting pressure is a key parameter, because too high pressure gives hard tablets and long dissolution time. Alternatively, if the pressure is too low the tablets can break down into smaller pieces. Thus, only a narrow range of tableting pressures can be used. Other process parameters include speed of tableting; homogeneity of mixtures; compressibility; arrangement of the dosing system; temperature and shape of the tablets; composition including liquid and moisture content; particle hardness, density, size, shape and distribution.

Tableting is done after the following process steps: *g*ranulation of the main components; sieving; post-addition of a range of additives (e.g., bleach, enzymes, salts, optical brighteners, polymer compounds, fragrances, pH regulators, dyes); sieving. For subsequent layers the same process steps are implemented. Since the moisture content is limited to 6–9%, the tablets are wrapped in a flow packer.

3.3.4 Liquid laundry detergents

Contrary to the European market and many other markets (China, Africa, India and Latin America) the liquid laundry detergents are becoming a new trend in the USA, where powder detergents cannot be easily found in department stores. Thus, liquids and powders coexist in laundry detergent markets globally with ca. 40–45% share each moneywise, with pod products being the rest [50]. At the same time the volume of powdered detergents (14 million metric tons per annum) is two-fold larger than for liquid detergents meaning that the latter are more expensive. The formulation of liquid laundry detergents is different compared to powder, not only in ratio of the ingredients (Figure 3.57) but also with respect to chemical composition.

For example, in addition to less expensive linear alkylbenzene sulfonate, the main surfactant in powdered detergents, liquid detergents contain alcohol ethoxylates, effective on challenging oily stains, as well as ether sulfates and amine oxides effective on grease.

Some ingredients, used for powder detergents, might not be stable in liquid formulations or may require an additional component.

For example, borax $Na_2H_{20}B_4O_{17}$ or calcium formate is added to ensure stability of enzymes (protease/amylase and lipase), polyethyleneimine ethoxylate is required as a complexing polymer instead of less expensive acrylic homopolymer complexing

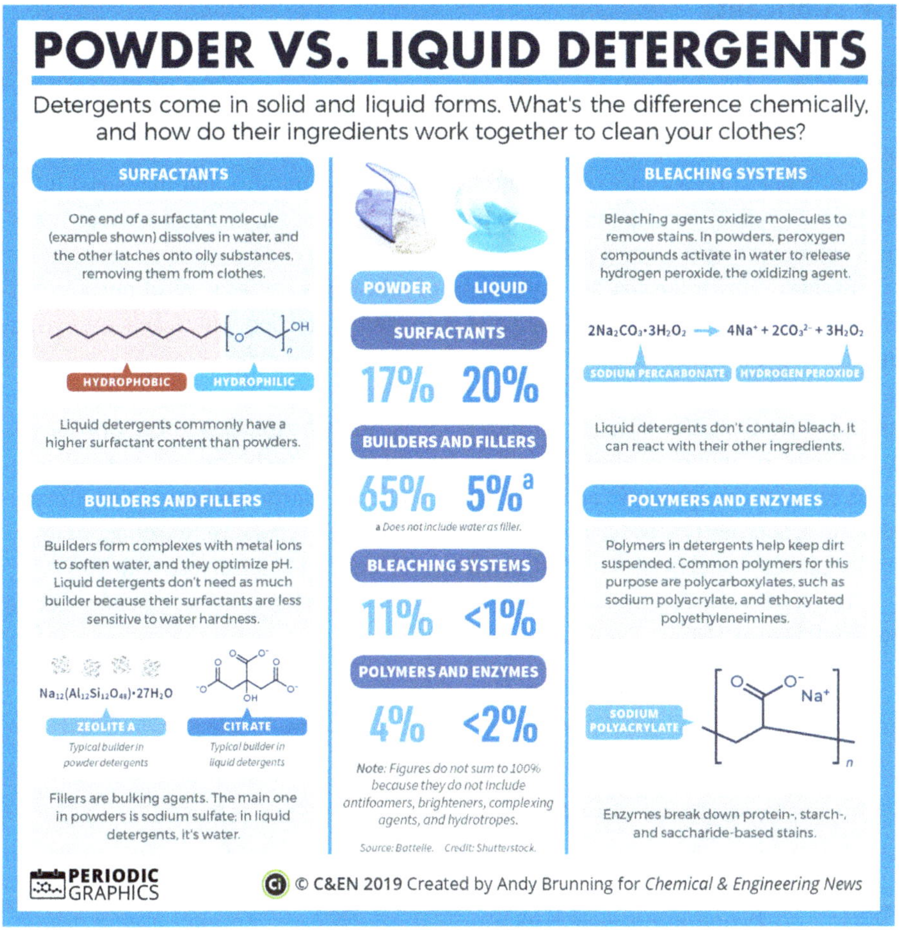

Figure 3.57: Comparison of powder and liquid detergents [50].

agents, and sodium citrate is applied instead of undissolved zeolite A as a builder. As an alternative to optical brighteners in powder detergents such as sodium percarbonate not stable in liquid, such additives as disodium diaminostilbene disulfonate are applied. The liquid laundry detergents also contain antibacterial materials, disinfectants, antifoaming and color-protecting substances along with perfumes.

3.4 Lubricants

3.4.1 Basics

Lubricants are substances reducing friction caused by interactions at the surfaces of adjoining parts. Lubricants can be subdivided into automotive and industrial lubricants and applied, correspondingly, to prolong lifetime of engines and various industrial equipment with moving parts.

Besides lubrication per se, application of lubricants serves several purposes including cooling, cleaning, suspending and corrosion protection. Lubricants act by separating surfaces with a liquid film.

Lubricants comprise a base fluid and an additive package. Base oils account on average for more than 95% of lubricant formulations. Some hydraulic and compressor oils are essentially only base oils (99%), while metalworking fluids, greases, or gear lubricants can contain up to 30% additives. These additives function by enhancing already-existing properties (viscosity, viscosity index, pour point, resistance to oxidation) or introducing new features (cleaning, suspending, anti-wear and anticorrosion ability).

The lube base oil is a complex mixture of hydrocarbons and is a product of oil refining (Figure 3.58).

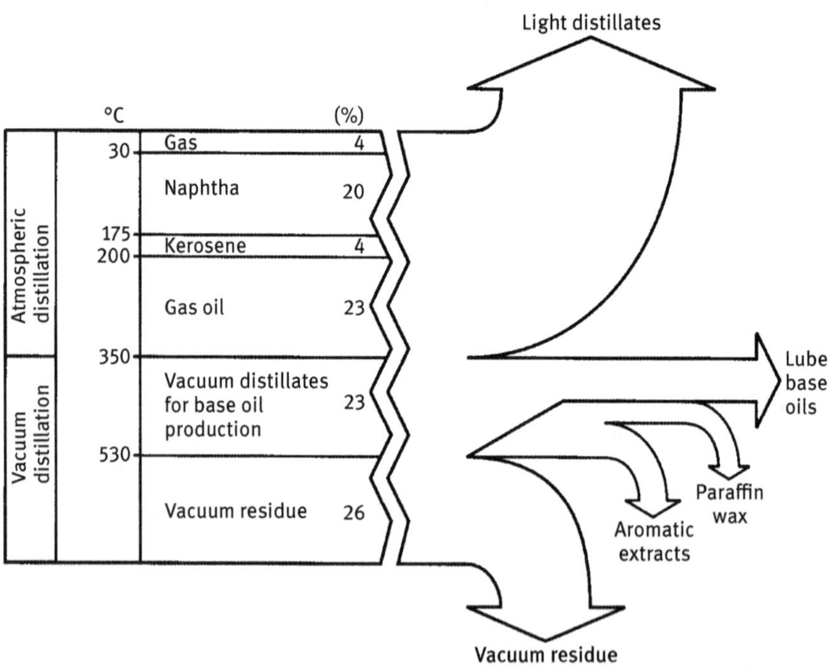

Figure 3.58: Products of oil refining.

Mineral and synthetic base oils are divided into several groups depending on the sulfur content, amount of saturated compounds and viscosity index (Table 3.16).

Table 3.16: Classification of lube base oils.

Group	Sulfur (wt%)		Saturates	Viscosity index
I	>0.03	and/or	<90	80–119
II	≤0.03	and	≥90	80–119
III	≤0.03	and	≥90	≥120
IV		All polyalphaolefins (PAOs)		
V		All stocks not included in groups I–IV		
		(pale oils and non-PAO synthetics)		

In recent years, these categories have been informally subdivided into Group I+, Group II+ and Group III+ . Groups 4 and 5 are synthetic oils, while mineral oils constituting mainly Groups 1–3 are obtained from petroleum crude. Mineral oils contain hydrocarbons with a molecular mass ca. 400–550. From the viewpoint of chemical composition, they include alkanes, cycloalkanes and aromatic compounds as well as some of the functionalized derivatives (e.g., low-molecular-weight oxidation products). Industrially available oils contain cyclic aliphatic hydrocarbons with several fused naphthene rings and polyaromatic molecules with alkyl chains.

Viscosity of lubricating oil is the most important parameter affecting heat generation, sealing effects, oil losses and cold start properties. Viscosity determines lubricant film thickness and flow properties. A certain high level of viscosity is needed for making lubricating films. At the same time, high viscosity results in high friction losses in the oil. Viscosity depends on pressure and shear rate for non-Newtonian fluids as well was on temperature (Figure 3.40), increasing as the temperature is decreased.

According to ASTM International, the viscosity is measured at 40 and 100 °C. The gradient between 40 °C and 100 °C is related to the viscosity index (VI), which is higher at lower changes of viscosity with temperature (Figure 3.59a).

A distillate oil from a naphthenic base crude exhibits VI lower than a distillate oil from a paraffin crude. Traditional mineral-based lubricating oils have VI from ca. 0 to 100, while more advanced hydroprocessing gives VIs well above 100.

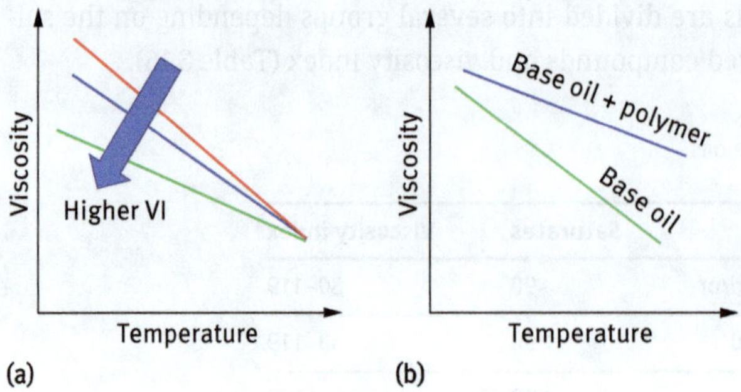

Figure 3.59: Viscosity dependence on temperature: (**a**) influence of viscosity index (VI) and (**b**) viscosity index improver.

3.4.2 Additives

Additives are needed to provide a lubricant with the desired performance (Figure 3.60).

Additives provide different properties to lubricants: detergents and dispersants are required, respectively, for cleaning and keeping dirt in suspension. Dispersants and detergents, which together account for ca. 45–50% of all lubricant additives, are used in engine oils, transmission fluids and tractor hydraulic fluids. Among various dispersants succinate esters, polymeric ones, succinimides or Mannich type compounds can be mentioned (Figure 3.61).

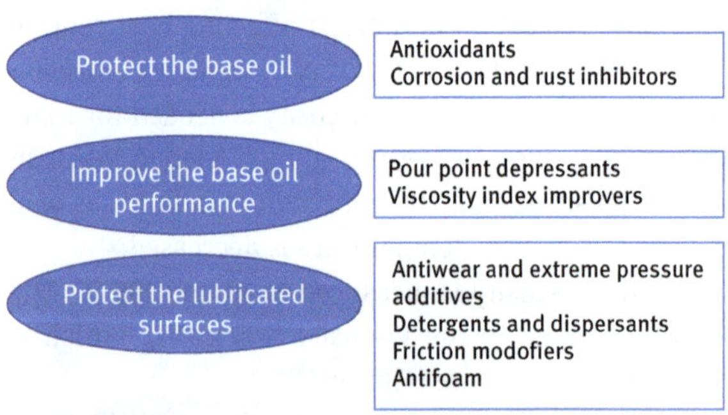

Figure 3.60: Function of additives.

Dispersants and detergents keep oil-insoluble resinous and asphalt-like oxidation products in suspension preventing their agglomeration as well as oil thickening and deposition of sludge and varnish on metal surfaces.

Dispersants differ from detergents in their basicity. Being metal-free, they possess low basicity and thus little or no acid-neutralizing ability. Dispersants have 4–15 fold higher weight than detergents and are more efficient in cleaning than detergents. The

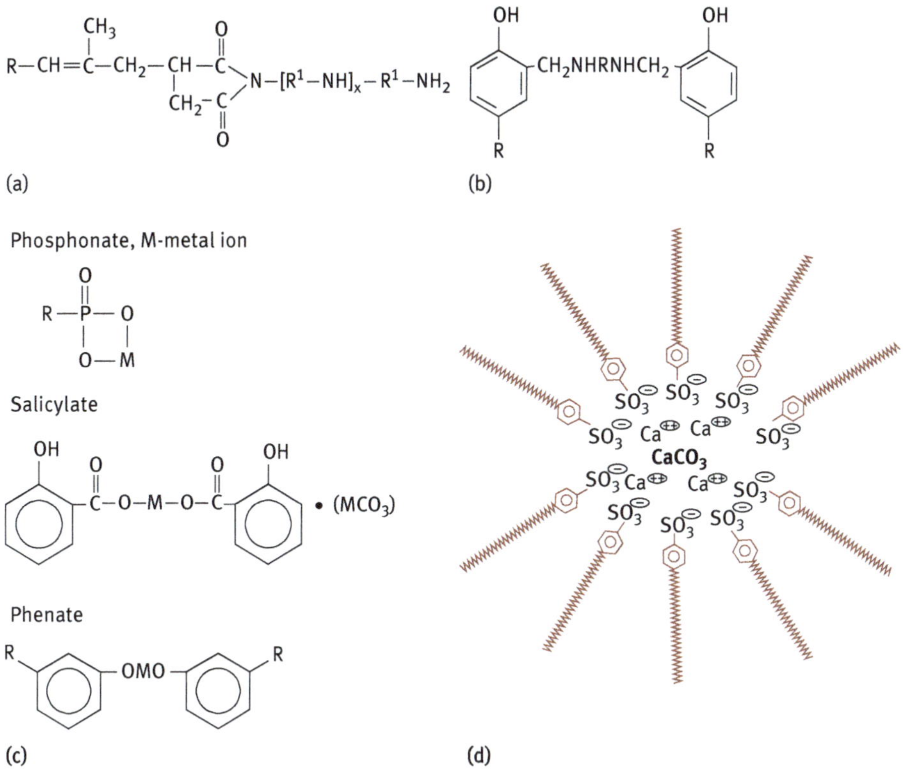

Figure 3.61: Dispersants: (**a**) succinimides and (**b**) Mannich types; and detergents: (**c**) phenates, salicylates and phosphonates and (**d**) sulfonates.

latter ones, especially basic detergents, contain such strong bases as metal hydroxides and metal carbonates and can therefore neutralize inorganic (H_2SO_4 and HNO_3) and organic acids.

Ash-less dispersants with improved dispersing properties are based on polybutenes with molecular masses of 500–3,000, for example, polyisobutylene succinic anhydrides. These dispersants have been developed to comply with the requirements of modern engines.

Antioxidants protect the base oil by delaying oxidation, which happens during operation of lubricants as a result of aging. Aged lubricants exhibit discoloration and a characteristic burnt odor. In fact, as a result of oxidation, resins (soluble in oil oxygen containing compounds), slightly colored relatively plastic lacquers, insoluble asphaltic compounds, acidic compounds and hydroperoxides promoting corrosion, are formed. Addition of antioxidants prevents formation of resins, lacquers and acidic compounds. Antioxidants act by destroying radicals (hydrocarbon and alkyl peroxy radicals) and hydroperoxides. The radical scavengers known as primary antioxidants interrupt the radical chain auto-oxidation by donating hydrogen to alkyl radicals and/or alkyl peroxy radicals. Examples include hindered phenolics

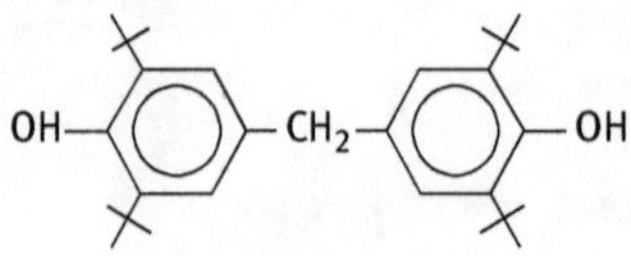

and aromatic amines

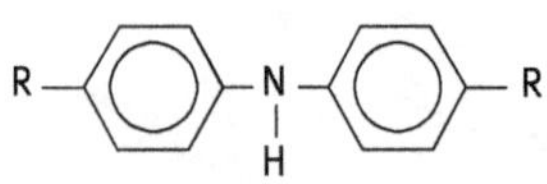

Secondary antioxidants are peroxide decomposers such as, for example, sulfur and/or phosphorus compounds, which form alcohols from alkyl hydroperoxides. The latter are generated in the radical chain during oxidation. Peroxide decomposes are oxidized in a sacrificial manner. Zinc dialkyldithiophosphate

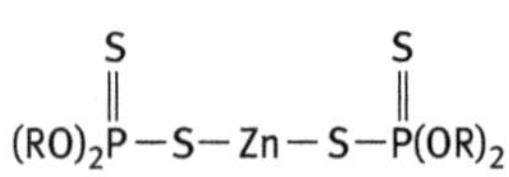

phosphites and thio-ethers are examples of various secondary antioxidants.

The last class of antioxidants is metal deactivators, which can be divided into chelating and film-forming agents. The former form a stable complex with metal ions, while the latter, either coat the metal surface, preventing metal ions from entering the oil or restrict access of corrosive species to the metal surface.

Another important class of additives is viscosity modifiers. They are polymeric molecules with solubility in base oil depending on the chain length, structure and chemical composition and temperature. Introduction of viscosity modifiers increases viscosity as well as VI. VI improvers are long-chain, high-molecular-weight polymers (olefin copolymers, polyisobutylene, hydrogenated styrene–isoprene and styrene–butadiene copolymers, polyalkyl methacrylates of the molecular weight ca. 10,000–250,000), which cause relative viscosity of an oil to increase more at high than low temperatures. In cold oil, the macromolecules are present in a coiled form; therefore, their impact on overall viscosity is minimal. Temperature increase leads to better solubility, expansion of polymer coils and higher viscosity (Figure 3.59b). The process is reversible because a temperature decrease brings back coil contraction. The use of viscosity improvers permits satisfactory lubrication in a much broader temperature range than with straight mineral oils alone.

Pour point depressants (e.g., alkylated naphthalene, phenolic polymers, ploymethacrylates, maleate/fumerate copolymer esters) enable lubricants to flow at low temperature. This is needed, as most mineral oils contain some dissolved wax, which at low temperature begins to separate, deteriorating flow characteristics. Removal of the residue waxes from the oil is somewhat expensive, and thus pour point depressants are applied as an economically attractive alternative. Two types of pour point depressants are used. Alkylaromatic polymers operate by adsorption on the formed wax crystals and prevent their growth and adhesion to each other (Figure 3.62). Other depressants (e.g., polymethacrylates) co-crystallize with wax also preventing crystal growth.

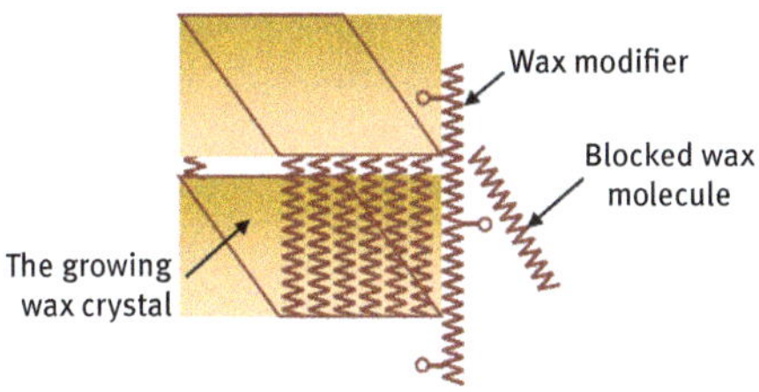

Figure 3.62: Mode of action of pour point depressants [51].

Apart from polyalkylated naphthalenes, pour point depressants are closely connected to several viscosity modifiers differing in application concentration and the type of monomers.

The foaming of lubricants is undesirable; therefore defoamers such as linear and cyclic polydimethylsiloxanes are added in 1–100 ppm concentrations.

To separate the lubricating system from water and prevent formation of relatively stable water-in-oil emulsions in many circulation systems (hydraulic, gear, turbine and compressor oils) good demulsification properties are required. Currently special poly(ethylene glycols) and other ethoxylated substances are used.

Some lubricants contain oil-soluble azo dyes for marketing, identification or leak detection purposes.

When the lubricating system runs in the area of mixed friction (e.g., under severe stress) additives are necessary to reduce wear, anti-wear (AW) and extreme pressure (EP). This is the case of metalworking and hydraulic fluids, engine oils and lubricating grease. EP additives act by forming protective films because of thermochemical reactions with the metal surfaces. Such films withstand extreme temperatures and mechanical pressures minimizing direct contact between surfaces. Some common phosphorus derivatives used as AW and EP agents are shown in Figure 3.63. They are different neutral and acidic phosphoric acid esters. Metal or amine salts or amides can be also used. The acidic form of these compounds is the most reactive one. Some AW /EP additives are also antioxidants, such as zinc dialkyldithiophosphates, zinc-bis(diamyldithiocarbamate) and the ash-less methylene-bis(di-*n*-butyldithiocarbamate). Oil-soluble organic sulfur compounds $R–S_x–R$ are also used to improve the EP properties of metal working fluids.

Friction modifiers work at temperatures where AW and EP additives are not yet reactive. They function by doing mechanical work (solid lubricating compounds, e.g., molybdenum disulfide, graphite, polyamide, etc.), forming adsorption layers (e.g., long-chain fatty acids and their derivatives), or tribochemical reaction layers (e.g., saturated fatty acids, phosphoric and thiophosphoric acid esters) or polymers (e.g., methacrylates). Some organometallic compounds (e.g., molybdenum dithiophosphates or dithiocarbamates, copper containing organic compounds) are also used.

Corrosion inhibitors are adsorbing on metal surfaces, protecting them from corrosion. They can be classified into two main groups: antirust additives for protecting ferrous metals and metal passivators for nonferrous metals. Since a polar group is present in those inhibitors they compete for adsorption sites with AW and EP additives also possessing polar groups. Rust inhibitors (amine succinates, alkaline earth sulfonates, derivatives of phosphoric acid) form a continuous film on the surface, thus preventing access of water. Among typical metal passivators also surface film-forming types (e.g., benzotriazole) can be mentioned. Other metal passivators such as N-salicylidenepropylenediamine or mercapto-thiadiazole derivatives act, respectively, as chelating agents and sulfur scavengers. Some metal passivators (e.g., zinc dithiophosphates and dithiocarbamates) have been already mentioned as they have multifunctional character.

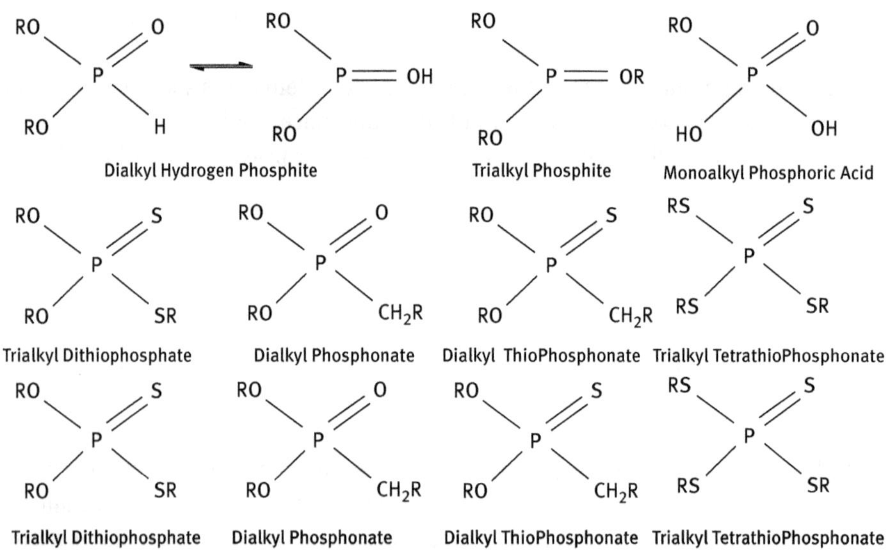

Figure 3.63: Common phosphorus derivatives used as antiwear (AW)/extreme pressure (EP) agents.

Besides all additives mentioned above there could be cases of several base oils, when more polar synthetic base oils can serve as additives for less polar ones, or vice versa. Lubricants can be made by blending as illustrated in Figure 3.64.

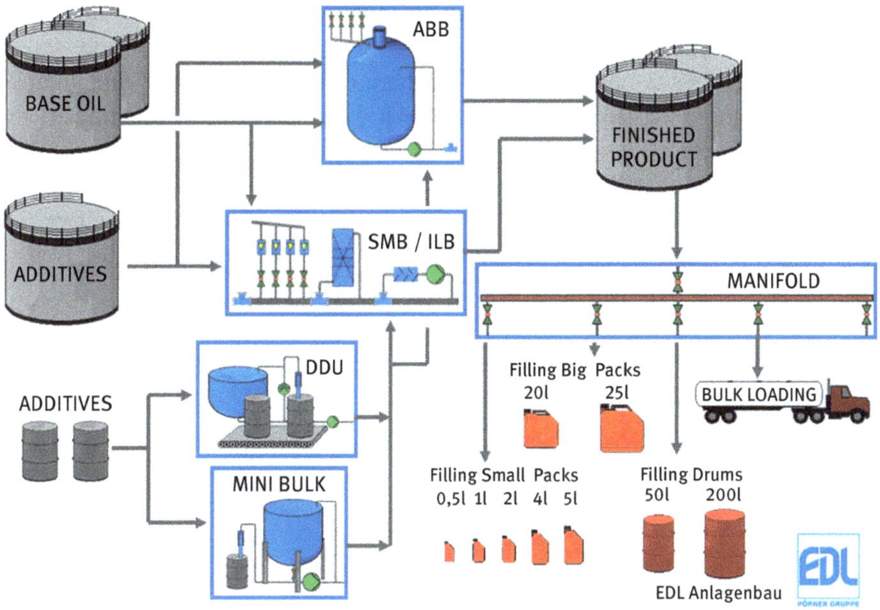

Figure 3.64: Lube oil blending plant. ABB, automatic batch blender; SMB, simultaneous metering blender; ILB, inline blender; DMB, dissolving mixing blender; DDU, drum decanting unit [52].

3.5 Ceramics

3.5.1 General

In general, ceramics are inorganic, nonmetallic products subjected during manufacture or use to high temperature (i.e., any temperature above red heat, ca. 540 °C). The overall classification of ceramics is given in Figure 3.65. Some ceramic materials (e.g.,

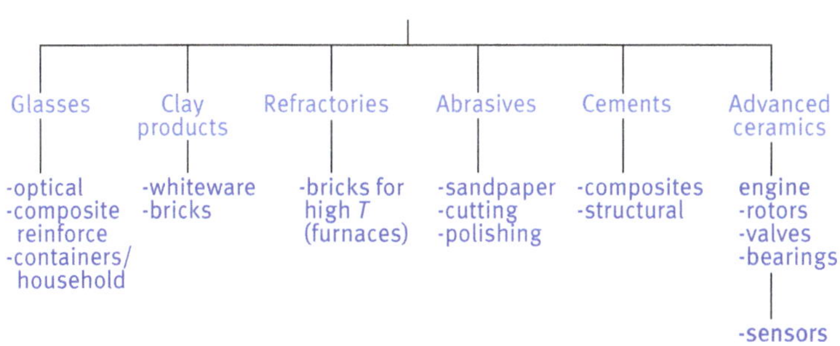

Figure 3.65: Characterization of ceramics.

certain glasses) can be easily formed, while other ceramics cannot be formed or cast and require special techniques.

Ceramics are metal oxides, borides, carbides, or nitrides, or their compounds. They have been used for centuries (e.g., glass or pottery), but have also found besides traditional applications, new uses in high-technology fields (e.g., electronics, computers, optical communication and medical implants). Ceramic products have a relatively high strength and thermal stability bearing at the same time low electrical conductivity and brittle fracture. A comparison between metals and ceramics is presented in Table 3.17.

Table 3.17: Comparison between metals and ceramics.

Metals	Ceramics
Crystal structure	Crystal structure
Large number of free electrons	Captive electrons
Metallic bond	Ionic/covalent bonds
Good electrical conductivity	Poor conductivity
Opaque	Transparent (thin sections)
Uniform atoms	Different-sized atoms
High tensile strength	Poor tensile strength
Low shear strength	High shear strength
Good ductility	Poor ductility (brittle)
Plastic flow	None
Impact strength	Poor impact strength
Relatively high weight	Lower weight
Moderate hardness	Extreme hardness
Nonporous	Initial high porosity
High density	Initial low density

3.5.2 Clay products and advanced ceramics

These materials include whiteware (crockery, floor and wall tiles, sanitaryware, electrical porcelain, decorative ceramics); structural clay products; refractory ceramics and advanced ceramics.

Clay, per se, is the oldest ceramic material and was used for making pottery or common bricks (Figure 3.66).

(a) (b)

Figure 3.66: Clay products: (**a**) pottery [53] and (**b**) bricks [54].

Silica powder and some alkali-containing minerals are added to make clay-based ceramic items. For instance, besides clay per se (50%) ceramic items can contain (25%) a filler (e.g., finely ground quartz) and (25%) a fluxing agent (e.g., feldspar, $KAlSi_3O_8$–$NaAlSi_3O_8$–$CaAl_2Si_2O_8$). Common bricks usually made of low-grade clays have the requirements of easy manufacturing and hard firing at temperatures as low as possible with minimum losses from cracking. Refractory ceramics with high silicon or aluminum oxide content are used to make firebricks for furnaces and ovens in various industries (steel-making, energy conversion, petrochemical industry, etc.). In some instances, they operate at rather high temperatures (1,500 °C) providing thermal protection of other materials. High melting points are thus absolutely needed. Besides silica, also alumina (T_m = 2,050 °C) and other oxides (e.g., MgO with T_m = 2,850 °C) fulfil this requirement. Advanced ceramics (Figure 3.67), which include silicon, boron or aluminum nitrides, silicon carbide (SiC), zirconia and alumina, have found applications as thermal barrier coatings especially in heat engines. A special feature of SiC is its inherent instability in oxygen requiring surface coating with a protective oxide for operation under oxidizing conditions.

Figure 3.67: Advanced ceramics [55].

Advanced ceramics in such applications offer, besides the possibility to be used at higher temperature and low density, also excellent wear and corrosion resistance, low frictional losses and option to operate without a cooling system. Brittle structure, difficulties in machining, as well as ease of void generation weakening an engine, are clear disadvantages. Reinforcement with fibers and whiskers opens up more applications for ceramics.

Ceramic armor made of very hard materials (Al_2O_3, B_4C, SiC and TiB_2) shatter the incoming projectile into pieces decreasing the penetration ability of the projectile (Figure 3.68). Ceramic materials cannot be used as stand-alone for armor applications because of brittleness. Therefore, an energy-absorbent material (metallic or polymeric composites) is used underneath improving the ballistic performance by preventing premature failure of ceramics.

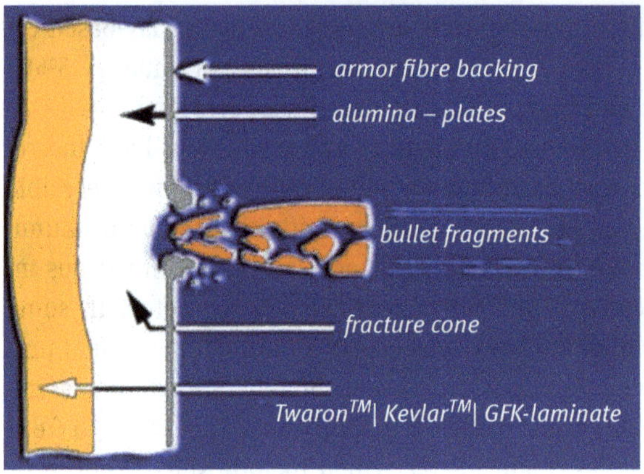

Figure 3.68: Armor ceramics principle [56].

Advanced ceramics can be used in electrical applications in capacitors, insulators, part of integrated circuit packages, piezoelectrics, magnets and superconductors because of their ability to securely hold microelectronics and provide efficient heat transfer having at the same time poor electrical conductivity. For electric applications, the thermal expansion coefficient of the advanced ceramic should match the corresponding values of the microelectronic chip and the electronic packaging material.

Fabrication of ceramic compounds includes milling and screening to the desired form; producing a slip by mixing particles and water; forming a "green" body by hydroplastic forming or slip casting; drying and firing. A more detailed sequence of operations is presented in Figure 3.69.

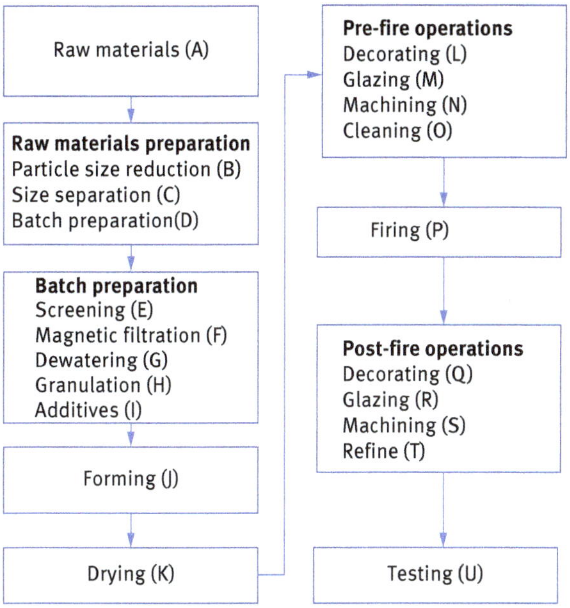

Ceramic type	Sequence of multi-operations to produce ceramic type
Cements	A-D/P/L/C/I/U
Structural clay bricks	A-D/J/K/P/U
Structural clay glazed pipe	A-D/J/K/P/R/T/U
Sanitary ware	A/D/E/F/O/M/P/U
Bone china (slip cast)	A/D-F/J/K/O/P/R/T/Q/T/U
Bone china (Haggard)	A/D-G/J/K/O/P/R/T/Q/T/U
Statists porcelain	A/D-F/J/K/N/M/P/ Q/T/U
Refractory fireclay (extruded)	A-D /J/K/P/ U
Refractory M_2O	A-D /J/K/P/ U
Wall tile	A/I/D-F/H/E/J/M/K/P/U
Glaze	A/I/D-F/M
Raw materials (washed clay)	A/C-F/K/C/U

Figure 3.69: Sequence of steps in making ceramics.

Figure 3.70 presents a flow scheme of a typical brick-manufacturing operation. As a raw material various kinds of clays mixed with loams, shales and sand are used depending on the product requirements and supply options. After breaking the clay using rollers and addition of water in a mixer, the mixture undergoes another step of rolling being fed thereafter through a horizontal pug mill. Extruded clay is cut thereafter to the required size on a wire-cutting table. After drying, the bricks are fired in a kiln, followed by glazing for the case of decorative bricks.

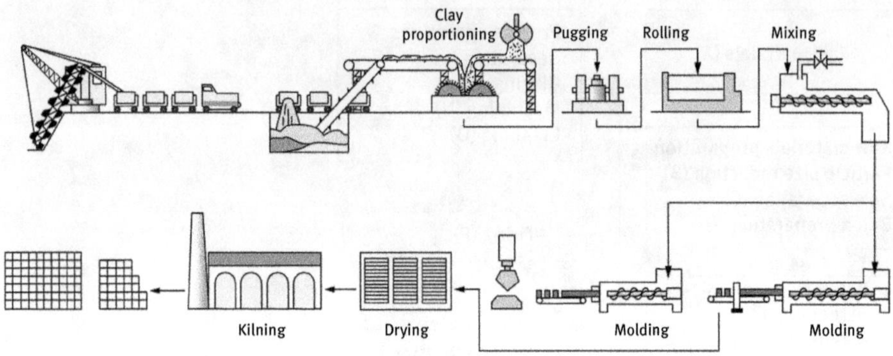

Figure 3.70: Manufacture of bricks [57].

Drying is very important for ceramic products. Removal of moisture should be rapid to avoid generation of stress capable of causing cracks and distortion. The basic principles of drying were discussed in Chapter 2. Drying methods are subdivided in convection and radiation methods. In the former methods, warm air is circulated around the ware heating the pieces and removing moisture. In radiation methods, heat is transferred as infrared or high-frequency radio waves.

Firing is done in kilns, which in general can be continuous (tunnel) and periodic (intermittent). In continuous kilns used in industry, the ware is loaded into cars and then passes through the kiln on a track or rail. The still-moving cars are unloaded after cooling, and then reloaded to start a new cycle. Tunnel kilns are refractory chambers sometimes reaching lengths of 90–100 m, which can be subdivided into three zones (preheating, firing and cooling). While tunnel kilns are used for firing of one type of products, periodic kilns can be applied for a range of products. Such periodic kilns operate on an intermittent schedule: loading, heating, cooling and unloading.

Traditional kilns using hot gases are heated either upward (updraft kiln) or downward (downdraft kiln) as illustrated in Figure 3.71. Clear disadvantages of such kilns are related to manual loading and unloading, long cooling period and need for reheating for the next firing.

Better air circulation in a downdraft kiln gives better fuel efficiency and more homogeneous heating. Despite lower fuel efficiency in updraft kilns resulting in hot and cold spots, they are more often applied because of simplicity to build and ship to a customer site.

Modern periodic kilns of elevator, shuttle or envelope types are presented in Figure 3.72.

In elevator kilns, several options can be realized. Either the kiln is lowered over the car or alternatively the car is elevated into the kiln, which is permanently fixed. An elevated kiln can also be moved horizontally over a series of cars. The firing temperature can be adjusted from 800–1,760 °C providing excellent temperature uniformity. Figure 3.72a shows a variant of a kiln with the hearth, which is raised and then

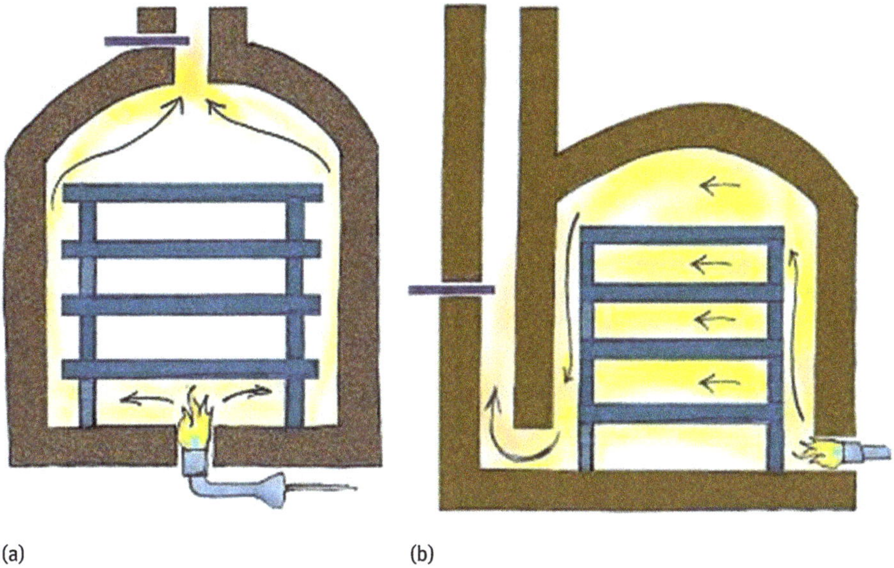

Figure 3.71: Periodic kilns: (**a**) updraft and (**b**) downdraft [58].

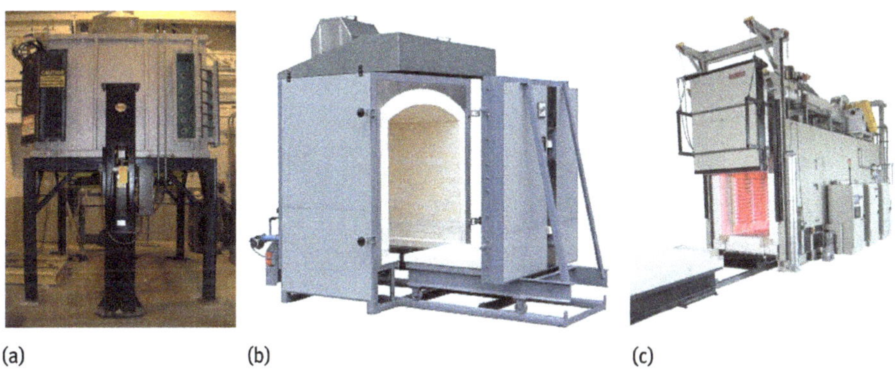

Figure 3.72: Modern periodic kilns: (**a**) elevator kilns [59]; (**b**) shuttle kilns [60]; and (**c**) envelope kiln [61].

lowered by a lifting device. Multiple cars can be also used. Such elevator kilns can operate with electric heating or natural gas burners.

Shuttle kilns (Figure 3.72b) are stationary, with one end having a movable door. The ware loaded onto a car is pushed on rails into the kiln. This type of design provides more options with floor layout and is economically preferred due to reduced duct work compared to an envelope kiln, which has its own advantages.

An envelope kiln (Figure 3.72c) is a kiln or furnace, which is moving over a stationary ware platform. The platform can have densely set ware without the risk of the load tipping as it is not moving. Two platforms can be used, one under loading (unloading) while the other undergoes firing. Envelope kilns require ca. 30% less floor space than shuttle kilns with a comparable size.

Kilns can operate not only under oxidizing atmosphere leading to burning carbon and forming oxides, but also under low-oxygen atmospheres resulting in reduction of multivalent ions to their lowest positive state, thus bringing color and some other changes to the ceramic product.

3.5.3 Glasses

3.5.3.1 General

Transparent and easily shaped glasses have silica as the main ingredient. The type of formed material structure depends on cooling with very slow cooling giving crystalline structure, while quicker cooling gives an amorphous structure with disordered and linked chains of silicon and oxygen atoms. The amorphous nature of glass is responsible for its transparency because presence of the crystal boundaries leads to light scattering, causing reflection.

Glass is thus an amorphous solid without long-range order being typically brittle and often optically transparent. It can be tempered to increase toughness and resistance to cracking. Dependence of specific volume on temperature is shown in Figure 3.73.

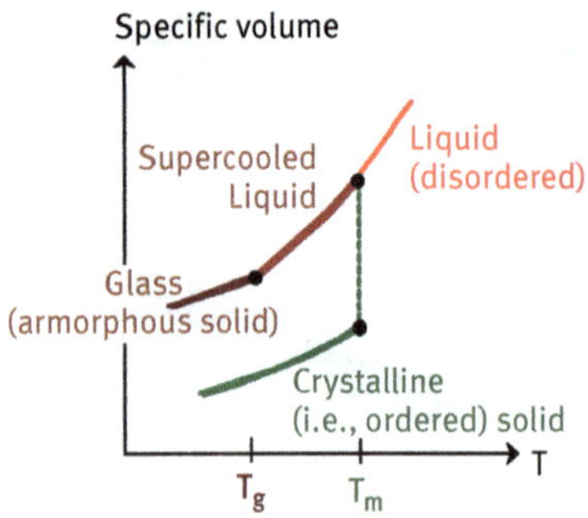

Figure 3.73: Temperature behavior of glass specific volume [62].

Crystalline materials crystallize at melting temp, T_m, and have abrupt changes in the specific volume at this temperature. In the case of glasses, there is no crystallization and specific volume varies smoothly with T. Arrangement of atoms in glass is random being more similar to liquids. Glass can be viewed, therefore, as a super stiff liquid. Glass transformation behavior is specific for a particular glass. The glass transformation region cannot be characterized by any single temperature. The glass-transition temperature T_g of a material reflects the range of temperatures over which this glass transition occurs rather than a single temperature.

The most frequently applied definition of T_g uses the energy release on heating in differential scanning calorimetry (Figure 3.74). In fact, different points (A, B, C in Figure 3.74) are used in scientific literature for calculation of glass transition temperature *Tg* resulting in substantial deviations between reported values. The most popular method of processing differential scanning calorimetry curves is to use the intersection of two tangents at the start of the corresponding endotherm (point A in Figure 3.74).

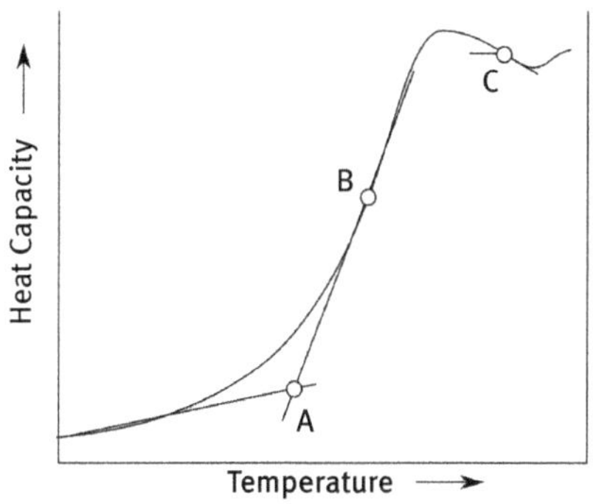

Figure 3.74: Temperature behavior of glass heat capacity [63].

Almost all industrially manufactured glasses are silicate glasses. An overview of different types of glasses is provided in Table 3.18. Glasses can have more complex composition than just silica, containing several types of oxides. Oxides, which form glasses by themselves, are termed network formers. Network modifiers are alkali-metal ions and alkaline-earth metal ions, while intermediate glass formers provide partial substitutes for network formers.

Alkali-metal or alkaline-earth oxides introduced in glass lead to breaking Si–O–Si linkages. These network modifiers are located at nonbridging oxygen sites decreasing silica network connectivity. As a result there is a very large decrease in viscosity, melting point and UV transmission. Introduction of modifiers also leads to a decrease in resistivity lowering chemical durability. Soda-lime glass is probably the most popular type of glass of this type. It also contains small amounts of alumina (Table 3.18). In general, addition of alumina to silica melts increases the resulting glass density and allows application of aluminosilicate glasses under conditions when a glass should withstand elevated temperature. The glass is further chemically strengthened by alkali-metal or alkaline-earth exchange. Application examples include airplane windows, lamp envelopes and cooking utensils to name a few.

Addition of boron to silica melts reduces the melt viscosity, which is, however, not as prominent as with addition of alkali. Borosilicate glass known as the Pyrex glass is used in laboratory equipment offering better temperature stability and chemical resistance. This glass is also less brittle than sodium glass. Other applications include ovenware, piping, glass fibers for reinforcement, textile or insulation fiber applications. Lead silicate glasses contain lead oxide to improve refractive index. The behavior of lead as a network modifier is related to its low coordination number with oxygen, varying between 2 and 4.

Inorganic glass-forming systems can contain besides those listed in Table 3.18 also Ge, Sn, P, Sb, S, Se or Te. Even minor components can substantially influence the ability to form glass and stability of the latter, which increases with increasing number of components.

Table 3.18: Composition of glasses in wt% [62].

Glass type	SiO_2	Na_2O	CaO	Al_2O_3	B_2O_3	Other	Characteristics and applications
Fused silica	>99.5						High melting temperature, very low coefficient of expansion (shock resistant)
Container (soda lime)	74	16	5	1		4MgO	Low melting temperature, easily work, durable
Fiberglass	55		16	15	10	4MgO	Easily drawn into fibers–glass–resin composites
96% Silica (Vycor)	96				4		Thermally shock and chemically resistant laboratory ware
Borosilicate (Pyrex)	81	3.5		2.5	13		Thermally shock and chemically resistant-oven ware
Optical flint	54	1				37PbO, $8K_2O$	High density and high index of refraction-optical lenses
Glass–ceramic (Pyrocream)	43.5	14		30	5.5	$6.5TiO_2$, $0.5As_2O_3$	Easily fabricated, strong, resists thermal shock-ovenware

Figure 3.75 illustrates that viscosity decreases with *T* increase and that presence of impurities lowers temperature of deformation T_{deform} at which the material is soft enough to deform and to work with.

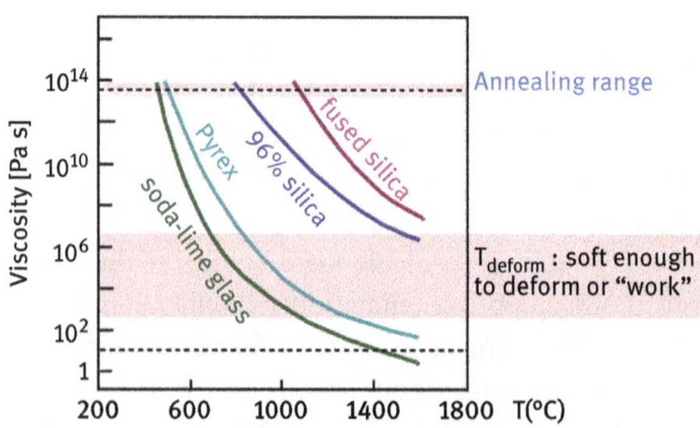

Figure 3.75: Dependence of viscosity on temperature [62].

3.5.3.2 Production of glass

The most widely used process for manufacturing architectural glass was invented by Pilkington in 1952. The process (Figure 3.76) starts with melting and refining when fine-grained ingredients are mixed to make a batch, flowing into a furnace heated up to ca. 1,500 °C. The raw materials for clear float glass are silica sand, sodium oxide from soda ash, calcium oxide (from, e.g., limestone), magnesia (dolomite) and alumina (feldspar). Other oxides are added to provide a color lint. NiO gives grey tinted glasses while iron oxide allows production of green tinted glasses.

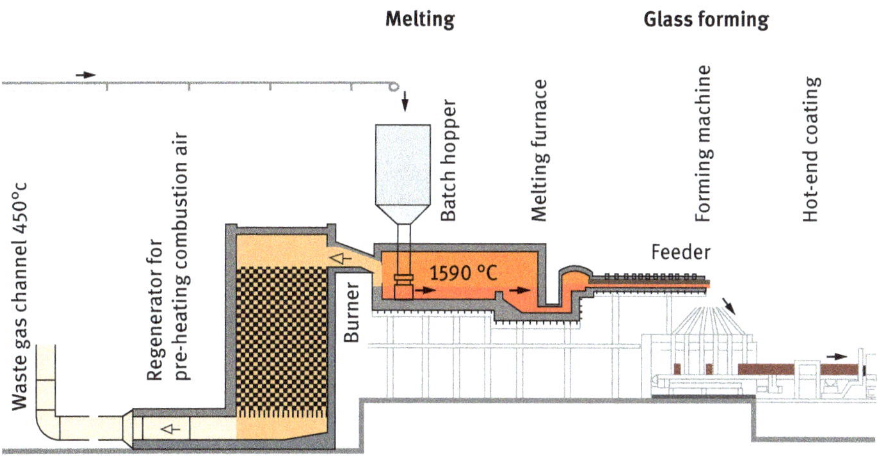

Figure 3.76: Technology for float glass production [64].

The broken or waste glass (i.e., cullet) is also added to the mixture in fairly large amounts helping reduce the melting point of the batch diminishing at the same time the overall energy consumption.

Mixed raw material is continuously pushed from the melting furnace to the float bath (Figure 3.77) passing onto the mirror-like surface of molten tin through the heating and fire polishing zones to the cooling zone. Glass leaves the float bath as solid ribbon at ca. 600 °C.

Coatings making significant changes in optical properties can be made when chemical vapor deposition is applied to the cooling ribbon of glass. Multiple coatings can be deposited in a few seconds as the glass flows beneath the coater.

Considerable stresses are developed in the ribbon during cooling of the glass. Annealing removes internal stresses caused by uneven cooling. The magnitude of high residual stresses depends on the cooling rate (at which glass is transformed from a liquid into a rigid state), coefficient of glass linear expansion, glass thickness and homogeneity. Compressive or tensile stress leads to anisotropic behavior of glass, which is otherwise amorphous and isotropic.

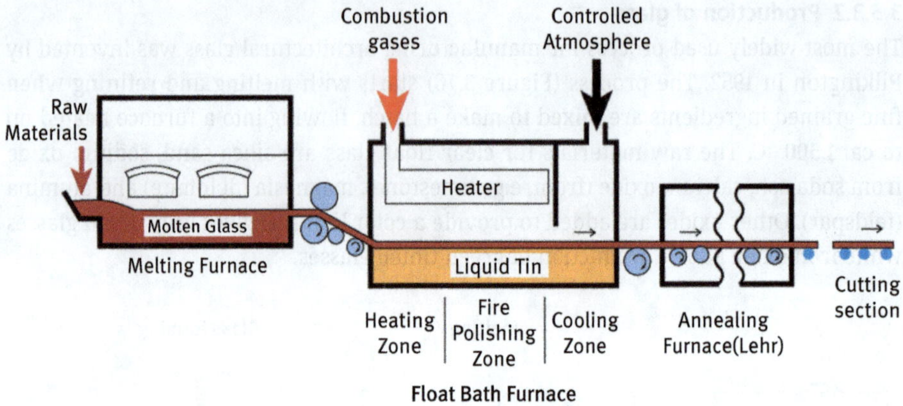

Figure 3.77: Float bath furnace [65].

Annealing furnace in the production line is followed by cutting ribbon to the required size.

Glass articles (bottles, dishes, optical lenses, television tubes, etc.) are formed by blowing, pressing, casting, drawing, rolling or spinning glass, cooling and setting the final shape. Such fabrication can be done either manually for small production or by machine for large scale manufacturing. For instance, when one end of the blow pipe is dipped in the molten mass of glass an operator blows vigorously from the other end of the pipe. As a result, the molten mass takes the shape of cylinder, which is then cut. In the casting method, the molten glass is poured in molds and is cooled down slowly. The molten glass can be also pulled by, for example, an iron bar dipped sideways. After lifting up the bar horizontally a sheet of molten glass is caught, followed by passing over a large rotating roller. The roller helps the molten glass to spread in the sheet. In pressing process, the molten glass is pressed into molds. The molten glass can also pass between heavy iron rollers giving a flat glass plate of uniform thickness. Another option is to pour the molten mass of glass on a flat iron casting table and then use a heavy iron roller. Finally, the molten glass can undergo spinning at high speed to a very fine size.

3.5.4 Cement and concrete

3.5.4.1 Cement

Cement (Figure 3.78) is used as a binder for construction materials. It binds, for example, sand and gravel (aggregate) to make concrete, which will be discussed later in this chapter, or is applied together with fine aggregate to give mortar. From the chemical viewpoint cements used in construction are usually inorganic materials based on lime (calcium oxide) or calcium silicate.

Figure 3.78: Cement [66].

Nonhydraulic cement does not set in wet conditions or under water. On the contrary, setting happens upon drying and reacting with carbon dioxide in air. After setting, it becomes chemically resistant. Lime (calcium oxide) is first produced from limestone (calcium carbonate) by calcination at temperatures exceeding 825 °C:

$$CaCO_3 \rightarrow CaO + CO_2$$

This is followed by slaking of the lime, i.e., reaction with water giving calcium hydroxide or slaked lime. After removal of water, carbonation starts, requiring exposure of the nonhydraulic cement (the slaked lime) to air:

$$Ca(OH)_2 + CO_2 \rightarrow CaCO_3 + H_2O$$

This reaction is slow, which is understandable because of low carbon dioxide partial pressure in air.

On the contrary setting of hydraulic cement (e.g., Portland cement), making it adhesive, occurs because of a reaction between dry ingredients and water. Hydraulic cements are made of a mixture of silicates and oxides with the four main components being belite ($2CaO \cdot SiO_2$), alite ($3CaO \cdot SiO_2$), tricalcium aluminate or celite ($3CaO \cdot Al_2O_3$) and brownnmillerite ($4CaO \cdot Al_2O_3 \cdot Fe_2O_3$). As a result of hydraulic hardening, calcium silicate hydrates (or calcium aluminate hydrates for calcium aluminate cements) are formed making cement that is not that water-soluble and thereby durable.

The production of Portland cement includes extraction and preparation of the raw materials, calcination of the raw materials mixture giving cement clinker, grinding and packaging (Figure 3.79). The feed to the clinker is a mixture of limestone and chalk while SiO_2, Al_2O_3, and Fe_2O_3 are introduced by adding clay and quartz sand. The

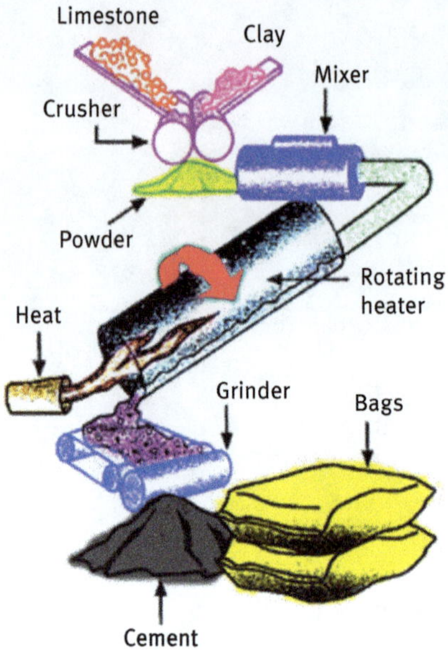

Figure 3.79: Production of hydraulic cement [67].

ratio of components is typically 75–79 wt% $CaCO_3$ and 21–25 wt% clay. An important issue in cement quality and uniformity is minor deviation of raw material composition over time.

In production, typically large clinker storage facilities (four to six weeks production capacity) are used compensating differences between production and dispatch related to, for example, plant interruptions and demand fluctuations. Such storage provides also additional clinker mass homogenization.

After mixing, the material passes through a preheater chamber consisting of a series of vertical cyclones before coming to the kiln phase. The heat integration is realized by using emitted hot gases from kiln in the preheating chamber. In the rotary kiln (i.e., a very large rotating furnace) at temperatures approaching 1,450 °C calcium carbonate is decomposed to calcium oxide followed by reactions of the latter and silicon dioxide forming, for example, calcium silicate. Heating of rotary kilns is done using either natural gas or coal. Downstream processes include cooling of clinkers by forced air and final grinding.

Such fine grinding can be done for the clinker alone or with addition of blast furnace slag, pozzolana or fly ash. Gypsum is added in a small percentage to control cement setting. Special grinding aids (e.g., glycols and ethanolamine) are introduced in small amounts (0.02–0.04 wt%) without deterioration of the water demand, setting and cement strength development.

Grinding systems can use ball mills and optionally include high-pressure grinding rolls to improve cement output and energy efficiency allowing 50–65% energy consumption compared to ball milling. This more energy efficient option includes comminution of the coarse feed under high compressive stress (50–350 MPa) in the gap between two counter-rotating grinding rollers. Stand-alone application of such rollers gives cement with a significantly higher water demand, and hence finish grinding with a ball mill is applied downstream high pressure grinding rolls. Vertical and horizontal roller mills can be used for cement grinding with also lower energy consumption (65–75% of ball mill grinding). Besides energy consumption, other parameters influence the selection of the grinding method including capacity, capital expenditure, maintenance costs, etc.

After grinding, cement is directly conveyed to silos and thereafter delivered to customers by trucks, railroad or ships. Only a small fraction is packed to 20–40 kg bags for individual customers.

Setting and hardening of cement happens upon exposure to water. The water/cement ratio for technical purposes is ca. 0.3–0.7. For workability water is added in amounts larger than needed for complete cement hydration. Additives can be also used to improve workability and decrease the water to cement ratio.

A mixture of cement with sand gives mortar, while addition of aggregates is done for the purpose of making concrete.

The exothermic nature of cement hydration influencing temperature of concrete should be considered in the construction of, e.g., tunnels to avoid cracks because of internal and residual stress. Heat release accelerating hydration under adiabatic conditions makes construction faster on one hand, while negatively influencing concrete workability on the other hand. The heat release can be regulated by controlling cement reactivity. Higher reactivity gives faster hardening with a faster release of the heat of hydration. The rate of hardening can be elevated by decreasing the size of different cement constituents and by changing the composition. In particular, for Portland cement higher amounts of tricalcium silicate and aluminate make the cement more reactive.

Presence of gases such as carbon dioxide can lead to their penetration inside the hardened cement paste and subsequent chemical reaction with calcium hydroxide, calcium aluminate or calcium silicate hydrates forming calcium carbonate. This positively influences the strength of the cement stone. At the same time carbonation can impair the steel reinforcement in concrete.

3.5.4.2 Concrete

Concrete is a composite material, comprising, besides cement and water, also coarse aggregates. Mixing aggregates with a lime-based dry Portland cement and water gives a fluid slurry, which is easy to pour and mold (Figure 3.80a). Asphalt concrete used for road surfaces is a material in which cement is replaced by bitumen. The cementing material can be also a polymer as in polymer concretes. The physical properties of the wet mixture or the final products can be improved by adding pozzolants, superplasticizers (special polymers) or reinforcing materials.

The latter give reinforced concrete with tensile strength, while pozzolans (referring to natural compounds of volcanic origin) represent siliceous or siliceous and aluminous materials, which react with calcium hydroxide. By themselves, pozzolans do not have cementitious properties,; however, the reaction mentioned above gives compounds with such properties. Superplasticizers act as dispersants preventing segregation of different particles, namely gravel, coarse and fine sands. They also behave as rheology improvers. Addition of superplasticizers brings more strength to concrete by decreasing water-to-cement ratio, at the same time not affecting the mixture workability. For instance, application of polycarboxylate ether based superplasticizers in a relatively low dosage (0.15–0.3% by cement weight) gives water reduction up to 40%. Aesthetic enhancement of, for example, driveways (Figure 3.80b) is achieved in decorative concrete by addition of certain components during pouring or after curing. A range of other additives can be introduced to cement in minor amounts including air-entraining additives to improve frost resistance, corrosion inhibitors, insecticidal agents, etc.

(a) (b)

Figure 3.80: (**a**) Ordinary [68] and (**b**) decorative concrete [69].

Reinforcement of concrete with steel bars makes otherwise brittle concrete much more stable, increasing the application scope. Fresh concrete (Figure 3.80a) can be shaped easily on a construction site, while some concrete products can be prefabricated in plants.

The nature of additives influences the way they are added to concrete. While solid additives are introduced together with the cement, mixing of the liquid additives with water is preferred. If a certain additive has limited time efficiency, it should be introduced immediately before use of concrete.

Concrete typically consists of ca. 25% of cement matrix while the rest 75% is made of aggregates acting as inert fillers and providing volume stability, wear resistance, ductility, stability to cracking, required thermal conductivity and thermal expansion, etc. Aggregates can have different origin (synthetic or natural), chemical nature (inorganic or even organic for special concretes), density or size. Tuff or calcined waste products can be used as aggregates for lightweight concrete, while such aggregates as e.g., baryte, ores, or slags are applied for heavyweight concrete, which then is utilized as a ballast or in shielding nuclear radiation. In general, the size of aggregates ranges from 0.01 to 100 mm. Particles above 4–5 mm are considered as coarse aggregates. Aggregates size distribution should ensure sufficient workability of fresh concrete with low water content.

Concrete aggregates should not influence hardening of the cement paste and should be stable during operation. This limits the content of some organic compounds or clays. A special type of concrete is mortar containing only sand (<2–4 mm). Such small size requires higher water-to-cement ratio compared to normal concrete. As a result, mortar exhibits lower compressive strength and volume stability. Water-reducing and expansion-producing additives as well as short reinforcement fibers (randomly oriented and homogeneously distributed) with diameter of several microns and length from few to 50–60 mm could be introduced to mortar.

Organic resins are applied in the case of polymer concrete giving several advantages such as high tension and compression strength, chemical stability as well as stability to temperature fluctuations around the water freezing point. Presence of polymers obviously influences thermal stability in a negative way, also making polymer concrete more expensive. In epoxy concrete, polymer is the only binding material. There are also options to apply polymers in a mix with cement and water or impregnate concrete after hardening with monomers for subsequent polymerization.

The water amount from the viewpoint of chemical reactions stoichiometry happening during hardening should be 20–25% of the cement mass; however, extra water is added to ensure workability and complete hydration. In principle, there are no special requirements regarding water apart from restrictions to use sea water, e.g., for reinforced concrete to avoid corrosion because of dissolved chlorides.

Thorough mixing of the materials ensuring uniform concrete can be done in a centralized way in a concrete production plant as well as on site (Figure 3.81a).

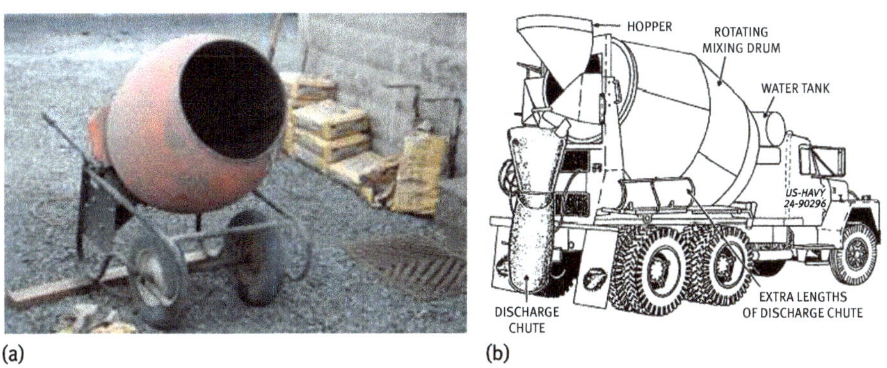

(a) (b)

Figure 3.81: Mixing equipment: (**a**) onsite [70] and (**b**) transit (truck) mixing [71].

Mixers generally consist of revolving drums with blades, stationary or revolving pans with blades rotating on vertical shafts, as well as horizontal drums equipped with spiral blades using a rotating horizontal shaft. Mixing time takes 30–120 s. Truck mixers (Figure 3.81b) are equipped with an inclined revolving drum operating at a speed lower than 2 rpm.

3.5.4.3 What can go wrong?

Finally, an interesting example of incompetent application of steel reinforced concrete is worth considering. A so-called metro bridge in Moscow was constructed in 1959 (Figure 3.82) as part of the subway (or underground) system and had to be completely rebuilt some 40 years later because of a wrong additive applied during its construction.

Figure 3.82: The metro station incorporated in a bridge across the Moscow river [72].

The original proposal was to use steel as a construction material; however, this concept was dropped in favor of reinforced concrete. The station located in the bridge was the longest in the subway system (272 m). The critical mistake was related to the fact that the reinforced concrete had to be used in the winter time because of a tight construction schedule. In order to prevent premature freezing of concrete prior to hardening, an inexpensive additive – sodium chloride was added. The latter is a notoriously known compound for decreasing the freezing point of water solutions. While the salt helped in a certain way during construction, it negatively influenced concrete and the reinforcement steel in the long term leading to corrosion. On top of that, the salt expanded in volume significantly during eventual crystallization. Already by 1963, it became clear that ca. 60% of steel was corroded. It took, however, another 20 years until the bridge was taken out of service and the metro station was closed down. The dissembling of the bridge was commenced in 1998 and after two years of reconstruction, first the bridge and then lately the metro station itself were reopened.

3.6 Catalysts

3.6.1 Basics

Catalysts accelerate the rates of chemical reactions while not being consumed. Catalysis is thus a kinetic process influencing not only the rates but also selectivity toward a particular product. In fact, catalyst selectivity is probably the most important characteristic of a catalyst, and it should be also sufficiently stable under operation conditions.

The dominating part of industrial processes using catalysts relies on solid catalysts, which are present in a phase other than reactants and products. The latter can be either in gas or liquid phases. There are also examples of using so-called homogeneous catalysts, when the catalysts are in the same phase, requiring their efficient separation from the products and unreacted substrates. This section will be devoted exclusively to heterogeneous catalysis or catalysis by solid surfaces.

Figure 3.83a shows the potential energy diagram of a catalytic process illustrating that initially there is a binding (Figure 3.83b) of the reactant to the catalyst (adsorption), followed by transformation of the catalyst-substrate complex and subsequent release of the product (desorption). The energy barrier between the catalyst–substrate and the transition state should be lower than the corresponding value of an uncatalyzed reaction.

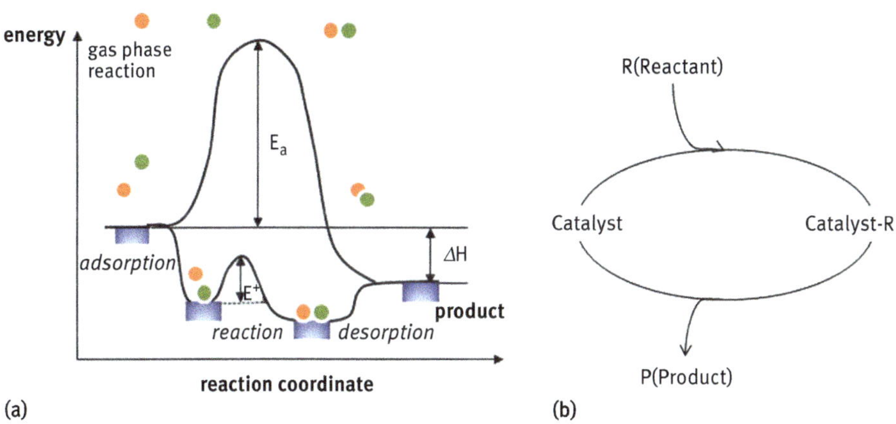

Figure 3.83: Reaction mechanism with substrate binding: (**a**) potential energy diagram of a gas phase reaction (upper curve) and a catalytic one (lower curve) and (**b**) catalytic cycle [73].

It should be also noted that although it is conventionally assumed that a catalyst participates in the process but remains unchanged at the end, there could be major changes in its structure and composition.

Potential energy diagrams for a catalytic and a noncatalytic reactions presented in Figure 3.83 indicate that in both cases the reactions should overcome a certain barrier, which is lower in the presence of a catalyst. These lower values of activation energy result in higher rates as clearly follows from the well-known temperature dependent expression for the rate constant:

$$k = k_0 \, e^{-E_a/RT}$$

where k_0 is the pre-exponential factor and E_a is activation energy related to the potential energy barrier. Catalysts accelerate chemical reactions, allowing them to be carried out at much lower temperatures and pressures, at the same time not influencing overall thermodynamics. As a result, a catalyst active in the forward reaction will also catalyze a reverse reaction.

Heterogeneous catalysis covers a wide length scale from a molecular scale (nm) of an active site to a catalytic reactor (m) scale (Figure 3.84). For the case of so-called supported catalysts the active particle with dimensions ca. 1–10 nm are located inside pores of a support material. Chemical composition, texture of materials, their shape and size as well as the pore structure are important for design of technical catalysts.

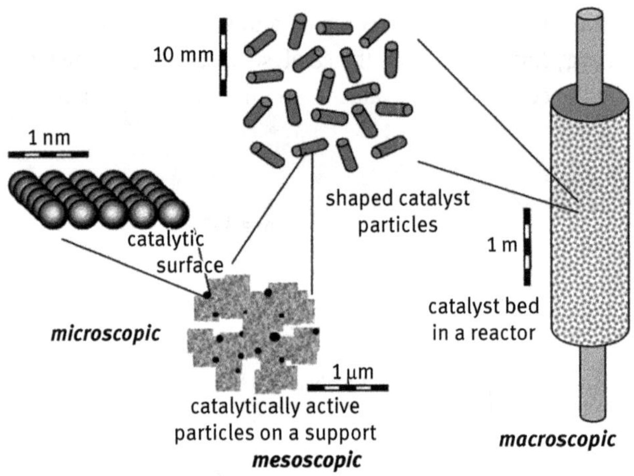

Figure 3.84: Length scales in catalysis [73].

In industrial practice, shaped catalysts (Figure 3.85) such as extrudates, pellets, tablets (mm scale) are used. Product engineering addresses therefore besides activity, selectivity and stability of catalysts, also mass transfer and heat transfer, porosity, mechanical strength, attrition resistance as well as a reactor type.

The shape and size of the catalyst particles should promote catalytic activity, strengthen the particle resistance to crushing and abrasion, minimize the bed pressure drop and diminish the fabrication costs. Small particle size is beneficial from the viewpoint of minimizing internal and external mass transfer. The corresponding pressure drop can be, however, very high highlighting a trade-off, which is made between a desire to have small particles exhibiting higher activity and large particles with

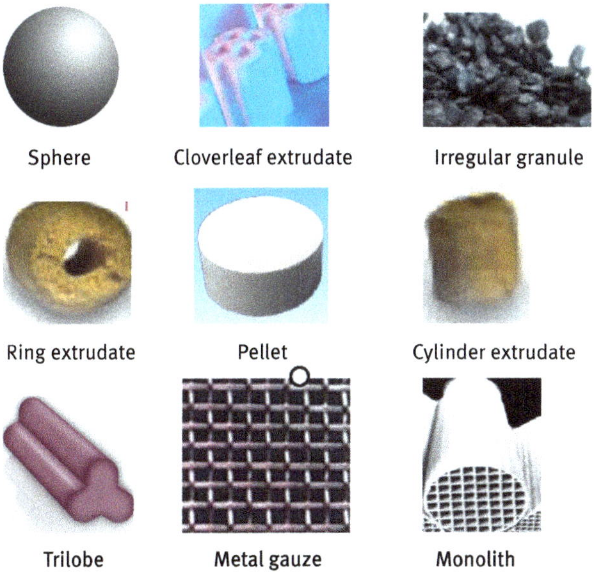

Figure 3.85: Various catalyst shapes [73].

lower pressure drop. The choice of the catalyst shape is closely related to the catalytic reactor type. Large particles (few mm) are applied for fixed-bed reactors, while slurry three-phase and fluidized-bed reactors require catalysts powders exhibiting good attrition resistance and easy filtration. Pressure drop can, in fact, be regulated by making special types of extrudates (e.g., rings, minilths with several holes, clover leaf extrudates, etc.) increasing, however, the complexity (e.g., manufacturing costs) and decreasing mechanical stability because of lower wall thickness. For off-gas cleaning applications monoliths are mainly used.

Mass transfer limitations inside a catalyst particle can be minimized by applying egg-shell type of catalysts with a reasonably thin catalytically active layer allowing obtaining activity similar to powders of a smaller size.

The choice of catalyst shapes is thus not straightforward and involves careful considerations of hydrodynamics, heat and mass transfer limitations, mechanical strength, as well as manufacturing methods and associated costs.

Composition of typical catalytic materials is demonstrated in Figure 3.86 showing that besides the active agent per se, catalysts can also include a support. Supports are needed to spread expensive materials, provide the mechanical strength, act as the heat sink/source, optimize bulk density and dilute an overactive phase.

Other functions (increase of surface area, porosity and cluster size optimization, improvement of activity, minimization of sintering and poisoning) can be also provided by a proper choice of a support, which is typically alumina, silica, titanium or active carbon to name the most important ones. Choice of such supports as alumina

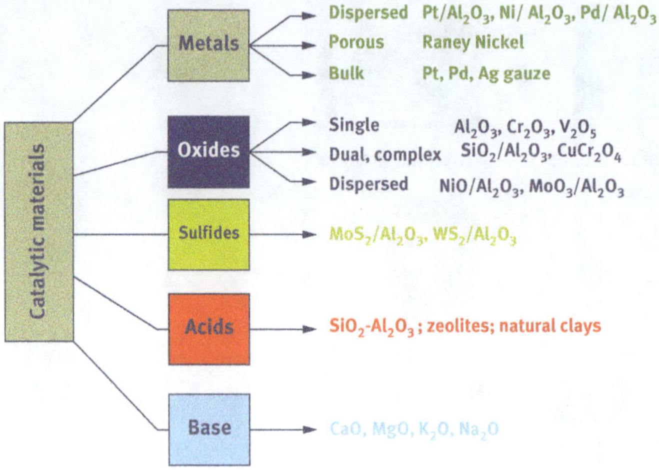

Figure 3.86: Catalytic materials [73].

and silica with high melting points is also justified by the fact that not only the active phase, but also the support can undergo sintering at temperatures substantially lower than the melting point.

Alumina is mainly preferred due to favorable bulk density, thermal stability and feasible price, being applied as γ-alumina phase in a number of important industrial reactions (hydrotreating, hydrocracking, hydrogenation, reforming, oxidation, etc.). Alumina as such is a Lewis acid, which should be considered when using it for certain reactions, because such activity can be either beneficial or detrimental. For example, platinum dispersed on alumina is applied for improving the octane number of gasoline by performing an overall reaction of alkane skeletal isomerization. In practice, platinum catalyzes dehydrogenation reactions of linear alkanes to linear alkenes, followed by skeletal isomerization on alumina and subsequent hydrogenation of formed branched olefins once again over platinum.

More thermally stable phase than γ-alumina, namely, α-alumina is used as a support for nickel in the case of natural gas steam reforming. Somewhat lower bulk density of silica than for alumina is translated into a larger reactor volume for the same active phase loading. Silica is more prone to sintering above 900 K than alumina and more volatile in the presence of steam and elevated pressures. It has at the same time an advantage of low acidity than alumina, and is also applied in a number of industrial processes (polymerization, oxidation and hydrogenation).

Amorphous or crystalline aluminosilicates can be used as supports and catalysts per se. In particular, crystalline zeolites are applied because of acidic properties. Carbon activated by physical or chemical means is often used as a support in synthesis of fine chemicals by different hydrogenation reactions. Other supports are used for some specific reactions. For instance, ceramic monoliths from cordierite are applied as supports for reactions requiring high flow rates and low pressure drop, while selective oxidations and steam reforming of natural gas relies respectively on titanium and magnesium aluminate.

Besides an active agent and a support, often various promoters are added in small amounts to improve activity, selectivity or stability. Textural promoters (oxides with high melting points Al_2O_3, SiO_2, ZrO_2, Cr_2O_3, CeO_2, MgO and TiO_2) inhibit sintering of the active phase. Structural promoters influence the chemical composition, generate lattice defects and change the electronic structure. The group designed as metal oxides in Figure 3.86 usually consists of a multicomponent mixture of various bulk or supported oxides. Typically, metal oxides are used in processes requiring higher temperatures.

In this textbook, the aim is not to cover in detail the basic principles of heterogeneous catalysis and catalytic materials as they are presented in detail in the easily available textbooks [73 and references therein] but rather to discuss catalysts from the chemical product technology viewpoint. Therefore, some industrially relevant catalyst preparation methods will be briefly considered.

3.6.2 Catalyst preparation technology

A more detailed discussion on catalyst preparation methods and industrial technology is available in a recent textbook [73]. The most relevant examples in the context of chemical product technology are reproduced here.

Large scale preparation of catalysts involving many parameters such as stirring rate, heating rate, heat transfer, flow pattern, filtration, washing, drying and calcination conditions can be different from laboratory-scale formulations. Industrially used methods are either kept as trade secrets or described in a very general way.

The catalyst formulations besides the active phase include various additives (Figure 3.87) to provide the required characteristics and performance not only in terms of catalysis per se, but also manufacturing, handling, pressure drop, mechanical properties, stability, etc.

An overview of the additives is given in Table 3.19. Materials with poor self-binding properties (e.g., sulfated zirconia or zeolites) require binders, introduction of which can influence the catalyst performance. The fillers are typically used for economical purposes. Generation of pores can be done using porogens. A compromise is required between formation of the pores and subsequently better mass transfer on one hand and the mechanical strength on the other as porosity increase results in a lower mechanical strength.

Some catalytically active phases should be introduced after shaping as they may not be able to survive, for example, shaping conditions. Impregnation of solid bodies will be discussed below as an example of such a strategy.

The main unit operations for shaping of industrial catalysts have been considered in Chapter 2. They include spray drying, tableting, extrusion and granulation, each

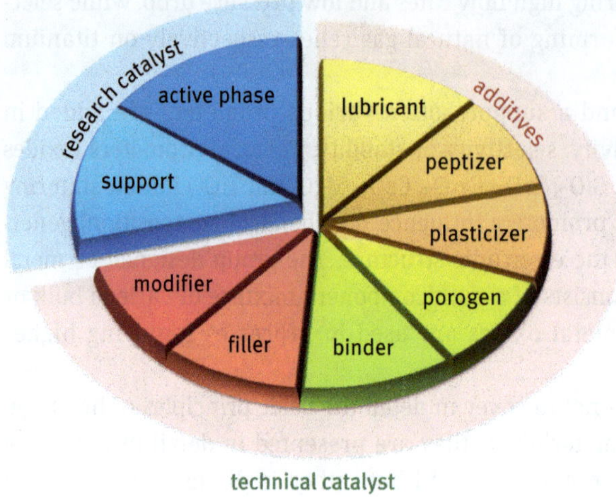

Figure 3.87: Composition of a technical catalyst [74].

Table 3.19: Overview of additives.

Additive	Function	Applicable materials	Examples
Binder	Increases mechanical strength upon hardening of technical body	Inorganic oxide with refractory properties	Alumina, aluminum phosphate, silica, silica-alumina, natural clays (attapulgite, bentonite, kaolin, sepiolite, montmorillonite), titanium, zirconia
Filler	Dilutes active phase to optimize content in technical body	Low-cost inorganic oxide with refractory properties	Natural clays (kaolin), alumina
Lubricant	Reduces friction during mixing and shaping operations	Oil or other viscous liquid	Ethylene glycol, glycerin, graphite, mineral oil, propylene glycol, aluminum stearate
Modifier	Enhances performance of a technical catalyst	Promoter, selective poison, co-catalyst, coke or metal traps, passivators	Metal or metal oxide precursor, clay, zeolite
Peptizer	Disperses particles to improve feed homogeneity prior to/ during shaping	Commonly a mineral or organic acid, can be a base	Acetic, citric, formic, hydrochloric, nitric, sulfuric or phosphoric acid
Plasticizer	Decreases viscosity of pre-mixtures to facilitate processing	Material dependent, typically containing polar and nonpolar functionality	Hydroxyethyl cellulose, polyethylene glycol, starch, sugar, water

Table 3.19 (continued)

Additive	Function	Applicable materials	Examples
Porogen	Increases interparticle porosity in technical body	Cheap material, which is removed by thermal decomposition during hardening	Carbon black, flax, sawdust, different size starch (potato, rice and corn)

Note: Reproduced with permission from [74].

having its own advantages and disadvantages. Tableting is done for powders that have enough plasticity, by applying high forming pressure. This operation results in shaped bodies with low pore volume and surface area. Spray drying and granulation are aimed at spherical particles of a size from 10 to 100 μm and 1 to 20 mm. In the latter case, the particle distribution is rather broad. Extrusion being, maybe, the most economical method, relies heavily on the knowledge of rheology, addressed in Chapter 2.

As an example of mechanical mixing and extrusion during catalyst preparation, the synthesis of sulfuric acid catalyst of the SVD type (on diatomite) is illustrated in Figure 3.88.

The active component in this catalyst is vanadium pentoxide, promoted with sulfates of alkali metals, potassium and sodium. The support is diatomite, which is a silica with some minor presence of other oxides such as Fe_2O_3 and Al_2O_3. The catalyst performance depends very much on the type of diatomite.

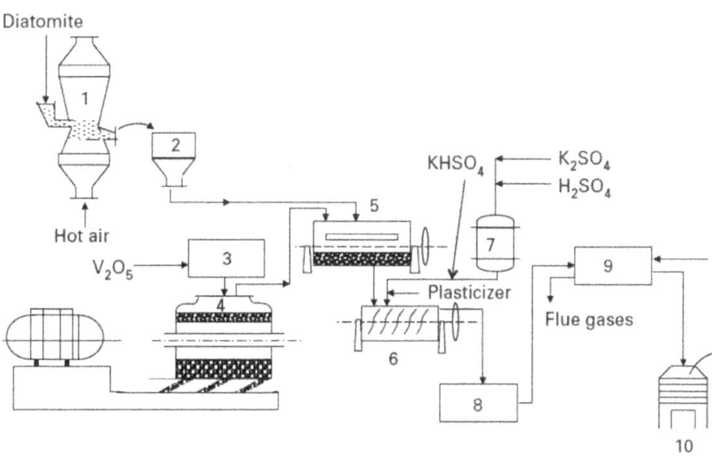

Figure 3.88: Production of sulfuric acid catalyst [73]: 1, dryer; 2, bunker; 3, grinder; 4, vibration milling; 5, ball milling; 6, extruder; 7, reactor for potassium bisulfate; 8, dryer; 9, calcination; 10, sieve.

The support is dried in a spray dryer at 120–150 °C, up to a humidity of 5–7%. The technical quality of vanadium pentoxide (80% of V_2O_5) is ground up, first to 2–3 mm (pos. 3) and then to ~ 200 µm (pos. 4). After mixing and ball milling (pos. 5) the solid phase that contains diatomite and vanadium pentoxide, it is extruded with the addition of potassium hydrosulfate (promoter), stearic acid (plasticizer) and graphite (porogen) directly to the extruder. Potassium bisulfate is prepared in the reactor (pos. 7) by contacting potassium sulfate with sulfuric acid. The extrudates contain 26–30% water and are thus dried at 120 °C (pos. 8) and calcined (pos. 9). During calcination, the following reactions occur.

$$2KHSO_4 = K_2S_2O_7 + H_2O; K_2S_2O_7 + V_2O_5 = K_2S_2O_7 \cdot V_2O_5$$

In this example, the supported catalyst was produced by a dry method of mechanical mixing, followed by calcination. This procedure leads to a catalyst when vanadium is present as a complex sulfated salt mixture, which is in the molten state during actual catalysis.

In general, catalysts preparation includes several steps, the most important being preparation of the primary solid (precipitation and impregnation) and subsequently, postprocessing (thermal treatment, chemical activation, etc.).

Precipitation and coprecipitation are mainly used for the production of oxidic catalysts and support materials. Metal salts or hydroxides are used as precursors. Typically, nitrates are preferred due to their high solubility in water, availability and a comparatively low price. Below, a preparation method of a catalyst for methanol synthesis, incorporating copper- zinc- aluminum oxides will be considered. Precipitation, which happens after the dissolution of the precursors, consists of nucleation and crystal growth. The catalytically active phase is typically metastable; thus, crystallization, which is enhanced at low temperatures, should be performed at conditions far from equilibrium.

Changing the conditions of precipitation may result in different phases, with varying catalytic activities. A particular case of the methanol synthesis catalysts (Figure 3.89) includes several steps, such as preparation of solutions, mixing, filtration, washing, drying, grinding, pelletizing, calcinations and final packaging. Water solutions of the corresponding salts of copper and aluminum, together with sodium carbonate, are fed into the reactor 5, where crystals of copper carbonate of the size 20–40 nm are form-mixed with aluminum hydroxide particles of size below 15 nm. After the addition of zinc nitrate, small crystals of zinc carbonate are generated, which are included in copper carbonate crystals, preventing the growth of the latter. After filtration (pos.6) from the mother liqueur, drying and calcination (pos.7) the crystals are grinded (pos.8), resulting in small particles. Tablets (pos.12) are made after the addition of graphite (pos.9) and additional grinding (pos.11). The amount of graphite in the final catalyst is below 2 wt%. Reduction (pos.15) results in the formation of copper crystallites that are incorporated in the matrix of Al_2O_3–$ZnAl_2O_4$ spinel. The presence of zinc is needed to diminish the dehydrogenation ability of the catalyst.

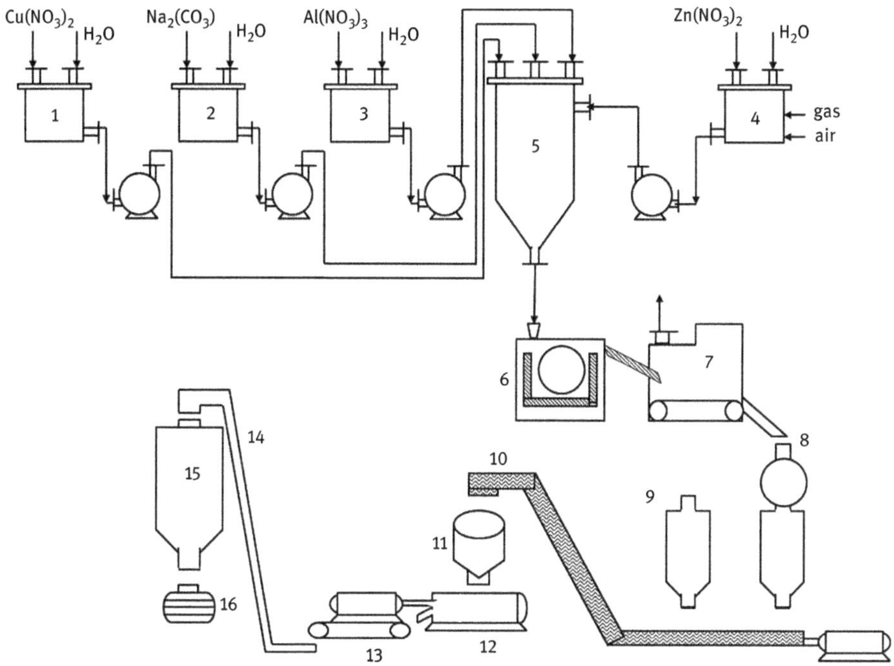

Figure 3.89: Flow scheme of Cu–Zn–Al oxide catalyst preparation: 1–4, vessels for the preparation of water solutions; 5, reactor; 6, filter; 7, drying and calcination unit; 8, grinder; 9, vessel with graphite; 10, mixing and conveyer; 11, additional grinding unit; 12, tableting machine; 13, 14, conveyers; 15, reactor for reduction (or oxidation); 16, vessels for the final catalyst (after [75]).

A special type of supports, which are widely used as catalysts per se, is solid acids – zeolites. These are crystalline aluminosilicates, containing pores and cavities of molecular dimensions. The complicated structure of zeolites is constructed from several secondary building blocks with different number of units that possess 6-, 8-, 10-, 12-membered rings of the so-called T atoms (Si, Al). Key properties of zeolites are size and shape-selectivity, together with the potential for strong acidity. In zeolites, SiO_4 and AlO_4 tetrahedra are linked by common oxygen atoms. Cations (such as Na^+) are needed to neutralize the charge difference between Al^{3+} and Si^{4+}. Brønsted acidity appears when Si^{4+} is replaced by Al^{3+} and the charge is balanced by a proton, localized in the vicinity of Al. Zeolites could thus have acidities comparable to mineral acids, which is a feature required for catalysis. Zeolites are also widely used as adsorbents for various purposes.

Aluminosilicate zeolites are formed by hydrothermal synthesis, typically under mild conditions (350–525 K) and autogenous pressure (above 373 K) using as precursors: (a) a silica source (the primary building units, i.e., SiO_2, of the framework); (b) an alumina source, which is an origin of the framework charge; (c) organic templates, (d) mineralizing agents (OH- anion, NaOH) and finally water (as a solvent), to form an aluminosilicate gel. Templates of the NR_4^+ type, such as tetra-alkylammonium hydroxide or alkylphosphoniums cations (R_4P^+) and organic complexes, can be used.

Important parameters for hydrothermal synthesis are the molar ratios (OH^-/SiO_2 and Si/Al), pH, temperature, the sources of silica and alumina, presence of contaminants, the order of addition of the reactants, aging and ripening, prior to crystallization, synthesis time, the nature of the organic templates, pretreatment of the reactants, inclusion of special additives, presence of seeds and stirring rate.

After crystallization, the autoclaves are quenched with cold water. The formed zeolite crystals are filtered, washed and dried. Washing is needed to diminish the high alkalinity created by the addition of NaOH. A flow sheet for the preparation of a sodium form of zeolites is given in Figure 3.90.

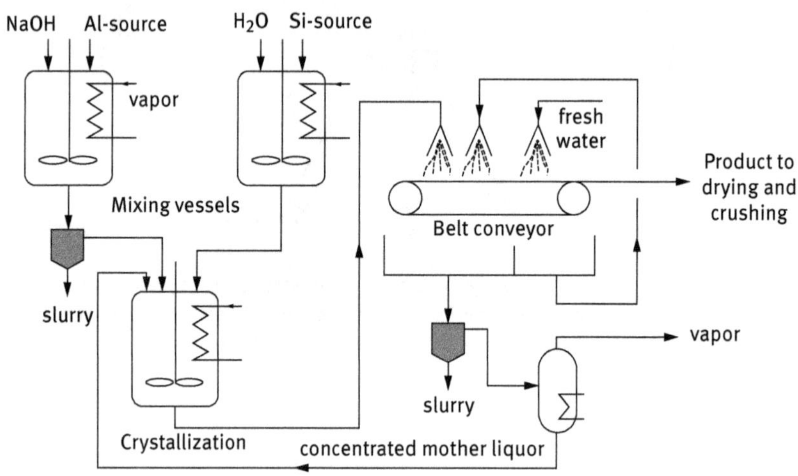

Figure 3.90: Preparation of a sodium form of zeolites (after [76]).

After drying at ca. 100 °C, calcination at ca. 500–600 C is performed. The duration (several hours) depends on the zeolite type. Calcination is performed to remove the organic template, giving the as-synthesized or the sodium form. Acidity is introduced by ion-exchanging sodium cations with NH_4Cl or NH_4NO_3, followed by drying and calcination, releasing ammonia and leaving behind the proton form of zeolite. It is also possible to make ion exchange with other ions. For example, the typical commercial catalysis for FCC (fluidized catalytic cracking) contains a mixture of rare earth-Y zeolite and SiO_2-Al_2O_3 amorphous aluminosilicate, providing the required mechanical stability. Small-pore ZSM-5 is applied as a cocatalyst, together with rare earth Y-zeolite, to increase the octane number of gasoline. The sodium form of zeolites Y is first produced, followed by an exchange of sodium with rare–earth ions.

The flow scheme for the synthesis of the FCC microspherical catalyst is given in Figure 3.91. In vessel 1 (Figure 3.91), sand, water and sodium hydroxide are introduced, followed by the preparation of water-glass at 98–110 °C. The latter is filtered to remove unreacted sand particles (pos.2) and transported to vessel 4, where the solution of needed concentration is prepared. Aluminum sulfate is first prepared in vessel 3, which is thereafter filtered and transported to vessel 5. In this vessel, a working solution of aluminum sulfate is prepared by adding sulfuric acid. This is needed, since otherwise the solution containing the required amount of alumina is too basic, which will result in generation of a zeolite with too large pores and a lower bulk density.

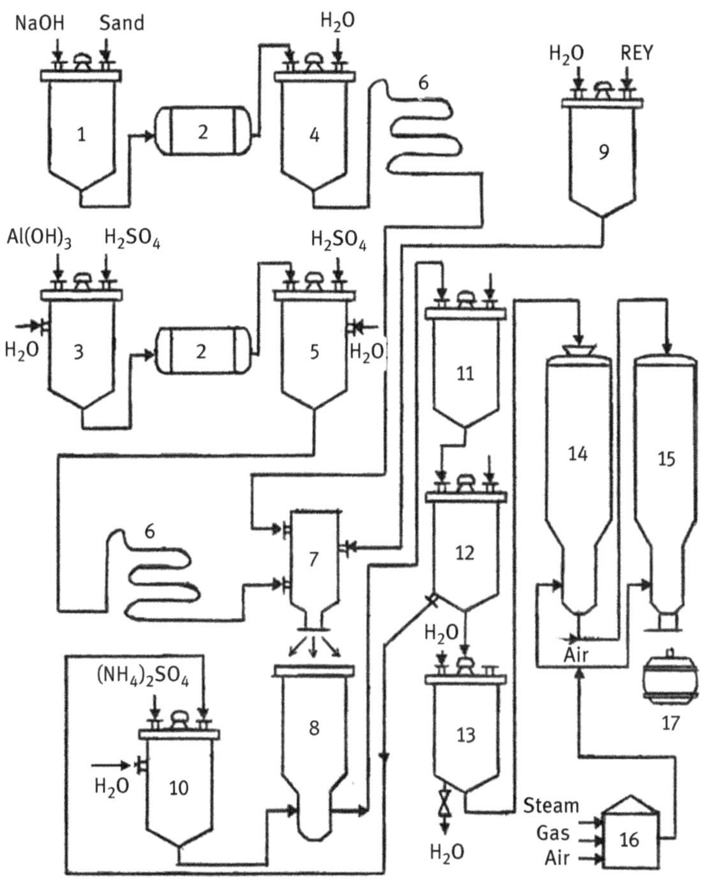

Figure 3.91: Microspherical catalyst synthesis, after [75]: 1, vessel for water-glass synthesis; 2, filter; 3, vessel for preparation of aluminum sulfate solution; 4,vessel for preparation of water glass solution; 5, vessel for the preparation of aluminum sulfate solution with sulfuric acid; 6, coolers; 7, mixer for the preparation of the sol of zeolitic aluminosilica gel; 8, column for the preparation of microspherical gel particles; 9, vessel for rare earth zeolite suspension; 10, vessel for water solution of ammonium sulfate; 11, vessel for gel syneresis; 12, vessel for gel activation; 13, vessel for washing out the microspherical gel particles from sulfate and other ions; 14, drying column; 15, column for calcination; 16, burner for flue gases generation; 17, storage. REY stands for a rare earth Y zeolite.

Vessel 9 contains water suspension of rare earth zeolite Y. Solutions from vessels 4, 5 and 9 are routed to a mixer-sprayer 7, where water-glass, aluminum sulfate and the rare earth zeolite Y suspension are mixed at the ratio required to assure pH 7.5–8.6. Microdroplets of size 2–5 nm leave vessel 7 and enter the column 8, which is filled (ca. 3 m) with transformer oil. The microdroplets passing through the column at ca. 10–12 °C form the gel. The large column diameter (1.5 m) is needed to prevent sticking of the gel to the vessel walls. A droplet passes through the transformer oil in 8–11 s, while 5–8 s are needed for coagulation. The coagulation rate increases with increase in temperature and the concentration of the parent solutions and depends on the pH and the presence of electrolytes. As mentioned above, an acid is added to the mixture to ensure the required coagulation rate. Passing through oil, a gel of microspherical particles is formed, containing 90% water and 10% dry substance. Syneresis (expulsion of a liquid from a gel) is done in vessel 11 at 40–50 °C for 6–24 h. Replacement of sodium ions by ammonium happens in vessel 12 at 50–80 °C for another 6–12 h. In order to prevent the removal of Al^{3+} and rare earth ions during washing (pos. 13), they are added in small amounts to the washing step, while sodium, ammonium and sulfate ions are washed out. Washed gel particles can be treated with water solutions of surfactants to prevent collapse of the gel structure during drying (pos. 14). Washing is followed by drying of the zeolite (pos. 14) and calcination (pos. 15). The flue gases needed for drying and calcination are generated by burning natural gas in the presence of steam and air (pos. 16). During drying, the flue gases enter the dryer from the bottom, while zeolite suspension falls countercurrently to the flue gases. Temperature along the dryer changes from ca. 315–325 °C at the top to 475–485 °C at the bottom. Such regime leads to a material with ca. 8–10% of water. Calcination can be done in a fluidized bed.

In supported catalysts, catalytically active compounds are attached to a support with a large specific surface area. This is done by deposition of the precursor on the support surface, followed by its transformation to the required active phase (metal, oxide, sulfide). The key parameters are: precursor type and concentration, nature of the support, presence of promoters, preparation conditions (pH, viscosity, solvent). In impregnation, depositing of the active phase on a support is done by contacting a certain volume of solution with a precursor and the support. Wet impregnation implies utilization of an excess of solution, which is eliminated by evaporation or draining. Conveyor type of wet impregnation has been used for large scale (Figure 3.92) when the support is loaded in buckets, which move down into the impregnation basin containing the catalyst precursor of constant concentration.

An alternative is incipient wetness or dry impregnation, when the impregnating solution is sprayed over the support (Figure 3.93) located in a rotating drum.

For both methods, the operating variable is the temperature, which influences both the precursor solubility and the solution viscosity, and as a consequence, the wetting time. The concentration profile of the impregnated compound depends on the mass transfer conditions within the pores during impregnation and drying. The maxi-

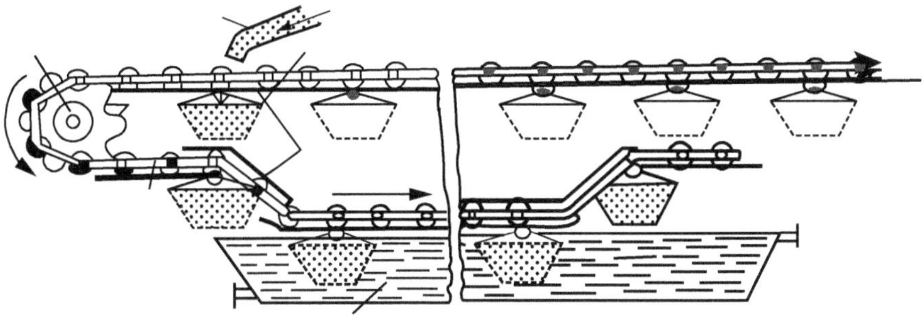

Figure 3.92: Conveyor wet impregnation [77].

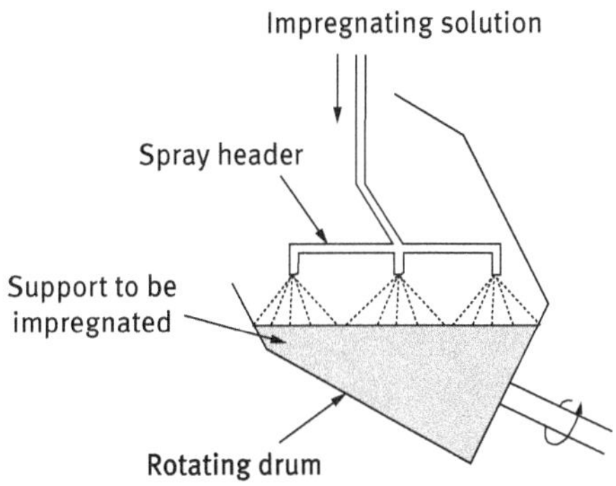

Figure 3.93: Incipient wetness impregnation (redrawn from [78]).

mum loading in these methods is limited to the solubility of the precursor in the solution. Several active components could be introduced in successive impregnations, with drying and even calcinations between these impregnations.

Drying, thermal decomposition, calcinations and activation are the steps, followed by impregnation. In the drying step, temperature, heating rate, gas flow and pressure, as well as drying time could be also altered to influence the resulting supported catalyst.

Technical arrangements of drying, calcination and activation should be done in a way that all the catalyst particles are exposed to the same conditions, which is not possible to achieve in fixed beds. Sometimes, activation of catalysts by, e.g., reduction of metal-supported catalysts or sulfidation of hydrodesulphurization catalysts is done at a plant site, thereby not being a part of the preparation procedure. The degree of activation (e.g., degree of metal reduction) should be examined, which might not be easy when a catalyst is placed in a catalytic reactor. Sometimes pre-activated catalysts are delivered to the plant site in order to minimize the startup time.

3.7 Pesticides

3.7.1 Overview

Pesticides are substances that control pests and are mainly intended to protect crops/ plans. Among them, the most common (ca. 40–50%) are herbicides to control undesired plants (weeds). Other pesticides include, e.g., insecticides (17%) and nematicides to kill insects and plant-parasitic nematodes, fungicides (ca. 10%) to kill parasitic fungi, repellents to keep insects and animals from certain objects or people and zoocides, for controlling harmful animals, such as rodents. While the benefits of using pesticides are apparent as they, for example, kill mosquitos that transmit malaria, kill bees, wasps and ants that cause allergic reactions, clear roadside weeds, control growth of algae, remove termites that can damage houses or kill rodents, there is a clear risk associated with their toxicity.

Some examples of pesticides are presented in Table 3.20, featuring both systemic and nonsystemic pesticides. The latter pesticides control pests or diseases through direct contact, without adsorption or penetration through foliage or leaves. Subsequently, nonsystemic pesticides stay at the exterior of a plant, targeting, for example, flying and chewing insects. Systemic pesticides, on the other hand, can penetrate plant tissues and translocate through the skin of fruits and vegetables.

Table 3.20: Examples of pesticides.

Name	Chemical structure	Function/specifics
Acifluorfen	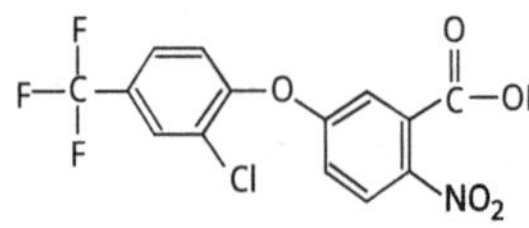	Contact herbicide; pre-emergence weed control; limited translocation; applied to foliage
Amicarbazone	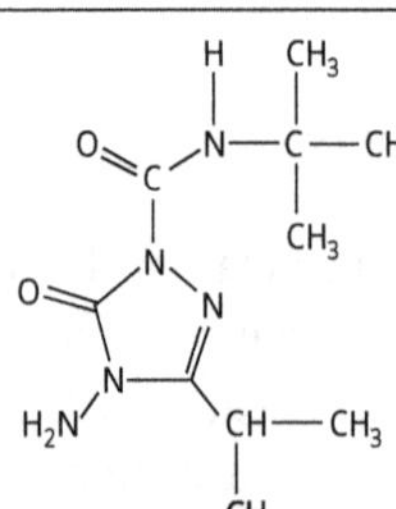	Herbicide to control broadleaf weeds in corn and soybeans

Table 3.20 (continued)

Name	Chemical structure	Function/specifics
Atrazine	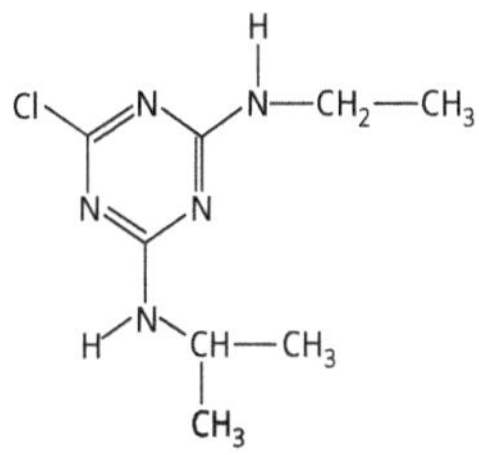	Herbicide, primarily on corn
Bentazone	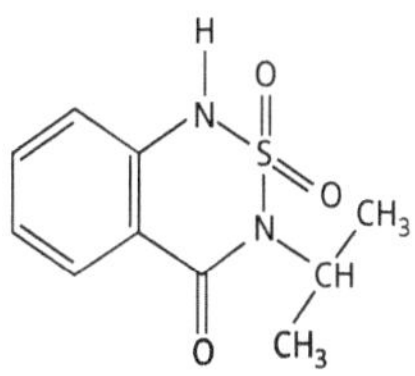	Kills broadleaf weeds in soybeans, rice, corn, peanuts, beans and peas by inhibiting photosynthesis; requires thorough plant coverage
Bromacil	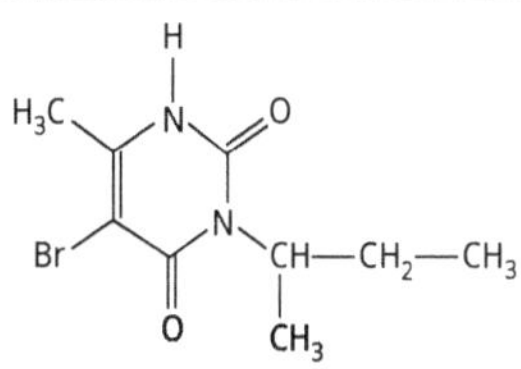	Pre- and postemergence herbicide for noncrop usage, pineapples and citrus varieties; carried to the root zone by moisture
Buprofezin	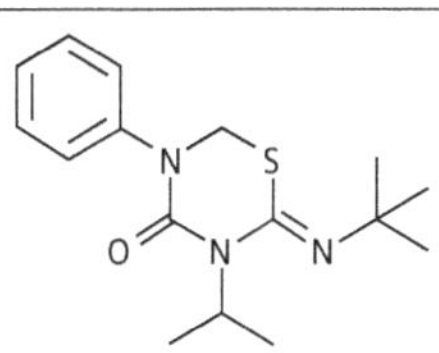	Insect growth regulator; inhibits chitin synthesis; Controls mealybugs, leafhoppers and whitefly on vegetable crops; nontoxic to humans, being toxic to aquatic organisms
Carbaryl	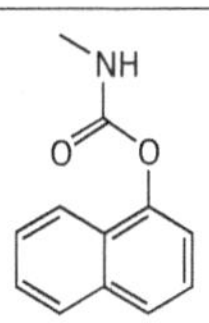	Insecticide, has very low mammalian oral and dermal toxicity; Very broad spectrum of insect control
Chlorfenapyr	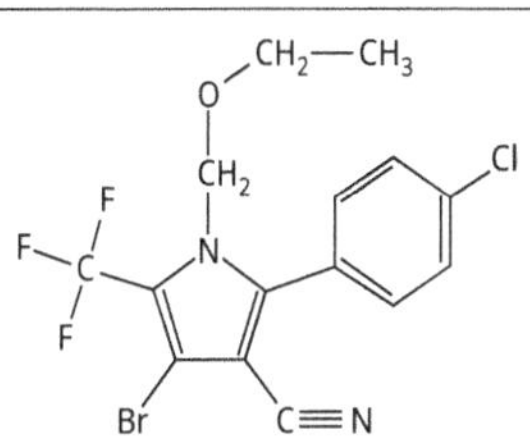	For cotton insect control; potentially, hazardous to birds

Table 3.20 (continued)

Name	Chemical structure	Function/specifics
Chlorothal	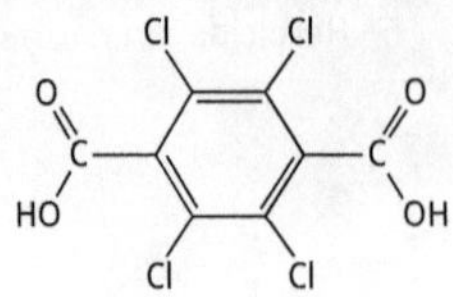	Crabgrass herbicide; effective against most grass and broadleaf weeds in vegetables
Chlorsulfuron	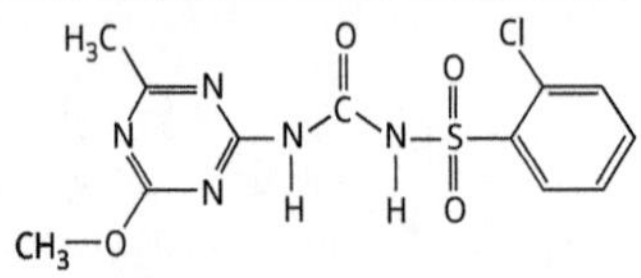	Herbicide for broadleaf weeds in wheat, barley, oats; very low mammalian toxicity
Cinnamaldehyde	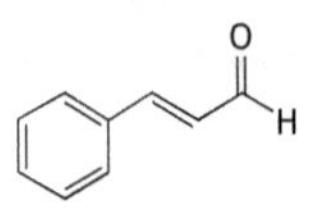	The dominant part of the essential oil of cinnamon bark; effective insecticide against mosquito larvae
Cyhexatin	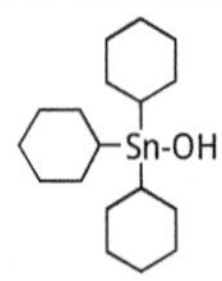	Acaricide, used against ticks and mites on, e.g., deciduous fruits (fall off the tree when ripe), citrus and greenhouse crops
DDT	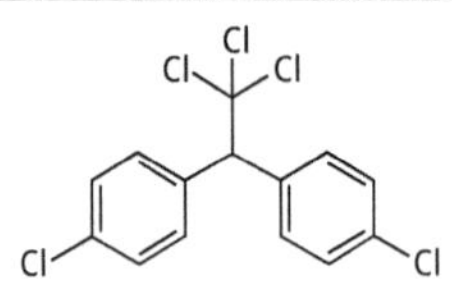	Toxic insecticide, still effectively used for malaria control in several Third World countries
Dicamba	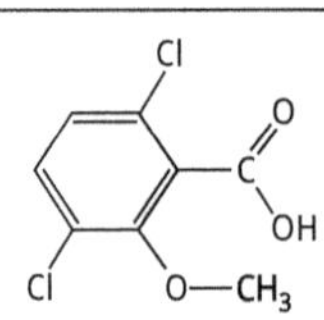	Growth hormone herbicide; moves from leaves to the terminal meristems of leaf, shoot and root; can be applied to soil, also moving in the transpiration stream.
Dichlobenil	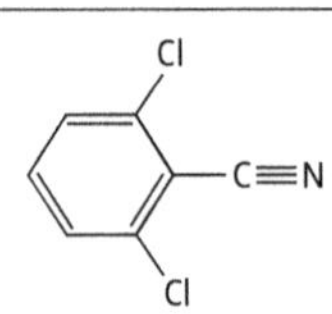	Herbicide
DEET, *N,N*-diethyl-*meta*-toluamide	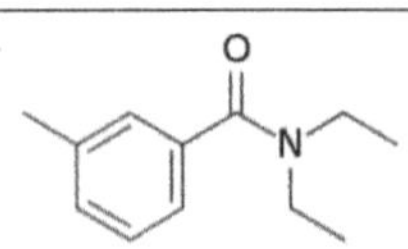	Insect repellent; provides protection against mosquitos, flies, ticks and many biting insects

Table 3.20 (continued)

Name	Chemical structure	Function/specifics
Diflubenzuron	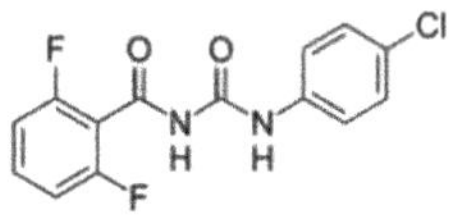	An insect growth regulator; used in forest management and on field crops; controls insect pests, particularly different types of moths
Dithiopyr	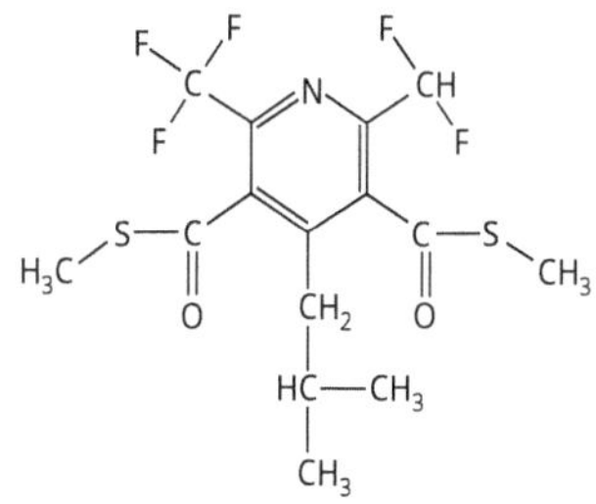	Used only on turf (sod) to control a wide variety of grass weeds
EPTC	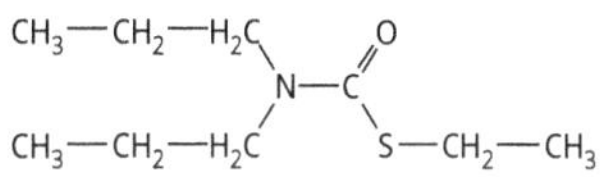	Herbicide, to control annual and perennial grassy weeds; volatile, must be incorporated in the soil at the time of application
Ethiprole	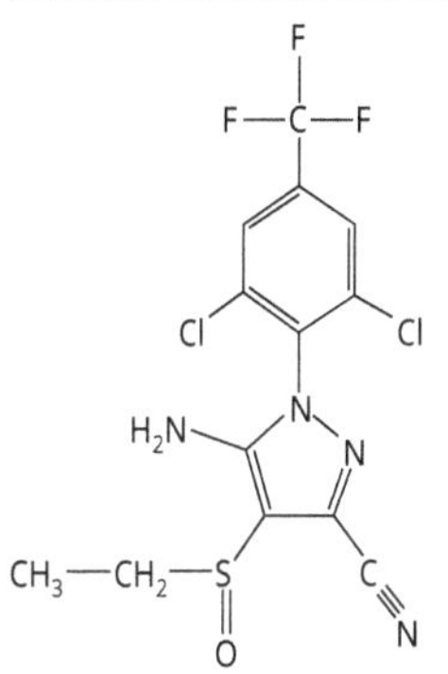	Active on chewing and sucking insects
Etoxazole	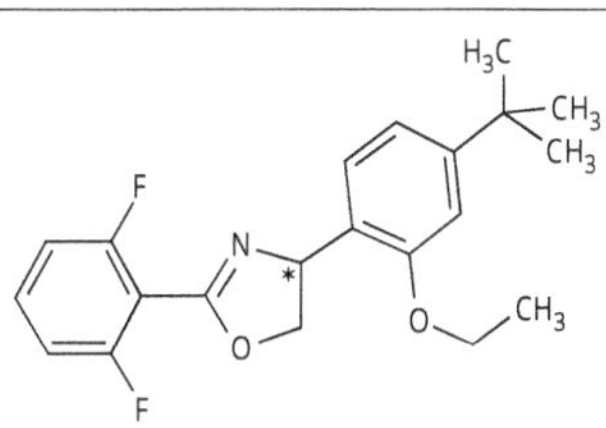	Systemic acaricide against spider mites; inhibits chitin synthesis; has some mammalian toxicity and toxicity toward fish
Eugenol	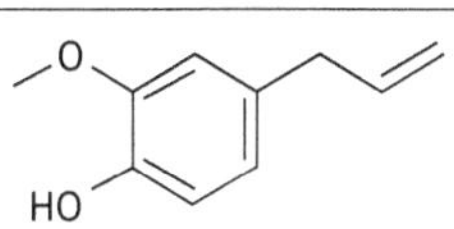	Bait to attract and collect orchid bees

Table 3.20 (continued)

Name	Chemical structure	Function/specifics
Fenvalerate	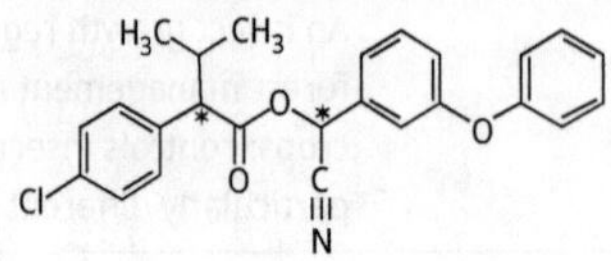	Synthetic pyrethroid (similar to natural compounds, produced by the flowers of pyrethrums) insecticide; photostable; moderate mammalian toxicity, while being toxic to bees and fish; can be applied in, e.g., emulsifiable concentrates, wettable powders, slow-release formulations and granules
Fipronil	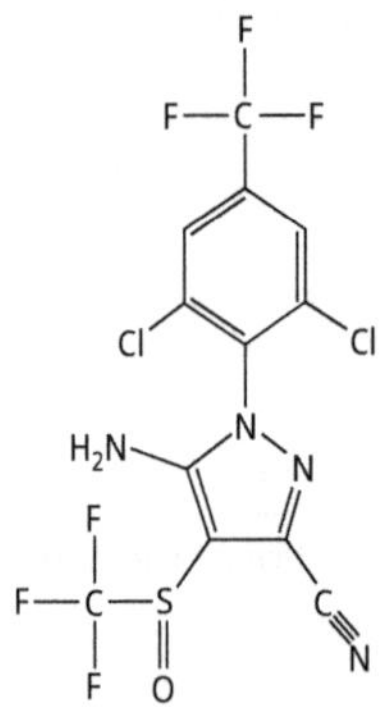	Insecticide to control many soil and foliar insects; is used on a variety of crops, including corn, on turf and for public health insect control
Flazasulfuron	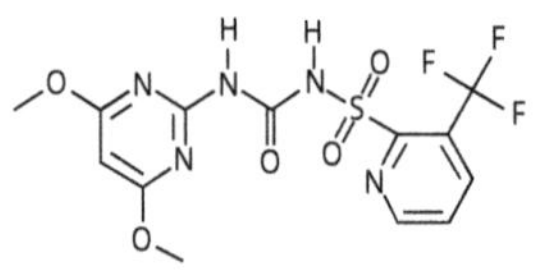	Herbicide; inhibition of the enzyme acetolactate synthase, cell division and weed growth
Fluazifop-butyl	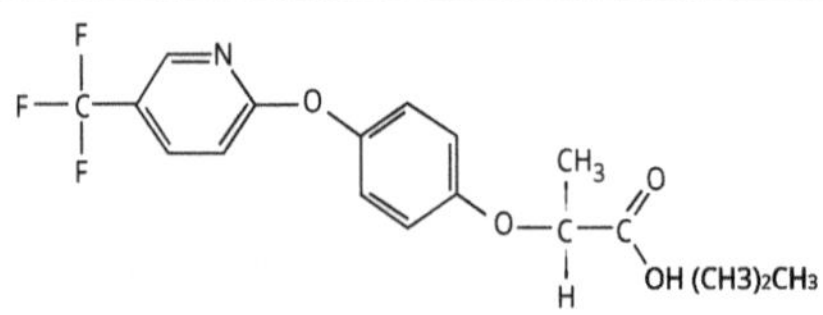	Postemergence herbicide; high activity against all grasses; good translocation; over-the-top application in broadleaved crops, especially cotton and soybeans
Glyphosate (Roundup®)	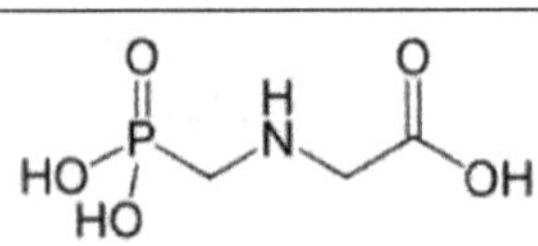	Nonselective, nonresidual, postemergence; Translocated, foliar-applied herbicide; several crops have been bioengineered to be tolerant to glyphosate

Table 3.20 (continued)

Name	Chemical structure	Function/specifics
Imidacloprid	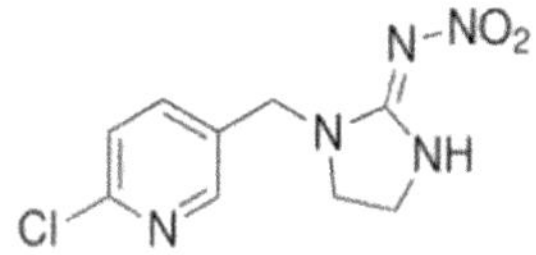	One of the most widely used insecticide; acts on the central nervous system of insects; can be applied by soil injection, tree injection, application to the skin of the plant, broadcast foliar, ground application as a granular or liquid formulation, or as a pesticide-coated seed treatment
Imiprothrin	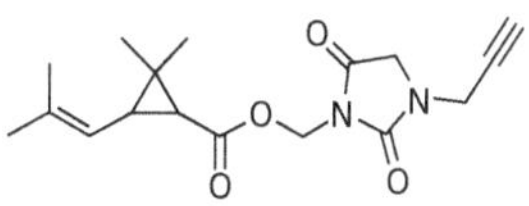	Synthetic pyrethroid insecticide, acting as neurotoxin; effective against, e.g., cockroaches, ants and spiders
Limonene	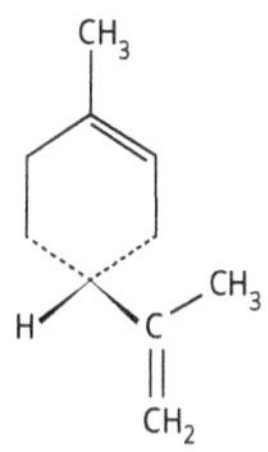	Extract from citrus peel; effective against pests of pets
Lindane	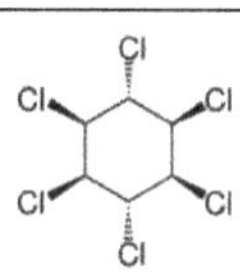	Only gamma out of five stereoisomers that has insecticidal properties; low-cost insecticide; has a negative temperature coefficient, i.e., becomes less toxic with increase in temperature
Methoprene	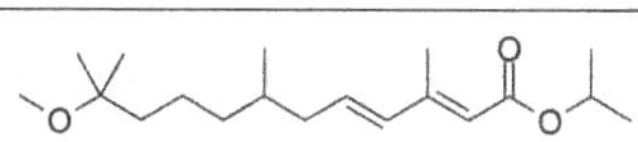	Mosquito growth regulator against second through fourth larval stage floodwater mosquitoes
Metolachlor	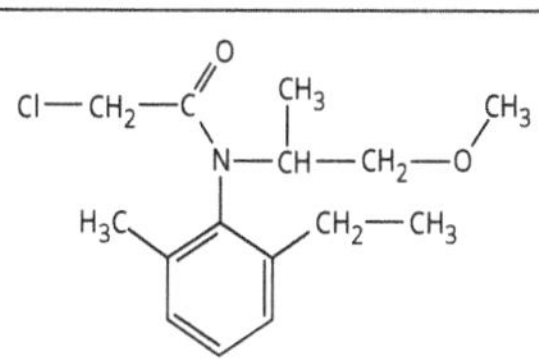	Herbicide for annual grass and some broadleaf weeds in corn, soybeans and peanuts.

Table 3.20 (continued)

Name	Chemical structure	Function/specifics
Metribuzin	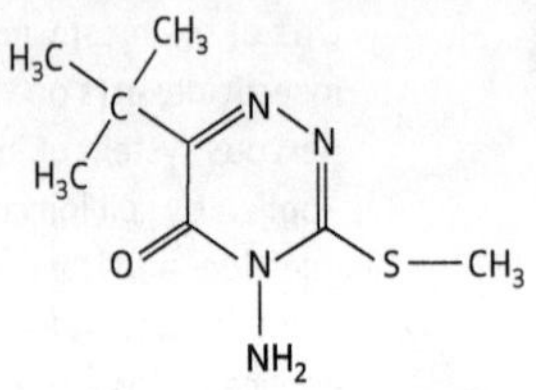	Herbicide that is used on soybeans, wheat, sugarcane and some vegetables
Metsulfuron-methyl	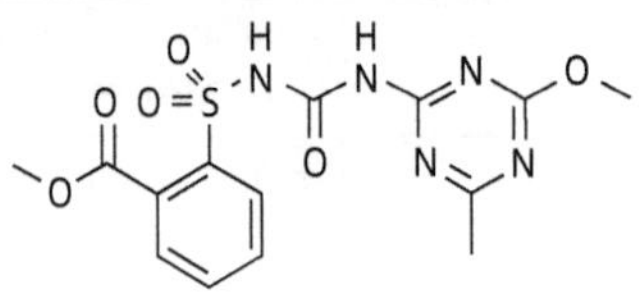	Systemic herbicide; kills broadleaf weeds and some annual grasses; low toxicity to mammals, birds, fish and insects
N-Methyl-neodecanamide	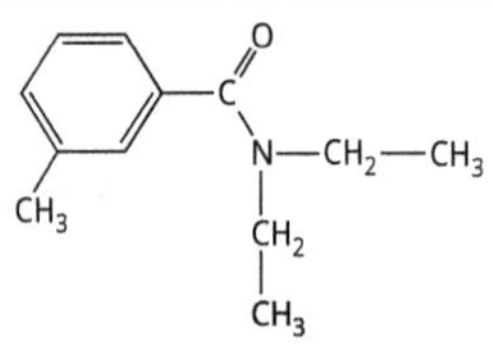	Repellent for cockroaches and ants
Norflurazon	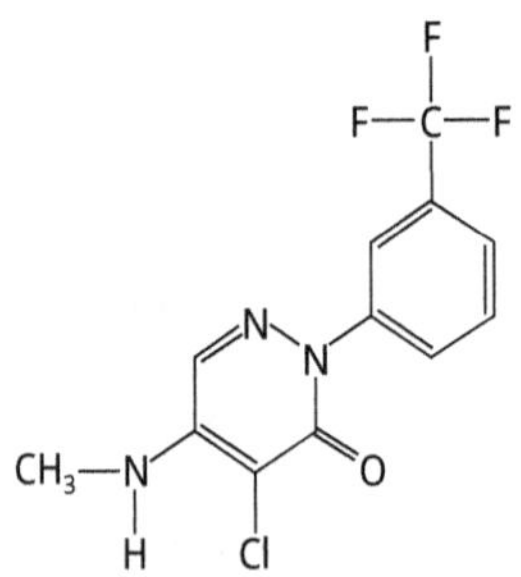	Pre-emergence herbicide for control of annual grasses and certain broadleaf weeds in cotton, soybeans, and vine and deciduous fruit crops; inhibits formation of chlorophyll in weeds
Paraquat	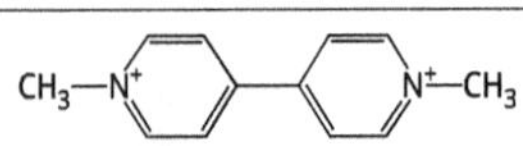	Contact herbicide; destructs cell membrane; Preharvest desiccants for seed crops, cotton, soybeans, sugarcane, and sunflowers
Piperonyl butoxide	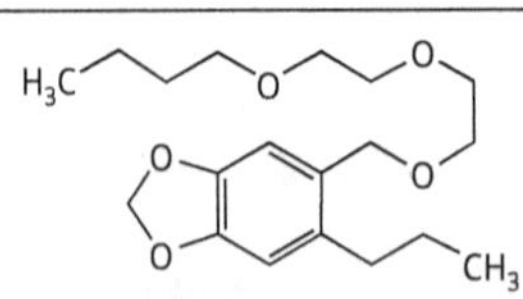	Synergist; does not have own pesticidal activity; enhances performance of certain pesticides such as carbamates, pyrethrins and pyrethroids

Table 3.20 (continued)

Name	Chemical structure	Function/specifics
Phenmedipham	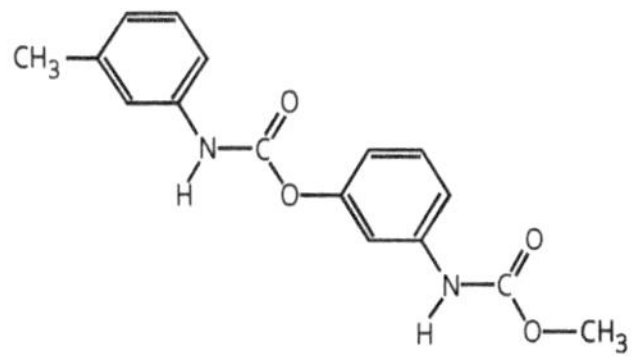	Postemergence herbicide in sugar beet and sunflower production
Propanil	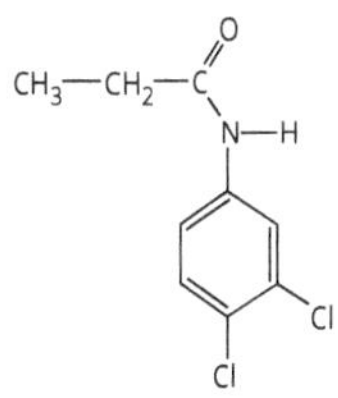	Herbicide used on rice fields
Propham	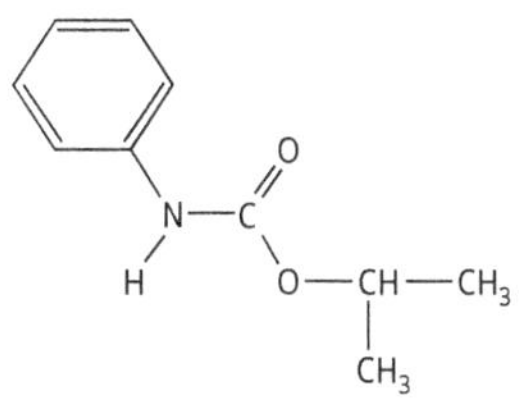	Herbicide, kills plants by stopping cell division and plant tissue growth
Pymetrozine	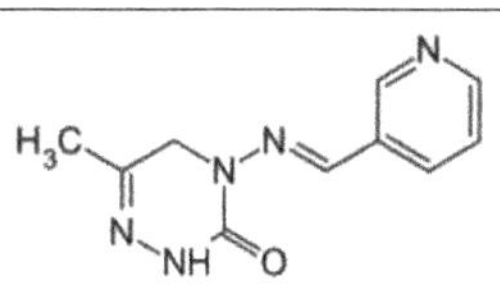	Controls aphids and whiteflies in vegetables, potatoes, tobacco, deciduous citrus fruit hops and ornamentals
Pyridaben	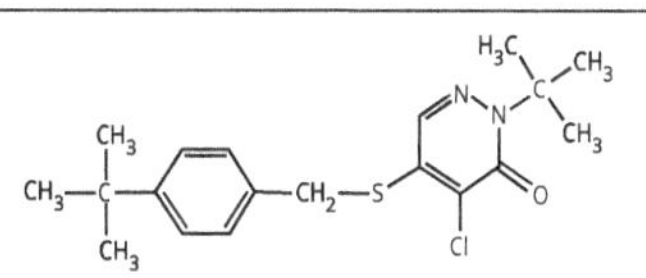	Contact acaricide; active against motile stages of mites; controls whiteflies.
Pyriproxyfen	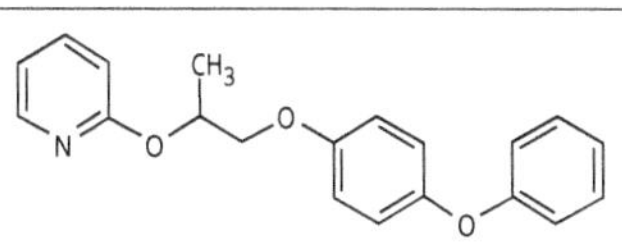	An insect growth regulator; protects cotton crops against whitefly; applied as aerosols, bait, carpet powder, foggers and pet collars
Sethoxydim	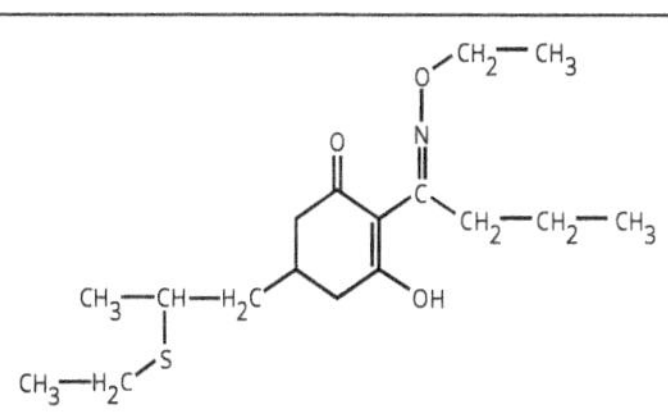	Postemergence herbicide; controls annual and perennial grass weeds in broadleaf crops; do not control broadleaf weeds; inhibit lipid biosynthesis

Table 3.20 (continued)

Name	Chemical structure	Function/specifics
Sulfometuron methyl	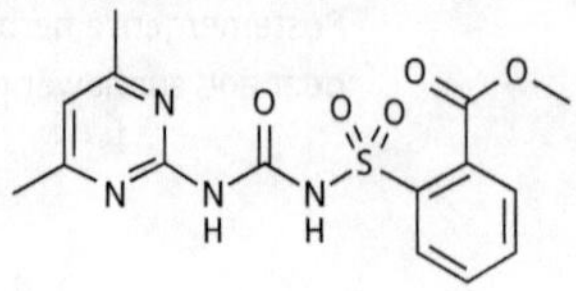	Herbicide, inhibits acetolactate synthase
Tetradifon	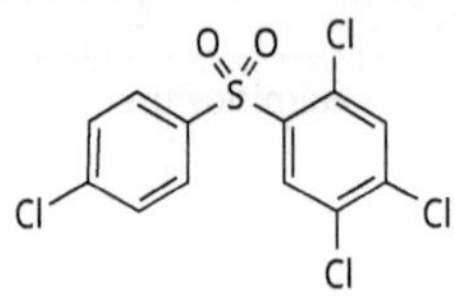	Acaricide, has low toxicity to insects
Trifluralin	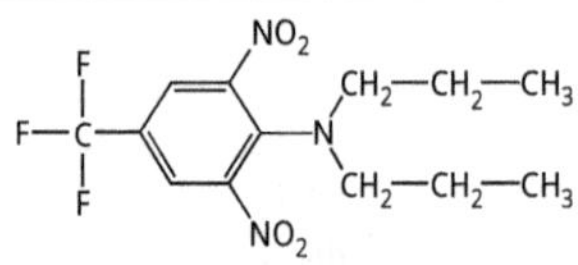	Pre-emergence selective herbicide; incorporated into soil; has low solubility in water
2,4-Dichloro-phenoxyacetic acid (2,4 D)*	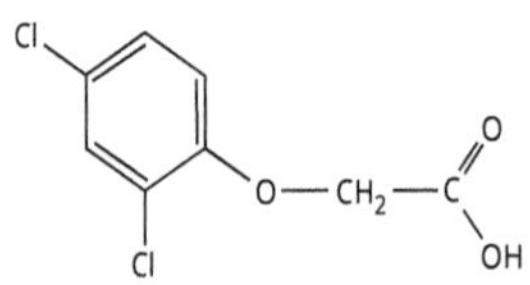	Highly selective for broadleaf weed; translocated; controls broadleaf weeds; applied on rights-of-way, turf and lawns, and in forest conservation programs; used in combination with a systemic herbicide picloram on pastures 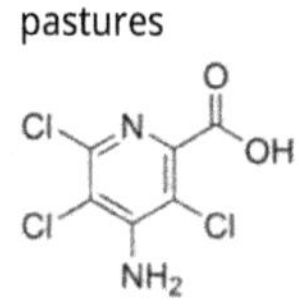

*The author spent 1-month internship at a factory making 2,4-D in 1985 as part of chemical engineering studies.
Note: Compiled from [79].

The main function of herbicides is to kill weeds (10% of all plant species), as the latter diminish the quality and quantity of agricultural production. This chemical for weed control has largely replaced mechanical methods of weed control. Herbicides are applied not only in farming areas, but also in roadsides, railroad embankments, fence lines, industrial sites, etc. to remove undesirable plants. Apparently, selective herbicides are required in farming areas, as the intention is not to harm crops, while nonselective herbicides remove all vegetation; they can be, for example, used in railroad embankments. In fact, selectivity in using a herbicide depends on the dosage and timing, as well as the placement, allowing otherwise a nonselective herbicide to reach the weeds but not the crop.

Herbicides can kill the plant parts because of a direct contact, thus requiring complete coverage of surfaces, or alternatively, by adsorption by roots or above-ground parts, followed by circulation of a herbicide within the plant system.

The timing of application obviously depends on the chemical nature of the herbicide, the crop and the weed, as well as, for example, the soil type and climate. Herbicides can be applied by spraying them to the soil or to crop stems without contact with the crop foliage or, alternatively, covering the whole area with the crop, per se.

Besides herbicides per se, in contemporary formulations, more and more so-called herbicide safeners are applied, which apparently enhance the metabolism of the herbicide by the crops.

Another development worth mentioning is the introduction of genetically engineered herbicide-tolerant crops (corn, cotton and soybeans) to withstand the action of nonselective herbicides like glyphosate (Table 3.20), allowing more efficient weed control. Application of genetically engineered crops and glyphosate, the most widely used herbicide in current agriculture, has also led to scientific controversies. Nevertheless, this chemical is marketed by many agrochemical companies in different solution strengths and with various adjuvants, which will be discussed below.

Besides control of weeds, an important issue is also the control of insects, including crop-eating ones, which is done by synthetic or natural compounds – insecticides, administrated by spraying, slow-release diffusion or using baits, etc.

Historically, not insecticides, but rather repellents have been applied to mitigate insect annoyance by, for example, spreading mud over the skin to repel biting and tickling by insects.

Notoriously known is the example of dichlorodiphenyltrichloroethane (DDT), originally developed as an insecticide to efficiently limit the spread of malaria, which even led to the Nobel Prize in Physiology and Medicine awarded in 1948 to P. H. Mueller. Lately, it is banned in most countries because of its persistence in the environment and due to human toxicity.

Some of the most popular insecticides, with their mode of action are given in Table 3.20.

In addition to synthetic insecticides extractives are also illustrated in this table and can be used for the same purpose. One example of such extractives is pyrethrum extracted from the flowers of a chrysanthemum. Application of pyrethrum results in immediate paralysis of insects, which could be recovered, unless pyrethrum is formulated with some synergists such as piperonyl butoxide. The role of synergists is to bind the oxidative enzymes, preventing them from degrading the toxicant. Similar to DDT, pyrethrum and some pyrethroids have a negative temperature coefficient.

Extracted from tobacco, an alkaloid nicotine, mimicking acetylcholine, is especially effective against aphids and caterpillars, while limonene, extracted from citrus peel, is effective against fleas, lice, mites and ticks, being nontoxic to warm-blooded animals. Other floral products applied to control various insects are eugenol and cinnamaldehyde (Table 3.20).

Besides insecticides mentioned in Table 3.20 also insect repellents are widely used. They are less efficient from the viewpoint of killing insects mainly discouraging insects rather than leading to their death. While DEET and some other insect repellents are used on humans, N-methylneodecanamide is applied to household floors and other surfaces to repel cockroaches and ants.

Finally, biorational pesticides or biopesticides should be mentioned. They are different from conventional pesticides, having fundamentally different modes of action, based on specific biological effects, rather than chemical poisoning. The active ingredient in a biopesticide, according to the European Environmental Agency definition, is a virus, fungus, bacteria or a natural product, derived from a plant source. Subsequently, lower risks regarding adverse effects on wildlife and the environment are expected. Such biopesticides that have a detrimental effect on specific targets, such as insects, weeds or plant diseases, are nontoxic to humans, domestic plants and animals.

3.7.2 Applications and formulations

Due to a number of reasons, including solubility or stability in water, challenges in handling and transportation, pesticides or active ingredients are not used in their neat form, but are applied together with a range of inert chemicals, in terms of biological activity of the latter. These chemicals include the carriers (organic solvents or mineral clays), adjuvants (stickers and spreaders) as well as other ingredients (stabilizers, safeners, dyes, performance enhances). Formulations for pesticides include, e.g., baits, granules, gels, dusts and emulsifiable concentrates. The latter might contain 30–50% active substance, 40–60% solvents (e.g., high-boiling mineral oils or water) and 5–10% emulsifiers. These special tensides (silicone-based ones, pyrrolidone derivatives, polymers and modified vegetable oils) influence the interactions of the formulations with plants. More specifically, as nonionic surfactants, polyethoxylated derivatives of C_8–C_{18} fatty alcohols, alkyl phenols, and dialkyl phenols, castor oil, alkyl amines, tristyrylphenol, sorbitan esters, EO/PO block copolymers and their alkyl phenyl ethers, silicone comb surfactants, alkylated polyvinylpyrrolidones are applied. Alkyl sulfates (C_8–C_{18}, preferably C_{12}), alkyl benzenesulfonates, e.g., calcium dodecylbenzenesulfonate, naphthalenesulfonate, dioctylsulfosuccinate are examples of anionic surfactants.

In most cases, an aqueous formulation is used for the spraying, prepared from emulsifiable or suspension concentrate, wettable powder, water-dispersible granules, etc. Spraying of insecticides and herbicides onto plants can be done either at the ground level (Figure 3.94) or from aircraft (Figure 3.95)

(a)

(b)

Figure 3.94: Spraying of pesticides at ground level (a) using a tractor [80] and (b) manual [81].

Figure 3.95: Spraying of pesticides from aircraft [82].

Surfactants mentioned above do not possess pesticidal activity per se; they, however, reduce or even eliminate spray application problems. Phytotoxicity of cationic surfactants prevents their utilization as stand-alone tensides; subsequently, anionic and nonionic surfactants are applied. These adjuvants are either added to the spray, tank prior the application, or are included in the pesticide formulation.

The role of surfactants is to wet the foliage and spread the pesticide evenly over a leaf, increasing the contact area with a pest (Figure 3.96)

For contact pesticides, anionic surfactants are the most effective. Nonionic surfactants, compatible with most pesticides, are often used with systemic pesticides, aiding in the penetration of pesticide sprays into plant cuticles. It should be noted that the critical micelle concentration of nonionic surfactants is usually two orders of magnitude lower than the corresponding anionic of the same alkyl chain length, which is beneficial from an economic viewpoint.

Other adjuvants used, in particular, with herbicides, include thickeners, evaporation and leaching inhibitors, buffers, antifoaming, wetting and spreading agents, ad-

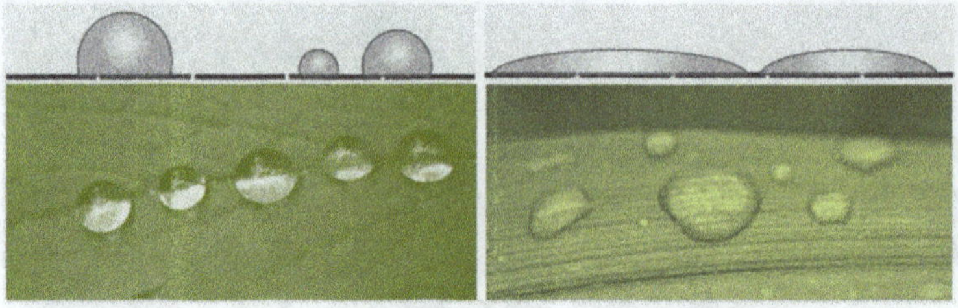

Figure 3.96: Surfactants increase the ability of the pesticide to spread evenly over the surface of a leaf [83].

hesion improvers, dispersants or drift controllers. The formulations can also include compounds, enhancing penetration, inhibiting phytotoxicity, improving activity or substances acting as markers.

Stickers or adhesion improvers diminish the amount of pesticide that can be washed away during irrigation or rain. Adjuvants of this type also decrease the evaporation of the pesticide, along with its degradation by sunlight.

Buffers and acidifiers are added to control pH and thus prevent degradation of pesticides that happen in alkaline conditions.

The role of drift control additives is to increase the average droplet size, as otherwise fine drops (<100 microns) drift away from targeted areas. Thickeners, increasing the viscosity of spray mixtures, not only control the drift but also slow down the evaporation. To prevent shear degradation of the antidrift agents, polymers that are stable at the high shear rates during spraying should be selected. Besides the droplet size, which with spraying using hydraulic nozzles is typically in the range 100–400 μm, impaction, adhesion, sliding, retention, wetting and spreading are the key parameters influencing the coverage of the target surface.

When a liquid drop is in contact with the leaf surface, it can adhere to the surface after flattening, spreading and finally forming a cap (Figure 3.96). This can happen either directly or after bouncing and fragmentation into smaller droplets, reaching the surface with a lower kinetic energy. The fate of the droplet depends on its mass, velocity and the distance between the leaf surface and the nozzle, which then defines the critical droplet diameter, below which adhesion is high. High surface tension decreases the critical droplet diameter. Increase in viscosity of the spray solution is beneficial for adhesion of small droplets; it has a minor effect on large drops. The surfactant concentration also has a minor effect on droplet adhesion per se, influencing, in any case, the size distribution of the droplets.

Application of large volumes of pesticides may result in coalescence of neighboring drops on the surface with the droplets, eventually reaching a critical value, beyond which, sliding and subsequent "drop off" can prevail. Retention of the droplets, being positively influenced by low droplet sizes and velocities, is significantly diminished at high wind speeds and smooth surfaces, more typical for plants grown under standard conditions (e.g., green/glass houses). On the contrary, retention on rough leaf surfaces of plants grown under the natural environment is better; thus, development of agrochemical formulations like selection of a surfactants and, in particular, understanding of spray retention should be done either on field-grown plants or at least considering the difference between plants grown under different conditions.

As mentioned above, agrochemicals can be systemic or nonsystemic. For the latter type, direct contact between the pest and the agrochemical is required as also mobility of the pest is important. Obviously, more dynamic pests do not need complete coverage of those areas of a plant that can be attacked by a pest. For more static pests, the liquid spray should be efficiently spread, preferentially achieving the maximum coverage.

One of the reasons that surfactants are added to agrochemical formulations in rather large concentrations for increasing wetting and spreading is the need to lower the contact angle of the droplets on, in general, unwettable hydrophobic leaf surfaces that are covered with crystalline wax of straight chain paraffinic (C_{24}–C_{35}) alcohols. An important issue with surfactants that should be carefully considered is their potential phytotoxicity. The local concentration of surfactants can be higher than the mean values, as evaporation from the edges is more rapid through the heat supplied by the dry air surrounding the leaf, subsequently leading to a higher concentration of the surfactant at the edge.

Nonionic surfactants, with HLB numbers between 7 and 9, are preferred as the wetting agents having two orders of magnitude lower is critical for micelle concentration than ionic surfactants with the same alkyl chain length. Considering that the agrochemical formulations are diluted before spraying, the impact of dilution will be less severe for nonionic surfactants because, even after dilution, the concentration of such surfactants can be beyond the one corresponding to *cmc*. A potential drawback with nonionic surfactants is the phase separation of surfactants with HLB numbers between 7 and 9 at high concentrations or temperatures. To mitigate the phase separation, small quantities of ionic surfactants may be added to the formulation. Overall, formation of the liquid crystalline phases above a certain concentration depends on the surfactant nature, the chain length of both hydrophobic and hydrophilic parts and the presence of other additives in the formulation (e.g. antifreeze agents and electrolytes).

As mentioned above, pesticides are categorized according to the target organism they control (Figure 3.97). Formulations as well as the application equipment and methods can significantly influence environmental and human exposure to a pesticide.

Application of emulsifiable concentrates (Figure 3.98a) as agrochemical formulations has the advantages of easy handling, transportation and storage. On the other hand, a high initial concentration of the active ingredient requires high precision in dilution, especially considering that the pesticides can be easily absorbed through the skin of humans or animals. Subsequently, should the animals' feed and water be contaminated by spray drift, the dilutions cannot be applied to premise treatment, di-

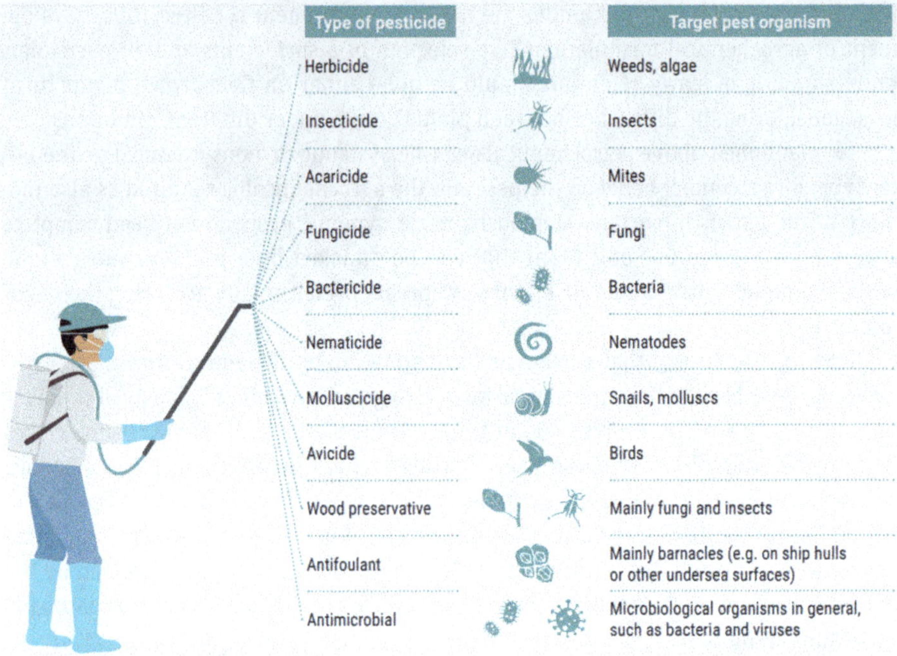

Figure 3.97: Categorization of pesticides [84].

(a) (b) (c) (d)

Figure 3.98: Formulations of agrochemicals: (a) emulsifiable concentrates [85], (b) aerosol container [86], (c) insecticide aerosol generator [87] and (d) granular [88].

rectly on livestock or poultry. A special case should be also taken when the concentrate is flammable.

Aerosols are commonly available in pressurized, sealed containers (Figure 3.98b) or as electric/gasoline-driven aerosol generators, releasing "smoke" or "fog" (Figure 3.98c). Utilization of such formulations, available in convenient to use and store quantities, can be still challenging because of the difficulty in applying them to the target site or pest, a potential risk of inhalation injury. Moreover, they are hazardous if punctured, overheated or used near an open flame.

Dry formulations (dusts, baits, granules, including water dispersible granule, wettable or soluble powders, dry flowables) are either ready to use and or concentrates that should be mixed with water, followed by spraying. Most dust formulations, containing low amounts of the active ingredients (<10 wt%) and a very fine, dry inert carrier, are ready to use. In some cases, dust concentrates with much higher loadings of active ingredients can be used, requiring mixing with dry inert carriers, prior to application.

Granular formulations, most often applied to the soil, are similar to dust formulations with larger and heavier granular particles (Figure 3.98d), with the active ingredient (1–15 wt%) either coated on the outer surface or distributed in them. The amount of active ingredient is relatively low, usually ranging from less than 1% to 15% by weight. Granular formulations, which apparently are slow releasing active ingredients, can also be applied from airplanes, with minimal risk of drifting or minor hazard to the applicator. The same is valid for other ways of application (e.g., simple seeders or fertilizer spreaders) as there is no spraying involved and, thus, little, if any, dust formation. The disadvantages are related to difficulties in uniform application, no sticking to foliage, insufficient release rates, in the case of poor raining and, especially, under drought conditions and, finally, potentially hazardous to nontarget species, like birds. A granular formulation, beside the active ingredients, includes dispersing, suspending and wetting agents, as well as a binder (e.g., lignosulfonate) and a filler. Granulation is done by, e.g., pan or fluid-bed granulation, extrusion, in the case of dry granulation, or by spray drying or granulation for the wet route.

Similar to dusts, wettable powders are commonly used with typically more than 50% of the active ingredient in the composition. They are mixed with water, rather than dissolved, and are applied as a spray. In this instance, only water penetrates the material (e.g., concrete or wood), while the active ingredient remains on the surface. Mixing, prior to application, should be rather efficient or otherwise, the powder will settle out. Other potential disadvantages include difficulties in mixing in very hard and alkaline waters, inhalation hazards, and wearing out of pumps and nozzles because of the abrasive properties of wettable granules. Clogging of nozzles can also happen.

In contrast to wettable powders, soluble powder formulations readily dissolve, avoiding all the disadvantages above, apart from inhalation hazards during mixing. A limitation of using such pesticide formulations is poor solubility in water for the dominant majority of pesticides.

Other agrochemical formulations include microencapsulated pesticides, when the active ingredient in the liquid or solid form is coated with a plastic. The coating, after spraying, breaks down slowly, releasing the active ingredient. While such release might depend on the weather conditions, being too slow has its apparent advantages related to inherent safety for the operators as well as longer effectiveness of the active ingredient. A potential danger is a special hazard to bees that may carry microencapsulated formulations to their hives.

In general, release of the active ingredient from a capsule depends on the diffusion, either through the microcapsule wall or a release behavior when the wall/coating is degraded, either by physical or chemical means (hydrolysis, thermal or biodegradation). The particle size, coating thickness and structure as well the type of the core and the concentration of the active ingredient control the release behavior. If the coating is thicker, it obviously leads to a slower release. The same is valid for larger capsules.

References

[1] http://www.meadex.co.uk/thermoplastic, accessed on 23.12.2017.
[2] http://www.stephensinjectionmoulding.co.uk/thermoplastics/, accessed on 23.12.2017.
[3] http://www.pvc.org/en/p/how-is-pvc-used/#PVCuse, accessed on 23.12.2017.
[4] http://www.sudarshan.com/perch/resources/plastics.jpg, accessed on 23.12.2017.
[5] https://media.licdn.com/mpr/mpr/shrinknp_800_800/AAEAAQAAAAAAAALSAAAAJDlhZDI5ODMwLTI2ZTktNDU1Yy1iMGJkLTc2Y2MyOGI1OGQxMg.jpg, accessed on 23.12.2017.
[6] http://www.gcsescience.com/plasticizer.gif, accessed on 23.12.2017.
[7] https://3.imimg.com/data3/PG/TB/MY-9020268/flexible-polyurethane-foam-250x250.jpg, accessed on 23.12.2017.
[8] http://polystyrenesheets.org/wp-content/uploads/2012/05/polystyrene-foam.jpg, accessed on 23.12.2017.
[9] http://www.novachem.com/expandablestyrenics/Documents/PROCESSING-EPS.pdf, accessed on 23.12.2017.
[10] https://nzic.org.nz/ChemProcesses/polymers/10G.pdf, accessed on 23.12.2017.
[11] https://www.slideshare.net/DesignWorldOnline/designing-with-adhesives, accessed on 23.12.2017.
[12] https://www.slideshare.net/ajaykottakota/adhesives-industry-analysis, accessed on 23.12.2017.
[13] https://1.bp.blogspot.com/-0aLRp_rTQLY/Ud6sElI6Y0I/AAAAAAAAAJg/YUM5NSAJpNM/s1600/Epoxy-Glue.jpg, accessed on 12. 03.2023.
[14] https://media.licdn.com/mpr/mpr/AAEAAQAAAAAAAAQZAAAAJDNlZDMzYTJlLTA5MDgtNDRhYy05YzYwLWIwYWE0NmQzMWM3OQ.jpg, accessed on 23.12.2017.
[15] http://www.technologystudent.com/images5/htglue1.gif, accessed on 23.12.2017.
[16] http://slideplayer.com/slide/10219257/, accessed on 23.12.2017.
[17] http://processsystems.sandvik.com/wp-content/uploads/2017/01/Dry-pastillation-of-hot-melt-adhesives.pdf, accessed on 23.12.2017.
[18] http://www.crystalandglassbeads.com/img/cali/glue/swarovski-crystal-adhesives.gif, accessed on 23.12.2017.
[19] https://commons.wikimedia.org/w/index.php?curid=91499337, accessed on 26.03.2023.
[20] http://www.ppgecoat.com/getattachment/bfb5044e-7da4-438f-9ae0-3532fb895936/anodic_cathodic_electro.gif.aspx, accessed 26.03.2023.
[21] http://www.herrindustrial.com/images/gallery/liquid_paint_spray_systems_full.jpg, accessed on 23.12.2017.
[22] https://unist.com/images/pfd/atomizationViaCoax.jpg
[23] http://www.freilacke.de/fileadmin/doc/index/Broschueren/Produkte/imc_EN.pdf, accessed on 23.12.2017.
[24] http://slideplayer.com/slide/3806689/, accessed on 23.12.2017.

[25] C. P. Izzo, Coatings and Finishes, Chapter 9, in Handbook of Materials for Product Design, Ed. C. A. Harper. McGraw-Hill, New York, 2001.

[26] https://nzic.org.nz/ChemProcesses/polymers/10D.pdf, accessed on 23.12.2017.

[27] http://www.kikalab.com/living-room-layout-ideas-with-flat-screen/attractive-living-room-layout-ideas-with-flat-screen-part-11-flat-vs-eggshell-paint-sheen/, accessed on 23.12.2017.

[28] https://knowledge.ulprospector.com/3230/pc-coatings-additives-defoamers-adhesion-promoters-uv-protection/, accessed on 23.12.2017.

[29] https://www.iot-gmbh.de/en/uv-systems/excimer/applications/matting, accessed on 23.12.2017.

[30] https://www.safinah-group.com/wp-content/uploads/2017/11/WROCCA-Rheology-Forum.pdf

[31] http://www.adhesionbonding.com/2012/03/18/high-performance-waterborne-coatings-2/, accessed on 23.12.2017.

[32] https://www.researchgate.net/publication/322666154_Review_on_Coating_defect_and_how_to_prevention_furthermore_include_ASTM_code_of_each_defect

[33] https://www.boatsnews.com/story/30666/choose-your-antifouling-paint-which-matrix, accessed on 2.04.2023.

[34] http://www.iloencyclopaedia.org/images/stories/enlarged/Part12/CMP_imgs/CMP040F3.jpg, accessed on 23.12.2017.

[35] N. Sekar, Direct dyes, in *Handbook of Textile and Industrial Dyeing, Principles, Processes and types of Dyes*, Woodhead Publishing Series in textiles, 2011, Vol. 1, 425–445. https://www.sciencedirect.com/science/article/pii/B978184569695550012X.

[36] https://i0.wp.com/textilelearner.net/wp-content/uploads/2012/01/chemical-structure-of-acid-dyes.gif?resize=400%2C205&ssl=1

[37] https://bdtextilemates.blogspot.com/2017/03/disperse-dyes.html

[38] M. S. Hossan, B. Ochiai Detailed study on rapid removal of cationic dyes using TiO2-poly(3-chloro-2-hydroxypropyl methacrylate) nanocomposite, J. Electrochem. Soc.2019, 166, B3240 DOI 10.1149/2.0381909jes

[39] https://i0.wp.com/textilelearner.net/wp-content/uploads/2012/01/Basic-steps-in-the-application-of-vat-dyes.png?resize=655,364&ssl=1

[40] D. M. Lewis, The Chemistry of Reactive Dyes and their Application Processes, in *Handbook of Textile and Industrial Dyeing. Principles, Processes and Types of Dyes*, 2011, 303–364. https://www.sciencedirect.com/science/article/pii/B978184569695550009X.

[41] L. Lin, W. Zhu, C. Zhang, Md. Y. Hossain, Z. B. S. Oli, Md. N. Pervez, S. Sarker, Md. I. Ul Hoque, Y. Cai, V. Naddeo, Combination of wet fixation and drying treatments to improve dye fixation onto spray-dyed cotton fabric, Sci.Rep., 2021, 11, 15403. https://doi.org/10.1038/s41598-021-94885-z.

[42] M. Peng, S. Wu, J. Du, C. Sun, C. Zhou, C. Xu, X. Hu, Establishing a rapid pad-steam process for bleaching of cotton fabric with an activated peroxide system, ACS Sust. Chem. Eng., 2018, 6, 8599. https://pubs.acs.org/doi/10.1021/acssuschemeng.8b00912.

[43] https://www.textileflowchart.com/2015/09/flow-chart-for-roller-printing-process.html

[44] http://www.washwise.org.au/_documents/Laundry%20detergent%20ingredients%20info%20sheet.pdf, accessed on 23.12.2017.

[45] Y. Yuan, T. R. Lee, Springer Series in Surface Sciences, in *Surface Science Techniques*, Ed. G. Bracco, B. Holst. Springer, Berlin, Heidelberg, 2013, Vol. 51, 3.

[46] http://surfactantschemistry.blogspot.fi/2013/07/laundry-detergents-formulation.html, accessed on 23.12.2017.

[47] E. Smulders, W. Rähse, W. V. Rybinski, J. Steber, E. Sung, F. Wiebel, *Laundry Detergents*, Wiley-VCH, Weinhem, 2002.

[48] M. McCoy, P&G and Henkel go head to head in the laundry aisle, Chem. Eng. News., 2017, 95(4), 15.

[49] https://stppgroup.com/products/detergent-plant-equipment/spray-drying-tower/

[50] https://cen.acs.org/business/consumer-products/Almost-extinct-US-powdered-laundry/97/i4

[51] http://www.stlehouston.com/2HoustonSTLE/2011-2012/Program/Lubricant%20Additives-Heverly-2012_2_8.pdf, accessed on 23.12.2017.
[52] http://www.edl.poerner.de/fileadmin/_migrated/pics/EDL_Blending_Plant_Overview_en.png, accessed on 23.12.2017.
[53] https://www.artsheaven.com/types-of-art-ceramics-and-pottery.html, accessed on 23.12.2017.
[54] https://diytips.co.technology/red-clay-bricks-better-cement-bricks/, accessed on 23.12.2017.
[55] http://global.kyocera.com/fcworld/first/images/process15_01.jpg, accessed on 23.12.2017.
[56] http://defense-update.com/wp-content/uploads/2012/01/armor-concept.jpg, accessed on 23.12.2017.
[57] http://www.iloencyclopaedia.org/images/stories/Part13/POT_imgs/POT10F15.jpg, accessed on 23.12.2017.
[58] https://www.thespruce.com/what-kind-of-kiln-is-it-2746127, accessed on 23.12.2017.
[59] http://www.swindelldressler.com/images/elevator-kiln1b.jpg, accessed on 23.12.2017.
[60] http://baileypottery.com/Portals/0/Images/Gas%20PRO/sh40whood.jpg?ver=2015-07-22-130900-600, accessed on 23.12.2017.
[61] https://www.keithcompany.com/includes/work/image_cache/c334c156e1fde269d18b553c077097ad.thumb.png, accessed on 23.12.2017.
[62] www.d.umn.edu/~rlindek1/ME2105/Ceramics_%20Properties%20_F2010.ppt, accessed on 23.12.2017.
[63] https://upload.wikimedia.org/wikipedia/en/1/16/Tgdscenglish.svg, accessed on 23.12.2017.
[64] http://de.verallia.com/sites/verallia.germany/files/images/graph_melting-and-forming-en.jpg, accessed on 23.12.2017.
[65] www.engr.mun.ca/~asharan/courses/5911_LECTURES/ch13.ppt, accessed on 23.12.2017.
[66] http://cdn.24.co.za/files/Cms/General/d/3688/85ca3d200327417a9a09f029cf247caa.jpg, accessed on 23.12.2017.
[67] https://essayforum.com/shared_files/uploaded/71868/285083_1.jpg, accessed on 23.12.2017.
[68] http://st2.depositphotos.com/2724605/5363/v/450/depositphotos_53633035-stock-video-labor-mixing-concrete-for-construction.jpg, accessed on 23.12.2017.
[69] https://www.duraamen.com/site_media/media/servee_images/Stamped-Concrete-Driveway.jpg, accessed on 23.12.2017.
[70] https://www.pobox355.com/wp-content/uploads/2011/08/mixer.jpg.jpg, accessed on 23.12.2017.
[71] http://constructionmanuals.tpub.com/14045/css/Mixing-Concrete-129.htm, accessed on 23.12.2017.
[72] http://kak-eto-sdelano.ru/kak-stroili-metromost-na-vorobevyh-gorah/, accessed on 23.12.2017.
[73] D. Murzin, *Engineering Catalysis*, 2nd edition. DeGruyter, Berlin, 2020.
[74] S. Mitchell, N.-L. Michels, J. Pérez-Ramírez, From powder to technical body: The undervalued science of catalyst scale up, Chem. Soc. Rev., 2013, 42, 6094.
[75] I. M. Kolesnikov, *Catalysis and Production of Catalysts*, Tekhnika, Moscow, 2004.
[76] http://www.tnw.tudelft.nl/fileadmin/Faculteit/TNW/Over_de_faculteit/Afdelingen/Chemical_Engineering/Research/Catalysis_Engineering/Education/Miscellaneous/doc/catcat03.ppt, accessed on 23.12.2017.
[77] I. P. Mukhlenov, E. I. Dobkina, V. I. Derjuzhkina, *Catalysts Technology*, Leningrad, Khimia, 1989.
[78] M. Campanati, G. Fornasari, A. Vaccari, Fundamentals in the preparation of heterogeneous catalysts, Catal. Today, 2003, 77, 299.
[79] https://ipmworld.umn.edu/ware-intro-herbicides, accessed 16.04. 2023.
[80] https://theecologist.org/2011/sep/01/consumers-and-farm-workers-risk-toxic-pesticides-sprayed-salad, accessed 16.04. 2023.
[81] https://www.ugaoo.com/blogs/garden-maintenance/how-to-spray-pesticides-on-plants, accessed 16.04. 2023.
[82] https://time.com/3721985/produce-pesticides/, accessed 16.04. 2023.

[83] https://www.researchgate.net/profile/Carsten-Bruehl/publication/341867348/figure/fig3/AS:898238982914056@1591168270547/Water-droplet-without-left-and-with-surfactant-right-Agricultural-surfactants-can.png, accessed 16.04. 2023.

[84] https://wedocs.unep.org/bitstream/handle/20.500.11822/40351/Pesticides_Ch2.pdf

[85] https://s.turbifycdn.com/aah/healthypetscom/merck-atroban-11-ec-emulsifiable-concentrate-insecticide-spray-1-pint-5.jpg, accessed 22.04.2023.

[86] https://cdn.mscdirect.com/global/images/ProductImages/1759557-21.jpg, accessed 22.04.2023.

[87] https://smhttp-ssl-60515.nexcesscdn.net/media/wysiwyg/solutions/products/Tomahawk_TMD14_Fog.jpeg, accessed 22.04.2023.

[88] https://www.lcicorp.com/-/media/files/lci/cards-and-heroes/applications/agrochem-powder-card.jpg?h=220&la=en&w=350&rev=82616449e74f435b8e22282151fda965&crop=1&hash=CEF342E6AA7184CD3DE6398B2F2BDCFE, accessed 22.04.2023.

4 Personal chemicals

4.1 Absorbent hygiene products

4.1.1 General

Disposable absorbent hygiene products for babies, menstruating women and people suffering from incontinence have transformed the lives of millions of people. Single-use diaper products will be considered as an example in the following section. Modern disposable diapers, designed for maximum leakage protection, are based on superabsorbent polymers, accounting for ca. 75% of the global demand. They keep the skin dry and allow achieving a more stable skin pH, with less dermatitis, compared to home-laundered cloth diapers. Introduction of disposable diapers eliminates the need for washing and laundering of cloth diapers, which, unless undertaken at high temperatures and in hygienic conditions, were not always able to kill the germs that cause infections.

The second largest segment (ca. 12% share) in the global superabsorbent polymer market is related to adult incontinence products. Urinary incontinence is rather common, affecting people in different life stages. More than 25% of women experience incontinence at some time in their life, in particular due to pregnancy and childbirth. For elder men, there is also a high risk of daily urinary incontinence, leading to prostatectomy or removal of the prostate. Incontinence can be experienced by both men and women due to, for instance, stroke or senile dementia. A current estimate is that more than half of female residents in long-term care facilities in Europe suffer from incontinence.

The segment of feminine hygiene products accounts for less than 10% of the global superabsorbent polymer market. Regular and intermittent loss of menses and other fluids in many countries are taboo and cause embarrassment. Product advertising, easy access to information, improvement in performance and availability of various feminine hygiene products for ordinary consumers have greatly increased the life of women.

4.1.2 Diapers

The word “diaper” originally referred to the cloth type rather than its use. Cloth diapers made of cotton were produced in the nineteenth century. In the twentieth century, disposable diapers were introduced. Of the 375,000 babies born every day, ca. 25% use disposable diapers, with large variations in terms of quantity used per day. For some children, their parents use just one diaper in a day. Consumption patterns in different markets depend on the geography, culture, religion, climate, etc. In the

https://doi.org/10.1515/9783110797961-004

EU, disposable baby diapers are the leading absorbent hygiene products, occupying 36% of the market and corresponding to consumption of over 21 billion units annually in the EU, and subsequently translating to waste, which on dry basis is equivalent to 3% of the EU municipal solid waste generation. Used diapers, disposed as household waste, are either incinerated in waste incineration plants or landfilled. The manufacture, distribution, use and waste management in the EU alone result in greenhouse emissions, equivalent to emissions from 1.5 million EU households [1].

In some countries on other continents, only a very small number of babies ever see a disposable diaper, while in other places, cloth diapers are used, requiring regular washing or even the use of open crotch pants is common.

Diapers (Figure 4.1) are aimed at different age categories (babies, children and adults). From the consumer's perspective, safety, absence of any leakage and skin irritation, overall comfort and ease of use, freedom of movement, gender specificity as well as environmental impact are important issues. As a chemical product, diapers should have sufficient absorption capacity, flexibility and resistance, fast adsorption rate and slow release.

Figure 4.1: Diapers [2].

In recent years, superabsorbent polymers have been introduced, allowing the production of thin diapers. Overall, in the last 30 years, the weight of an average disposable baby diaper has decreased by ca. 50%. Superabsorbent polymers (SAPs) are able to absorb and retain large volumes of water and aqueous solutions, keep the urine inside the diaper, even under pressure, allow liquid transfer through a swollen layer of gel, reduce the risk of leakage and ensure a dry skin. Previously, superabsorbents were made from chemically modified starch, cellulose and other polymers such as like poly (vinyl alcohol) and poly(ethylene oxide). These polymers are hydrophilic, having high affinity for water. At the same time, when mildly cross-linked, either chemically or physically, these polymers only swell in water but not dissolve. Present day superabsorbent polymers are based on partially neutralized, lightly cross-linked poly(acrylic acid).

The polymers are first manufactured at low solids levels, followed by drying and milling, giving granular white solids. Superabsorbents can take water more than 100 times their weight because of their swelling to a rubbery gel. Absorption liquid can hardly be released mechanically. The full absorption capacity of a diaper requires an efficient liquid transport from the entry point to the rest of the diaper. Large amounts of superabsorbents in modern diapers imply significant swelling, which means that the swollen superabsorbent should be permeable for additional liquid through the gel mass. Otherwise, because of low permeability, the gel is blocked for further liquid, leading to leakage of the diaper. Permeability depends on the particle size distribution, porosity, interconnection between pores and the ability of the swollen particles to retain their shape under pressure.

A modern diaper consists of a cover stock or a permeable cover layer, a distribution system, an absorption layer (absorbent core) and an impermeable outer layer (impermeable back sheet). The cover and distribution layers that are needed to distribute the baby's urine over the absorbent core are usually hydrophilically treated nonwoven fabrics. The absorption core (Figure 4.2) consists of a mixture of cellulose fluff and superabsorbent granules.

A fluff is needed for good liquid distribution, which by itself cannot effectively store the liquid. A decrease of the fluff amount, from 80 to 90% in the past to the current level of ca. 50%, allowed diapers to be made thinner, while at the same time, increasing its core density. This increase comes with an increased amount of superabsorbents in the absorbent core, requiring more efficiency. Diapers for adult care obviously need a higher absorption capacity and the ability to reduce odors.

In addition to surface adsorption, uptake of liquids is due to absorption, which occurs because of the physical filling of the open space of a porous media. The driving forces for such filling are capillary pressure as well as diffusion of the liquid into the absorbent material itself. A range of parameters influence the absorption, including wettability, radius, length and volumes of the pores, surface tension and viscosity.

Superabsorbents are rubber-elastic materials that have polyelectrolyte networks (Figure 4.3). These hydrophylic polymers use acrylic acid of high purity as the basis and are manufactured by the cross-linking radical copolymerization of partly neutralized acrylic acid in water with a cross-linker containing two or more groups that can undergo polymerization.

Monomeric acrylic acid is first converted to sodium acrylate (55–75%) by partial neutralization with sodium hydroxide. Cross-linking in superabsorbent polymers is done through bulk (core) cross-linking during the polymerization stage or by surface cross-linking. For core cross-linking, organic molecules with two or more double bonds that have an ability to take part in polymerization are added in small amounts. They are then incorporated into the polymer backbone (Figure 4.4).

Polymerization starts after oxygen removal from the water solution and the subsequent introduction of radical initiators. Deterioration of the product is avoided by

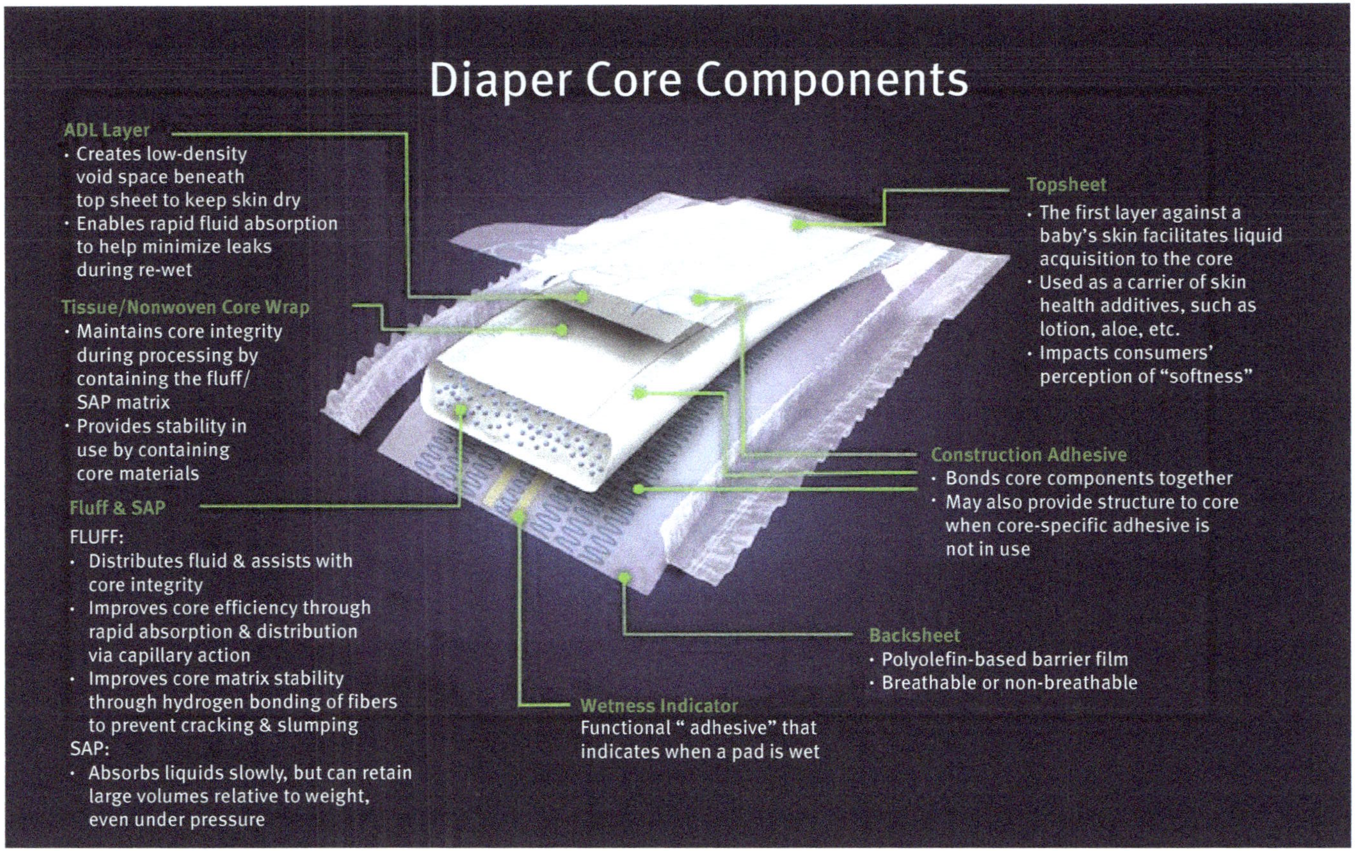

Figure 4.2: The structure of the diaper showing the composition of the diaper core [3].

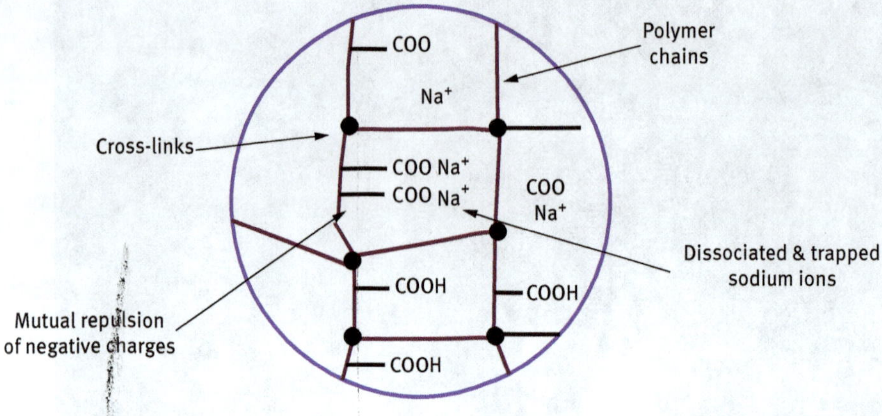

Figure 4.3: Mechanism of water absorption in superabsorbents [4].

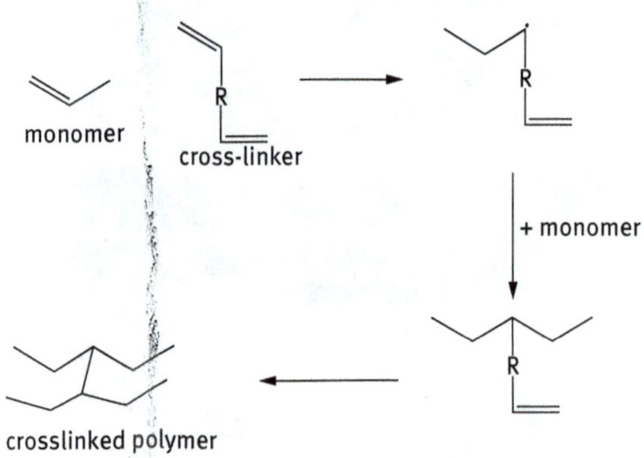

Figure 4.4: Illustration of cross-linking [4].

heat removal. Important issues are also high conversion and prevention of side reactions, which might increase the residual acrylic acid monomer content.

Absorption capacity of diapers strongly depends on the degree of neutralization or a charge on the polymer chain. Higher absorption is associated with a higher charge and a lower ionic strength of the swelling liquid. Other parameters influencing the absorption capacity, when superabsorbent granules are put in contact with a liquid, are density and homogeneity of the polymer network as well as cross-linking.

Cross-linking between polymer chains allows the formation of a three-dimensional network, which is important for making superabsorbents of the required characteristics. Higher cross-linking density gives a lower swelling capacity, which is then translated into an increased gel strength and lower absorption.

If the cross-linker is too reactive, it will be consumed early during polymerization, and the polymer chains formed at the later stages might not be incorporated into the network. During the application of superabsorbents, these polymers (so called extractables) will be extracted. This apparently should be avoided. Alternatively, if a cross-linker is not reactive, there will be difficulties in the formation of the three-dimensional network.

The concentration of a cross-linker is also important in optimizing the product performance. When the amount of a cross-linker is too low, it will result in extensive swelling, a high level of extractables, a tacky feel and a low strength against pressure. On the contrary, with too high amounts of cross-linkers, a downside will be a low swelling capacity with, however, less extractables and low tackiness.

As cross-linkers compound with at least two polymerizable double bonds, compounds with a double bond that can be polymerized, and a reactive functional group as well as compounds with two functional groups that are reactive toward monomers can be used. Some examples are given in Figure 4.5. Multivalent (e.g. aluminum) salts can be also applied.

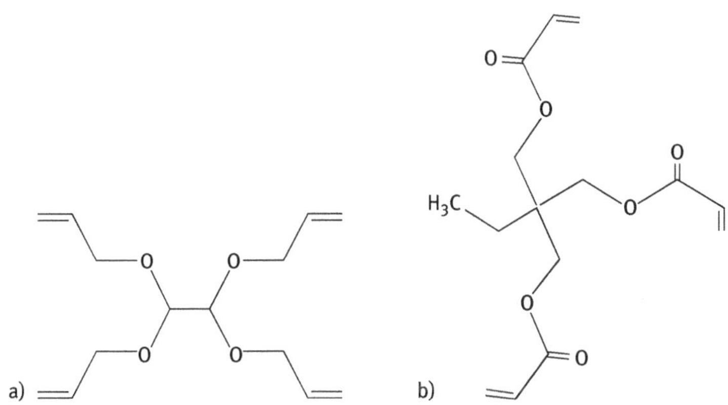

Figure 4.5: Some examples of cross-linkers: (**a**) tetraallylethoxy ethane and (**b**) 1,1,1-trimethylolpropanetricrylate.

Polymerization of acrylic acid in aqueous solutions depends on the solution pH and the degree of the carboxyl groups' neutralization. Commercial superabsorbents have a degree of neutralization between 50 and 85 mol %, which can be improved by partial neutralization after polymerization.

Polymerization typically gives a gel, which is not dissolved, but swells in water. Thereafter, the gel is disintegrated, dried, milled, and sieved to the desired powder grain size.

Another option is to apply surface cross-linking of superabsorbent polymers after drying (Figure 4.6). Such treatment is needed to achieve a cross-linkage density at the surface, higher than inside the particles. This improves the liquid distribution inside the diaper. Without such surface cross-linking, the liquid coming to the surface is hindered from penetrating further into the diaper, resulting in leaking of the diaper. As a consequence, a baby becomes wet. Surface cross-linking allowed the development of ultrathin baby diapers, with a low ratio between fluff and the superabsorbent. Overall, this approach improves not only the swelling capacity of superabsorbents by decreasing the core cross-linking but also the absorption against pressure or load when a baby sits on a diaper. The cross-linker for surface coating has at least two functional groups reacting with the carboxyl groups on the polymer backbone. Such cross-linking treatment is followed by curing through heating.

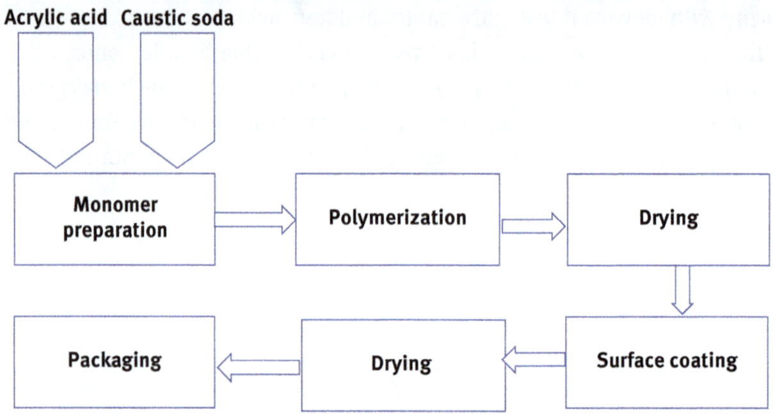

Figure 4.6: Flow sheet for the manufacturing of superabsorbents with surface coating (modified from [5]).

A grain size between 0.15 and 0.8 mm used in diapers is achieved using the following procedure. Coarse milling in an attrition or a knife mill is followed by more fine milling in roll crushers, giving the final grain size. After sieving, larger particles are sent back to milling, while the fine fraction is recycled back. This can be done, for instance, by adding it as a powder to the monomer solution or by mixing it with the polymerized gel particles and water.

The final powder of the required size is then supplied to companies making hygiene products where it is blended with the fluff to form the absorbent core of a diaper.

Overall, the standard design of disposable baby diapers is based on a layered construction using different nonwoven fabrics and various materials, including SAP, cellulose pulp, films, elastics, adhesives and other minor synthetic elements [6]. The diaper components are cut, shaped and bonded to build up such a multilayer structure.

The process starts by cutting a nonwoven fabric into two strips of material, cuffs that act as barriers to avoid fluid leaks. The topsheet (Figure 4.2), being highly permeable to minimize the time that the fluid is in contact with the skin of the baby, is glued to the cuffs. An acquisition and distribution layer (ADL) is glued onto the upper central part of the topsheet. The ADL (Figure 4.2) stores the fluid temporarily before it is transferred and distributed through capillaries to the absorbent core, which is the main storage for fluid in a diaper. An appropriate blend of fibrous fluff pulp and SAP, encapsulated between two nonwoven layers is used for constructing the core, which is then glued to the ADL topsheet layers. The backsheet, as an outer layer, seals the diaper and is mainly breathable for the protection of the baby's skin. Gluing the backsheet to the whole construction is followed by cutting the diapers, folding and packaging.

Some newer diapers are fabricated with "glueless" methods, reducing the impact of CO_2 release during manufacturing by ca. 10%.

4.2 Pharmaceutical products

4.2.1 Administration routes

In this section, a few main systems for administering pharmaceutical compounds to humans or animals through various routes (Figure 4.7), with an aim to achieve therapeutic effects will be discussed. The dosage forms can be solids (tablets, powders, granules, etc.), liquids (e.g., solutions, suspensions and emulsions), soft (e.g., capsules, patches and gels) or gaseous (gases and aerosols).

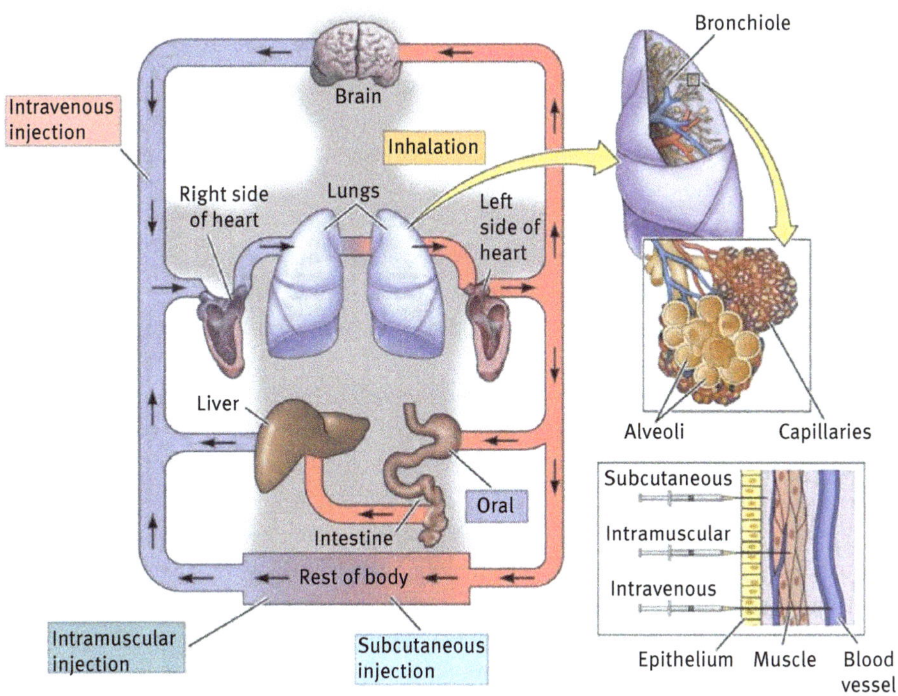

Figure 4.7: Routes for administrating drugs [7].

The choice of formulation takes into consideration not only physicochemical properties of a drug, but also body effects, comfort, age, presence of other deceases and overall well-being of a patient, as well as the price, traditions, insurance and so on. Moreover, a decease can be acute, long term or chronic, of different severity, requiring local or systemic treatment, either by the patient or by healthcare professionals.

As shown in Figure 4.7, the delivery can be oral, ophthalmic, vaginal, through lungs, nasal cavity, rectum or by injection.

Oral delivery has an advantage from the point of view of a patient, being, in general, very convenient and noninvasive. Moreover, manufacturing of traditional oral delivery systems, such as tablets or various types of capsules, suspensions, elixiris, etc. is well established, allowing the best case production of drug delivery systems with controlled-release profiles. From the production viewpoint, tablets allow large-scale production at lowest costs, are easiest and cheapest to package and ship, have simple product identification and, overall, could be rather stable unless some crystal phase transformations or a formation of more stable crystalline structures from amorphous phases happen during storage. From the utilization point of view, such delivery forms are the lightest and most compact, have excellent dose precision and the least content variability, are easy to handle and can contain, otherwise incompatible active ingredients.

There could be, however, also some disadvantages in the use of oral delivery routes related, for instance, to low solubility and permeability. Dissolution is subsequently the rate-limiting step in the delivery of a drug from a tablet to the systemic circulation. Other disadvantages include premature degradation, interactions with food, potential manufacturing difficulties, bitter taste or sensitivity to oxygen, thus requiring encapsulation, entrapment prior to compression or coating, irregular adsorption or inability of unconscious patients (Figure 4.8) to take a dose.

Overall, for oral delivery with e.g. tablets and capsules (Figure 4.9), the following should be fulfilled: maintenance of optimum therapeutic drug concentration in the blood with minimum fluctuation; predictable and reproducible release rates for extended duration, elimination of side effects, frequent dosing and wastage of drug.

Somewhat related to oral delivery from the viewpoint of drug administration, are systems for nasal and pulmonary delivery. Inhales of various types, one of which is shown in Figure 4.9b, contain solid particles or drug suspensions to treat illnesses such as asthma and chronic obstructive pulmonary diseases.

These drug delivery systems should be able to comply with stringent requirements: ability to form an aerosol and stability during its formation, biocompatibility, targeting of specific sites in the lung, predetermined release profile and degradation within an acceptable period of time. Inhalation can be for a wide range of substances of different size, offering a possibility of delivering high drug concentrations directly to the disease site and minimizing the risk of systemic side effects. In particular, this noninvasive, needle-free delivery method reduces side effects for gastrointestinal tract problems, avoiding first-pass metabolism in the liver and allowing for a smaller

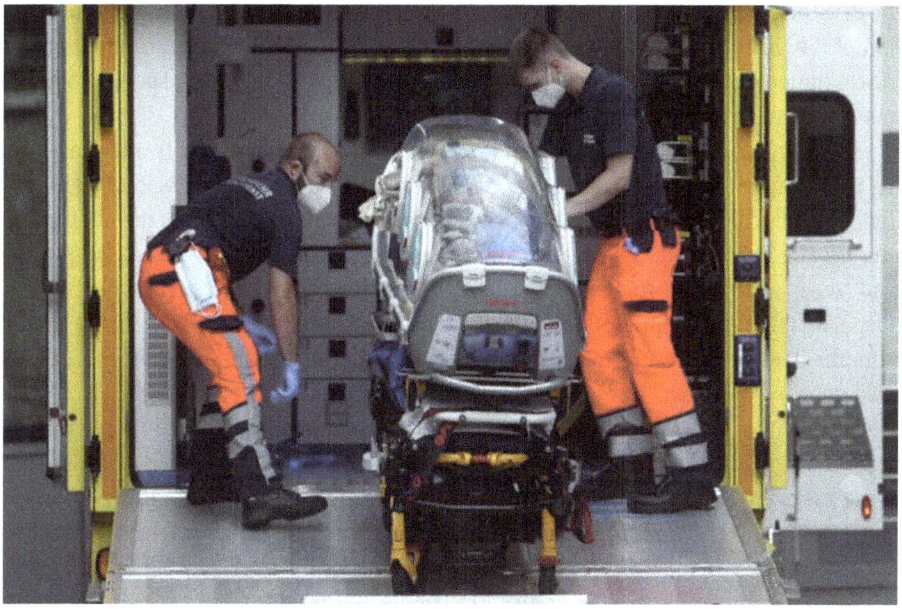

Figure 4.8: German emergency personnel load into their ambulance, the Russian opposition figure Alexei Navalny on a stretcher on August 22, 2020, at the Berlin's Charite hospital, where Navalny will be treated after his medical evacuation to Germany, following suspected poisoning (image from [8]).

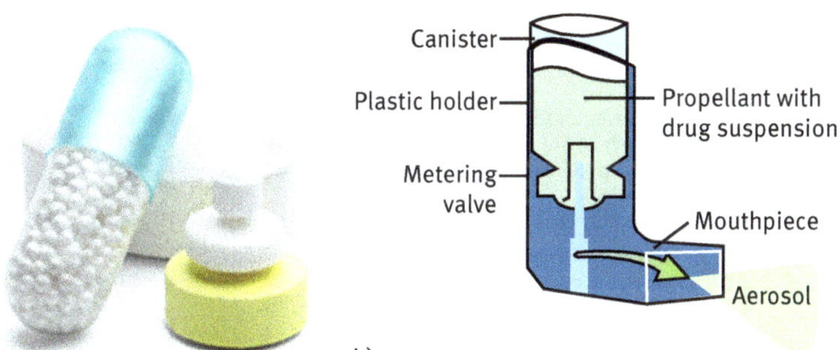

a) b)

Figure 4.9: Some systems for oral drug delivery: (**a**) tablets and capsules [9] and (**b**) inhaler [10].

dose than conventional oral delivery. On the downside, poor delivery efficiency, high dose variability, difficulties to disperse powders 1–5 μm in diameter into aerosol, stability problems and difficulties in obtaining the optimum particle size should be mentioned. The size requirements stem from the practical aspects of pulmonary drug delivery, namely that too small size will result in exhaling, while if the particles are too large, they can affect the oropharynx and larynx. Besides the size and shape of the particles, other pharmaceutical factors affecting aerosol deposition are aerosol velocity, density and physical stability. Aerosol deposition in the lungs and efficiency of the drug depends not only on the design on the inhalation device but also on the creation of the turbulent flow in the inhaler. Such flow relies on the ability of a patient to inhale the powder at a sufficiently high inspiratory flow rate, posing special requirements for inhalation devices intended for children.

Intravenous delivery has many advantages, including bioavailability and maximizing incorporation of degradable drugs, rapid response and control of blood concentration. This way of administration bypasses the first-pass effect (first-pass metabolism or presystemic metabolism) when the drug concentration is substantially reduced. Significant losses are related to absorption in the digestive system and metabolism in the liver. The latter can happen to an extent that only a small amount of the drug comes from the liver to the rest of the circulatory system, thereby essentially diminishing bioavailability of the drug.

Intravenous delivery is, however, invasive, influencing very much the quality of veins if there is a need to make intravenous infusions for a long period of time. This is the case, for instance, of oncological deceases requiring many months of chemotherapy. Moreover, trained personnel are needed for intravenous infusion and there are always risks related to possible toxicity because of incorrect dosing and sterility. Traditional delivery systems include Intravenous (IV) bags for infusion (Figure 4.10). Some of the engineering issues related to the application of intravenous bags are shown in Figure 4.10a.

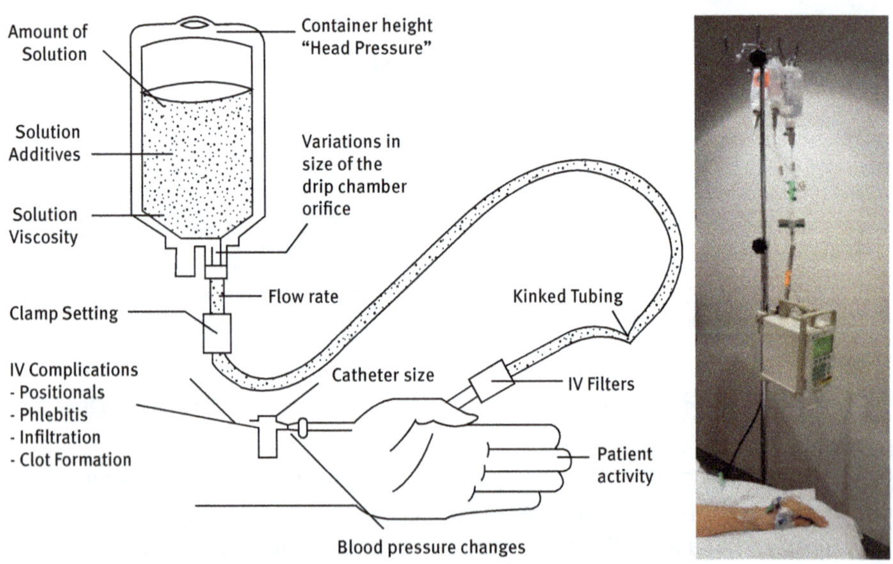

a) b)

Figure 4.10: Application of IV bags: (**a**) schematics [11] and (**b**) visualization (photo of the author).

Another method for administration of drugs is intramuscular or subcutaneous injections (Figure 4.11).

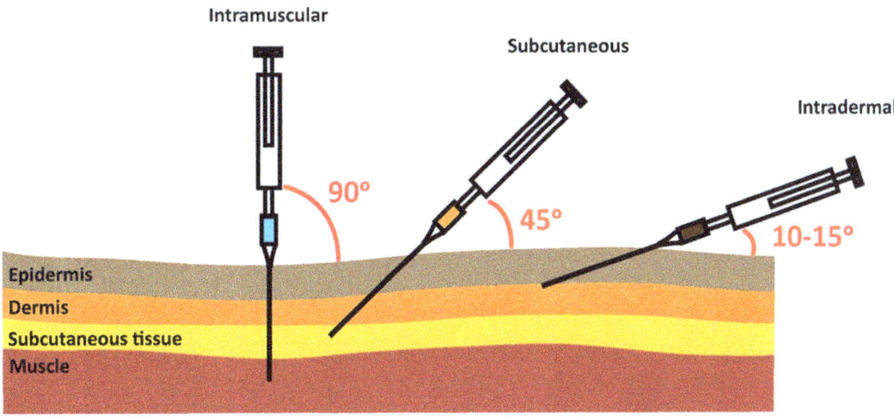

Figure 4.11: Types of injection [12].

Advantageous of the intramuscular route, also bypassing the first-pass metabolism, are related to a possibility of administrating the drug by a patient and in larger volumes than via subcutaneous injection. At the same time, this method is invasive and might result in patient discomfort, irritation and inflammation, and requires some training.

Home infusion therapy recently evolved as a cost-effective way of the intravenous or subcutaneous administration of drugs to, especially, oncological patients at home. The components needed to perform home infusion include, besides the drug, the equipment and supplies (Figure 4.12). Such approach is typically applied on the second day of a chemotherapy protocol; it is mainly used to help in reducing the risk of infection resulting from strong chemotherapy and does not require administration of the main drug to treat cancer, for example, a monoclinical antibody.

On-body drug delivery devices, including a housing with an interior space, a needle having retracted and deployed states, an injector to move the needle between these states and a reservoir disposed within the interior space, are currently under development and implementation in clinical practice. Such delivery devices are wearable, disposable and of single use (Figure 4.13). Design and placement of the on-body injector for pegfilgrastim, developed by Amgen to fight infection in patients undergoing chemotherapy is shown in Figure 4.13. It demonstrates a battery-powered electromechanical drug delivery system that is activated upon complete filling, followed by the automatic deployment of a rigid needle into the

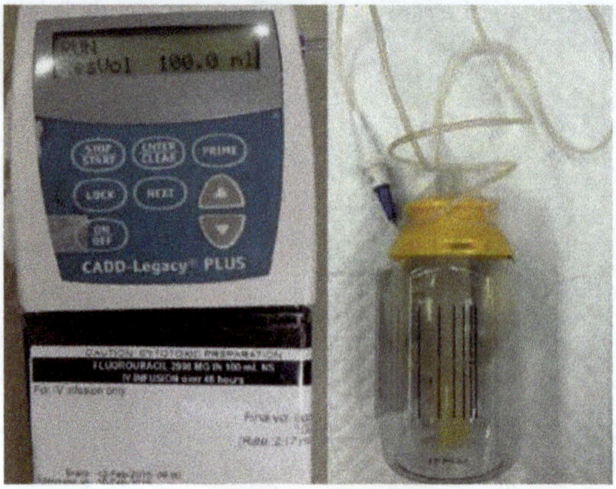

Figure 4.12: Portable infusion devices for ambulatory home chemotherapy [13]. Copyright by the Hong Kong College of Radiology.

subcutaneous tissue. While the rigid needle is retracted, the soft cannula remains in the subcutaneous tissue until the on-body injector is removed.

Cartridges with auto injectors have been developed for reducing signs and symptoms in patients suffering from moderate to acute rheumatoid arthritis (Figure 4.14).

Whatever is the drug delivery method, it should ensure optimum therapeutic drug concentration in the blood, with minimum fluctuations, elimination of side effects, avoidance of frequent dosing and wastage of drug. The release rates should be also predictable and reproducible. Selection of the drug delivery route depends on the physicochemical properties of a drug (e.g., molecular size, chemical stability, solubility, pK_a, crystallinity and particle size) and the environment (pH, temperature and concentration). Moreover, biological interactions with a drug (sensitivity to first-pass metabolism, degradation and side effects) and the desired pharmacological effect (local or systemic, immediate or sustained, etc.) should be considered.

Different release profiles are illustrated in Figure 4.15. Apparently, the concentration of the released drug should not exceed the upper limit of the therapeutic range, as otherwise the drug becomes toxic. Too fast release of a drug can be also unwanted, as after a short while, the concentration becomes too low for a drug to be effective. In the case of immediate release, dosage delivery begins right after ingestion of a pill.

Contrary to immediate release, modified-release dosage is used in oral dose formulations (pills and capsules). Slower and steadier release into the bloodstream has the advantage of less frequent intake of drugs. Controlled (sustained)-release medication is provided over an extended time. In fact, sustained release maintains drug release over a sustained period but not at a constant rate. An interesting case is the zero-order release when the rate is independent of the drug left in a tablet or capsule.

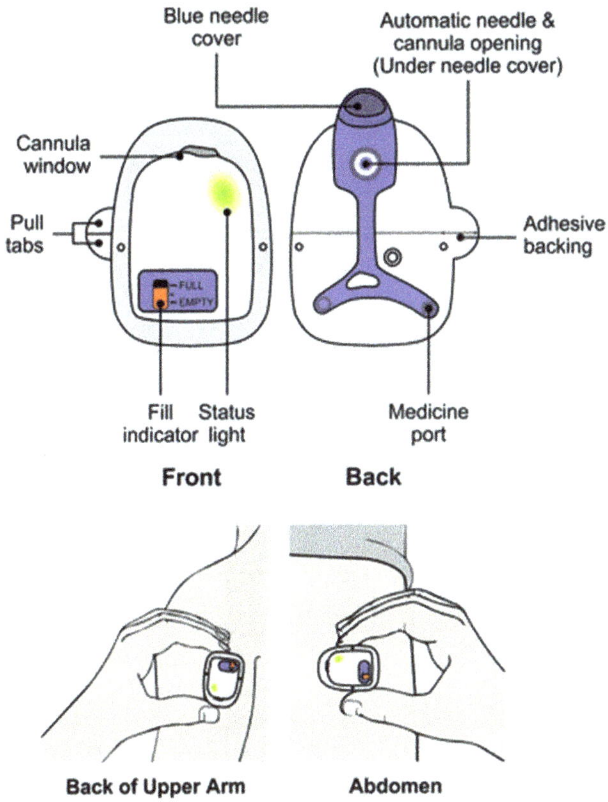

Figure 4.13: Design and placement of the on-body injector [14].

These cases correspond to so-called extended-release variants (Figure 4.16) of the modified release dosage. Other possible dosage forms rely on delayed release (Figure 4.16) or targeted release. In the latter case, the dose is delivered to a specific target in the body.

4.2.2 Advanced systems for controlled drug delivery

For such systems, release rates can be independent or dependent on environmental factors in a controlled manner, and are regulated by the design of a delivery system with external trigger (e.g., pH, etc.). Other requirements include inertness, biocompatibility (absence of inflammation), biodegradability, minimal aggregation or inactivation. There are several drug delivery systems operating with different principles. First, systems based on dissolution will be discussed. For the dissolution-based delivery (Figure 4.17), in the case of matrix systems, a drug is homogeneously dispersed in a polymer matrix and

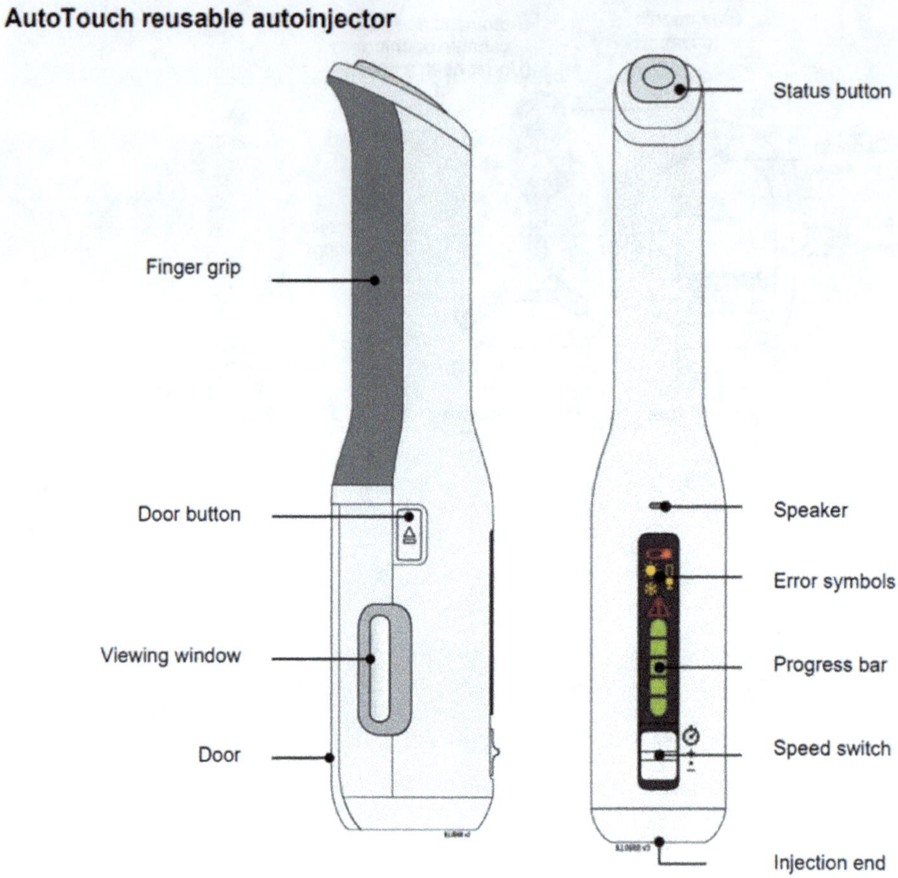

Figure 4.14: AutoTouch autoinjector [15].

is released as the polymer matrix dissolves or degrades (Figure 4.17a). The amount of drug decreases with time because the size of the matrix also decreases; thus the drug release is of nonzero order. In reservoir systems (Figure 4.17b), the drug core is surrounded by a rate-controlling membrane.

Dissolution of the polymeric material in an aqueous environment is thus needed (physical mechanism) or alternatively, (bio)degradable polymers can be used (chemical mechanism). In the case of a chemical mechanism, a drug is dispersed in a degradable polymer matrix and cannot be readily released. Such release happens as the matrix degrades. Biodegradable polymers, being hydrophobic and water insoluble, break down into biocompatible monomers. Among such polymers, the most widely used ones are poly(glycolic acid), poly(lactic acid) and poly(lactic-co-glycolic acid) (Figure 4.18). Glycolic acid ($HO–CH_2–COOH$) and L-lactic acid ($HO–CH(CH_3)–COOH$) copolymers are also applied in the manufacture of synthetic biodegradable surgical sutures, clips, etc.

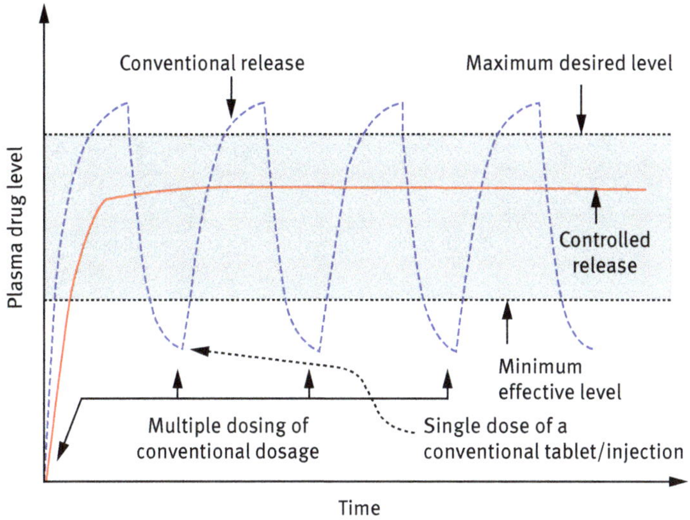

Figure 4.15: Drug release profiles (reproduced with permission from [16]).

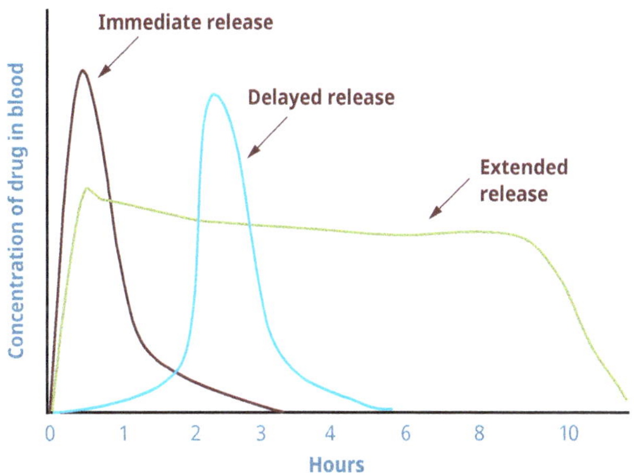

Figure 4.16: Concentration versus time for different release profiles [17].

Drugs loaded onto a certain carrier can be also released by diffusion. In reservoir drug delivery systems (Figure 4.19), this can happen through nonporous, homogeneous and dense polymer (solution–diffusion) membranes between polymer chains. Alternatively, diffusion occurs through microporous membranes, with pores filled with the liquid (water for hydrophilic drugs or oil for hydrophobic drugs). In monolithic devices, release of a drug dispersed in polymer matrix without any membrane is controlled by diffusion out of this polymer matrix, which can be nonporous or microporous.

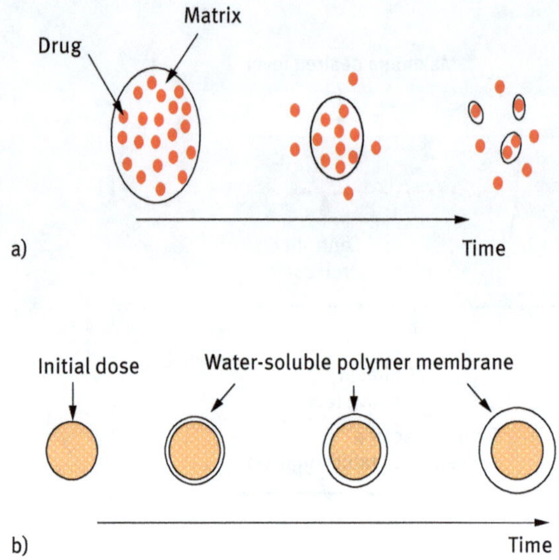

Figure 4.17: Drug delivery systems based on dissolution: (**a**) matrix and (**b**) reservoir.

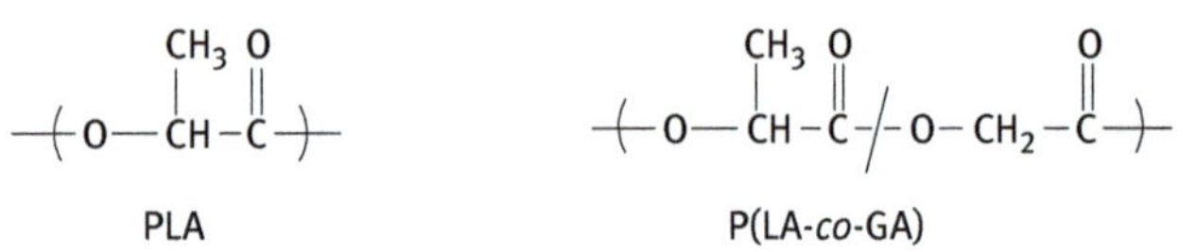

Figure 4.18: Structures of some biodegradable polymers.

In dissolution and diffusion-based drug delivery systems, a range of variables influence the release kinetics, including drugs powder size, molecular weight, solubility and loading; polymer properties (composition, structure, surface coating and hydrophobicity); device geometry and application of stimulation such as mechanical, magnetic or ultrasonic. For instance, when a polymeric matrix contains drug powder and magnetic beads, the release of a drug is accelerated when an oscillating external magnetic field is applied.

Environmentally responsive systems can also react to changes of other external conditions, such as temperature or pH. As an example, pH-responsive gels can be mentioned, which are protonated or deprotonated, depending on the pH. As a result, swelling behavior is different. Chemically, different gels will respond differently to pH. Since biological systems have regions with distinct pH, drug delivery systems can be designed for specific pH, stimulating release in desired locations. An illustration is given in Figure 4.20 when a hydrogel system (a copolymer of acrylic acid, N,N' dimethyl acrylamide, and *tert*-butyl acrylamide cross-linked with diaminocaproyl-diaminoazobenzene) is delivered specifically to the colon.

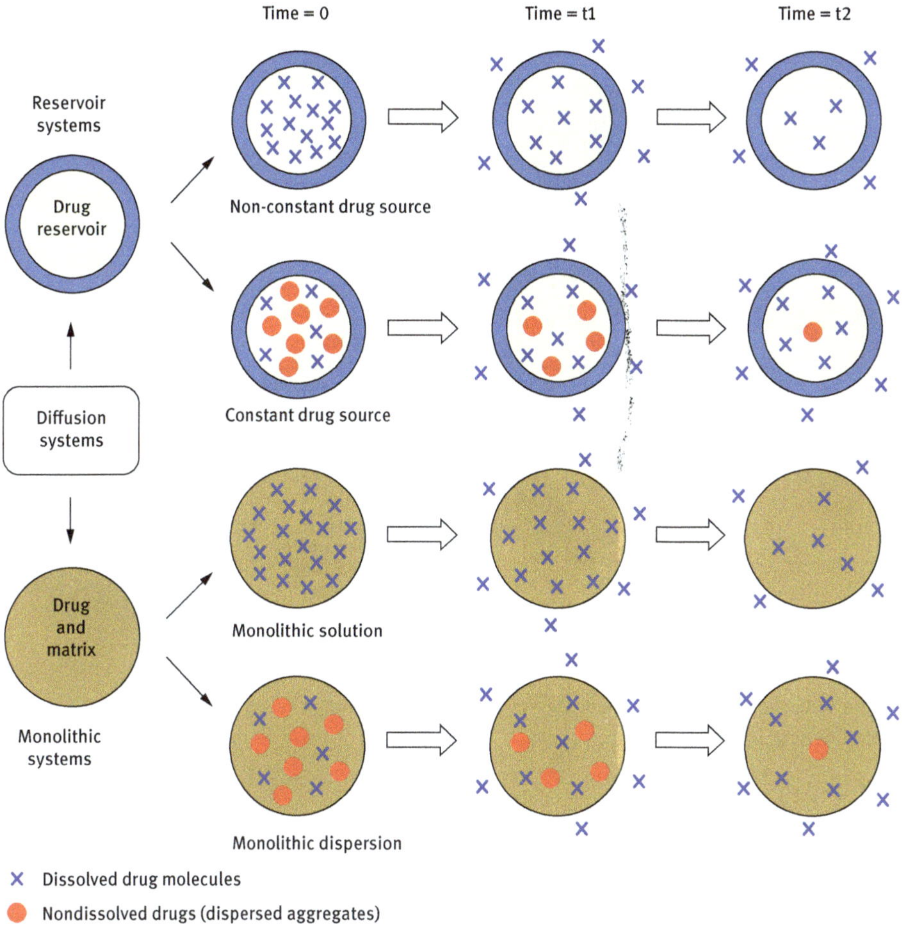

Figure 4.19: Diffusion-based drug delivery systems: reservoir and monolithic systems (reproduced with permission from [16]).

Osmotic and swelling-controlled systems are based on penetration of water. Swelling-controlled release systems are initially dry and when placed in the body, absorb water or other body fluids and swell. Rigid tablets with a water-permeable jacket having one or more laser-drilled small holes are used in osmotic systems. Osmotic pressure is created when water is drawn in through a semipermeable membrane, which, however, prevents drug from moving out. Water infusion leads to swelling of a gel and release of drugs through the holes (Figure 4.21).

A summary of the methods used to produce nanocapsules and nanoshperes of polymers containing drugs is presented in Figure 4.22. Oil-in-water (O/W) emulsions are used for lipid-soluble drugs, while water-in-oil (W/O) emulsions are used for drugs soluble in water.

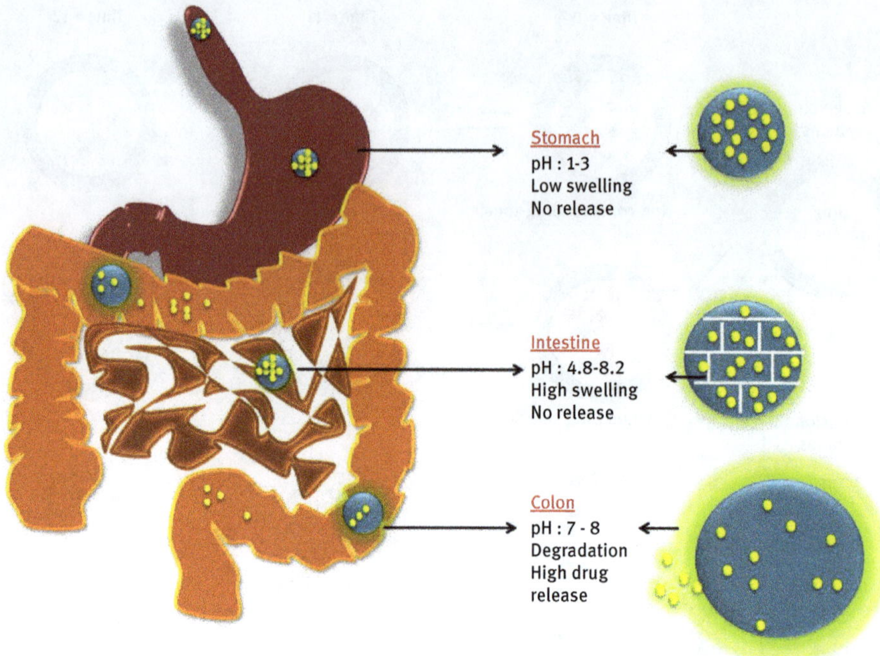

Figure 4.20: The response of the hydrogel system as it passes through various pH in the gastrointestinal tract [18].

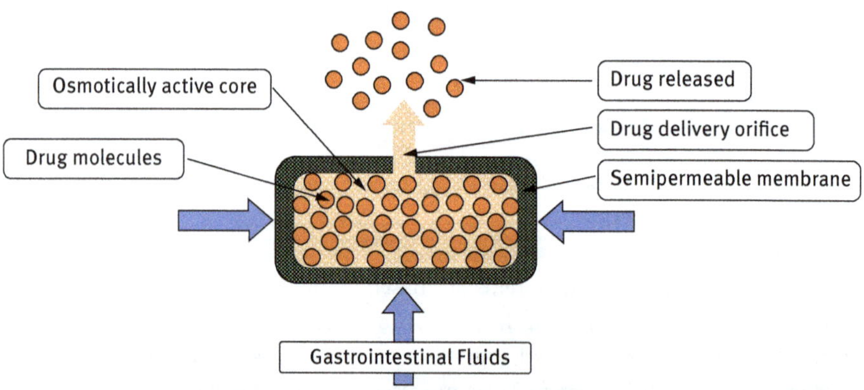

Figure 4.21: Osmotic-controlled one-chamber system with drug dispersed in an expanding gel [19].

Let us consider, as an example, methods based on microencapsulation. For O/W emulsions, a drug is dissolved or suspended in a polymer solution. Then, an aqueous solution containing an emulsifier is added. For the case of solvent evaporation, the system is evacuated to remove the solvent. Alternatively, the solvent is extracted by quenching with water. Water-in-oil-in-water (W/O/W) multiple emulsion method for water-

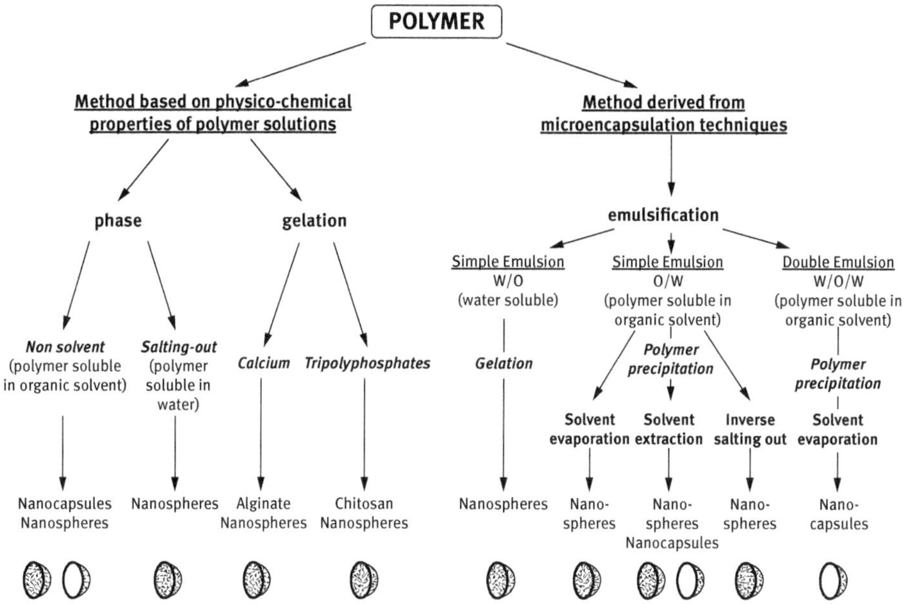

Figure 4.22: Summary of methods for the preparation of nanospheres and nanocapsules from a polymer [20]. W/O, water-in-oil; O/W, oil-in-water; W/O/W, water-in-oil-in-water.

soluble drugs (e.g., proteins and peptides) includes the formation of W/O emulsions containing a drug and a buffer as an aqueous phase and a polymer dissolved in the organic phase. To this emulsion, an aqueous solution with an emulsifier is added, resulting in W/O/W emulsion. Solvent evaporation or solvent extraction, described above, can be used to make the micro/nanospherical product.

In general, besides gels, micro/nano-capsules and -spheres, there could be systems having other structures and morphologies (Figure 4.23), and include, for example, polymer–drug conjugates, polymeric micelle and liposome systems.

Other promising microparticulate systems are highly porous materials, based on mesoporous silica or porous silicon. They exhibit a narrow pore size distribution in the nm range, in combination with high specific pore volumes. The significant attention devoted to them recently is related to a potentially homogeneous incorporation of drugs with high loading and subsequently, a controlled release. Porous silicon and mesoporous (pores in the range of 2–50 nm) materials such as MCM-41 (2D hexagonal) or SBA-15 (3D cubic), with a large surface area, can be used for storing hydrophobic drugs. They can deliver large dosages of poorly water-soluble drugs without premature release. Clear advantages of these materials are their cost competitiveness, chemical inertness, thermal stability, resistance to mechanical stress and pH as well as biocompatibility. The presence of abundant silanol groups allows anchoring organic functionalities by covalent bonding, and can be used in tuning drug loading and its subsequent release. Even if the

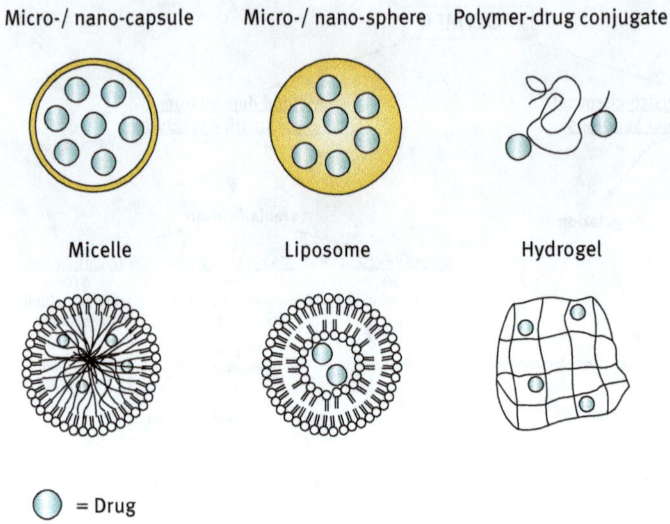

Figure 4.23: Different structures and morphologies of drug delivery systems [21].

pore size is the governing factor in drug release, surface functionalization helps to control drug release for certain external stimuli (chemical, pH, temperature, etc.).

4.2.3 Manufacture of pharmaceutical forms: tablets

First, a certain formulation should be developed, providing a combination of active ingredients with the appropriate excipients. The latter are inactive ingredients used to dilute, protect, stabilize and control the release of the active ingredient. Moreover, they can be used as fillers or to control color. Some ingredients for pharmaceutical tablets are presented in Table 4.1.

Table 4.1: Examples of excipients for pharmaceutical tablets.

Component	Function	Examples
Fillers/diluents	Increase size and weight of the final dosage form, ensure appropriate size and convenient volume	Microcrystalline cellulose, sucrose, calcium phosphate, dextrin
Binders	Increase powder adhesion and ensure appropriate granule formation and compression	Pregelatinized starch, hydroxypropyl methylcellulose
Disintegrants	Promote breakdown of aggregates	Sodium starch glycolate

Table 4.1 (continued)

Component	Function	Examples
Flow aids	Reduce interaction between particles	Talc
Lubricants	Reduce interactions between particles and surfaces of the processing equipment	Magnesium and calcium stearates
Surfactants	Promote wetting	Sodium lauryl sulfate, polysorbate
Modified release agents	Influence the release of active ingredient	Hydroxypropyl methylcellulose

The active ingredient should preferably be amorphous in a metastable form rather than crystalline. Solvent mixtures and/ or rapid solvent evaporations are used to precipitate drugs. Alternatively, in spray drying a solution of lactose and an active compound is applied. Fxreeze-drying is also a viable alternative for heat-sensible compounds, resulting, however, in mixtures of crystals and amorphous structures.

The solid dosage forms for oral delivery are different types of tablets (chewable, multilayer, with modified release, etc.), powders or capsules (hard or soft gelatin), while different types of aerosols (metered dose or dry powder) are used for inhalation.

A general scheme for drug product manufacture is shown in Figure 4.24, illustrating how the drug and the excipients are combined into the dosage form.

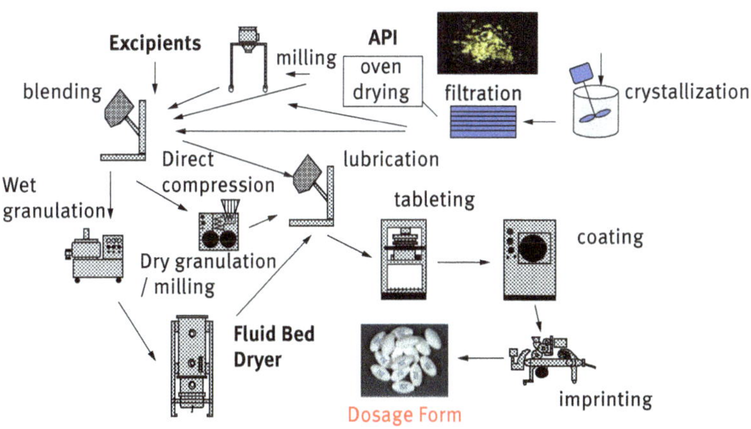

Figure 4.24: Manufacturing of dosage forms from API (active pharmaceutical ingredient) [22].

Overall, among the factors influencing tablet dissolution, the following can be mentioned: particle size of the drug and its solubility, type and concentration of excipients (disintegrants, binders, and lubricants), method of manufacturing and the granulation compactness.

The majority of the active pharmaceutical ingredients cannot fill the matrix of a tablet press, because of a substantial amount of fines and overall inhomogeneity, resulting in strong intraparticle friction. Subsequently, various additives are added, with their roles mentioned in Table 4.1. Additives, for example, are needed to improve flow behavior, while binders promote adhesion of particles, enabling preparation of the granules and maintaining integrity of the final tablet. Disintegrants are required for efficient breakup of the tablets to smaller particles, which is important for drug dissolution and bioavailability. Upon exposure to aqueous solutions, disintegrating agents must swell or expand; subsequently, they should be hydrophilic and have a high water uptake. Apparently, the size of constituents of a tablet should not differ too much. Some excipients, with their functional properties, are presented in Table 4.2.

Table 4.2: Examples of excipients in pharmaceutical tablets.

Agent	Function	Concentration (%)	Comments
Starch	Binder, also acts as a disintegrant	5–25	The most common in the past, requires time-consuming paste preparation; disintegration by wicking (absorbing draw off liquids) rather than swelling
Pregelatinized starch	Binder	0.1–0.5	Easier to prepare than starch because of solubility in cold water
Acacia	Binder	1–5	Requires paste preparation
Polyvinyl-pyrrolidone	Binder	2–8	Soluble in ethanol and water; available in a range of molecular mass/viscosity
Hydroxypropyl cellulose	Binder	2–8	Low viscosity grades preferred
Gums – agar, guar	Disintegrant	<5	Swells in water; limited amounts are used; otherwise, viscous gels can be formed
Sodium starch glycolate	Disintegrant	2–8	Swells in water
Sodium alginate	Disintegrant	4–6	Swells in water; forms less viscous gels than gums
Talc	Glidant*	1–5	Often used; fine powder
Fumed silica	Glidant	0.1–0.5	Small particle size
Sodium lauryl sulfate	Glidant	0.2–2	Anionic surfactant and lubricant
Magnesium lauryl sulfate	Lubricant	1–3	Wetting agent
Stearic acid	Lubricant	0.25–2	Hydrophobic

Table 4.2 (continued)

Agent	Function	Concentration (%)	Comments
Magnesium stearate	Lubricant	0.25–2	Hydrophobic
Sodium stearyl fumarate	Lubricant	0.5–2	Less hydrophobic than stearates
Light mineral oil	Lubricant	1–3	Hydrophobic
Glyceryl behenate	Lubricant	0.5–4	Acts as a controlled-release agent

*A glidant is a substance that is added to a powder to improve its flowability.

Let us consider as an example, Erlotinib,

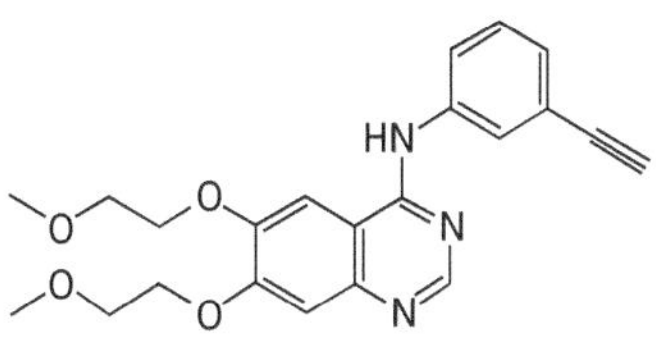

sold under the brand name Tarceva, among others, which is a medication used to treat non-small cell lung cancer and pancreatic cancer. The tablet composition is presented in Table 4.3.

Table 4.3: Composition of Erlotinib hydrochloride formulation of 150 mg strength [23].

Composition	%
Intragranular components	
Erlotinib hydrochloride form A	36.4
Lactose	23.1
Microcrystalline cellulose	6.7
Sodium carboxymethylcellulose	1.1
Water	q.s.*
Extragranular components	
Microcrystalline cellulose	22.7
Sodium starch glycolate	8.0
Magnesium stearate	2.0
*Uncoated tablet mass***	100

*q.s. (quantum satis) is a Latin term meaning the amount which is enough.
**After coating with Opadry white, the coated tablet mass increases from 450 to 472.5 mg/tablet.

According to [23], the manufacturing process for Erolotinib includes sieving sodium carboxymethyl cellulose, microcrystalline cellulose and lactose; adding Erlotinib hydrochloride and mixing everything in a granulator; adding water to granulate; sieving wet granules to de-agglomerate; drying in a fluid bed dryer; sieving the granules, microcrystalline cellulose (extragranular) and sodium starch glycolate; and mixing with the sieved granules in a blender, followed by sieving magnesium stearate and mixing with the blend in a blender. Other unit operations for this particular case are compression of the homogeneous blend under monitored humidity conditions on a rotary tableting machine, coating of the tablets and finally packing the tablets in blisters or any other suitable packaging material.

For this and other examples, the following unit operations are thus used during manufacturing: size reduction (milling), blending, dry and wet granulation, drying, tablet compression and coating. An overview of compressed tablet manufacturing is given in Figure 4.25.

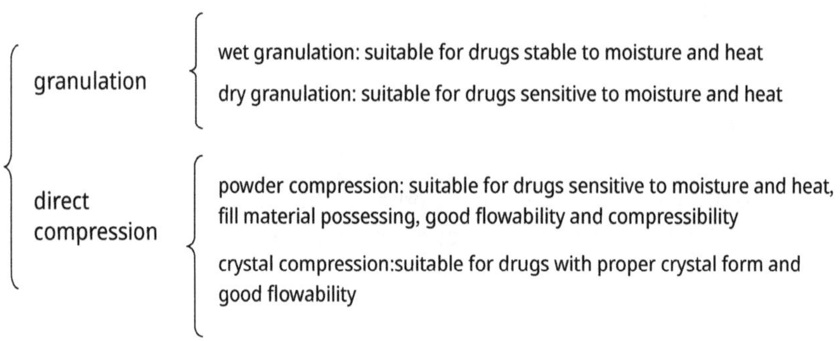

Figure 4.25: Compressed tablet manufacturing.

Powder compression is suitable for drugs that are sensitive to moisture and heat, and for drugs that have good flowability and compressibility. Compressed tablets, being the most popular dosage form, offer ease of administration, precise dosing and good bioavailability. As follows from Figure 4.25, either direct compression of powder blends can be used or suitable granulations are prepared, followed by compression. In direct compression, an admixture of the necessary powders undergoes compression with tablet compression presses. Excipients are added in a granular form. In case, direct compression is not possible because of, for instance, poor compressibility or flowability, wet or dry granulation is done. Wet and dry granulations are suitable for drugs that are, respectively, stable or sensitive to moisture and heat. In the case of dry granulation (or slugging), the ingredients are blended and compressed on a heavy-duty tablet press, giving large hard tablets of ca. 28–50 cm. These slugs should thereafter be milled and recompressed.

Granulation, in general, is conducted to enhance the flow properties, prevent segregation during tableting or storage, diminish dust production and avoid or minimize toxic dust, improve compression characteristics and overall appearance of the final product. The aim of granulation is thus to:

- ensure uniform distribution of ingredients and uniform size, as well as sufficient physical strength,
- allow good disintegrating properties and lubrication to reduce die-wall friction,
- provide near-spherical shape for reproducible flow and constant tablet weight.

Moreover, sufficient amounts of fines must be formed to fill the empty spaces between coarse granules, for better compression.

In wet granulation, fine powder particles are agglomerated with the aid of a suitable nontoxic granulating fluid such as water, isopropanol or ethanol, as well as their mixtures.

In spray granulation, the liquids, containing solids (i.e., solutions, suspensions or melts), are sprayed into a fluidized bed system with immediate evaporation. The starter core formed in this way can be further coated by applying other liquids that would also evaporate. Different arrangements of wet granulation are possible (Figure 4.26).

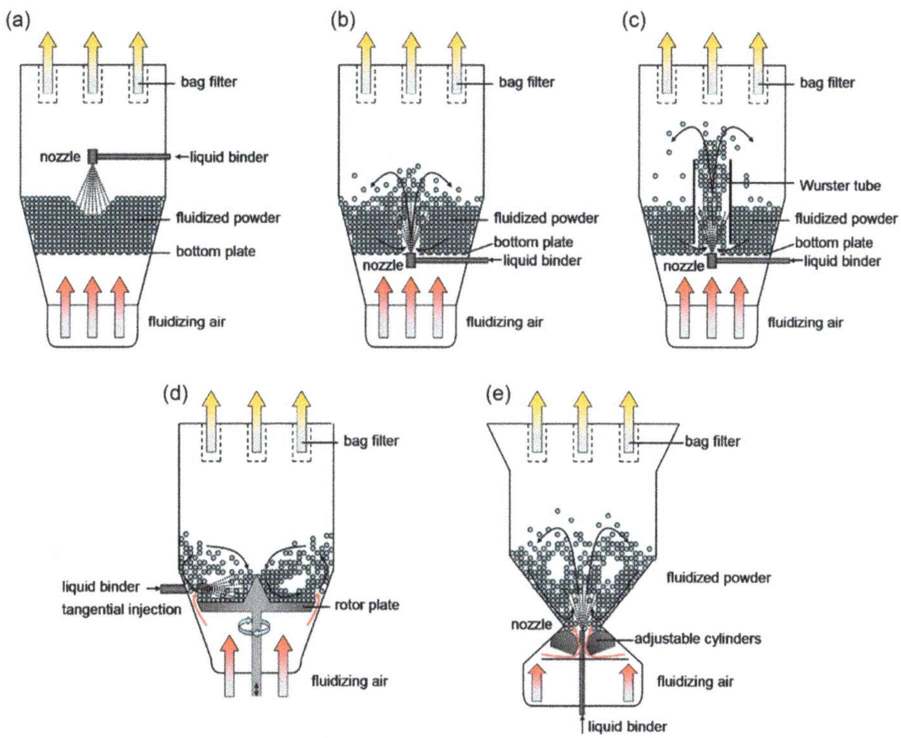

Figure 4.26: Representative configurations of wet fluidized bed granulators and coaters: (**a**) top spray granulator, (**b**) bottom spray granulator, (**c**) Wurster granulator, (**d**) rotor granulator and (**e**) spouted bed granulator (reproduced from reference [24] with permission).

Tablet presses could be either single-punch or multistation rotary presses. Basic principles of tableting were discussed in Chapter 2. Tablet properties are influenced by the properties of the primary particles, the way they are processed, the nature and properties of additives and excipients, temperature, humidity and posttreatment. As mentioned above, agglomeration, prior to tableting, influences the particle size distribution and can thus improve the tableting properties, which are also influenced by precompression, tableting speed, deaeration and time of exposure to pressure.

On a multistation rotary press, with one station consisting of one die, one upper punch and one lower punch, more than 60 dies are assembled in a rotating circular table. The operation of a rotary press is illustrated in Figure 4.27. In rotary tablet presses, a more homogeneous force distribution is achieved than in essentric presses, as the forces act on both sides. In the rotary press, the pressure increases gradually, thus exposure to pressure is longer, which in turn allows better removal of air from the mixture; otherwise, it leads to expansion of air after the pressure is released, destroying the tablet. Reduced heating of a tablet at the surface is beneficial for heat-sensitive constituents of a tablet.

Compressed tablets can be divided into single-layer (uncoated), layered and coated tablets. Single-layer tablets are used in a variety of applications such as swallowing for systemic use, chewing before swallowing, or quick dissolution for fast therapy (angina), etc. Layered tablets that contain two or more layers are applied when there is a need to separate chemically or physically, incompatible ingredients during processing and storage. Alternatively, such tablets are applied when an initial immediate dose release from one layer is needed, followed by a prolonged-release dose from a second layer. An inert layer is applied in-between. This action dosage is used for the treatment of chronic conditions when repeated dosing is required.

Preparation of such tablets requires, for instance, coating of the second dose in the inner core while the immediate-release dose is located in the outer shell. Coating can be also used for protection from oxygen, CO_2, water, or light, masking bad taste, providing aesthetics or assisting in controlling the drug release rate.

Subsequently, coating can be functional or nonfunctional.

The following coating methods are used: sugar coating, film coating, compression coating and fluid-bed or air suspension coating.

Sugar coating, being a multistage process aimed at masking the taste and increasing the bulk, consists of several coating steps. It starts with sealing the tablet cores with a thin coat of a food-grade water impermeable polymer shellac (purified resinous secretion of the insect *Laccifer lacca*), cellulose acetate phthalate and polyvinyl acetate phthalate, to protect the core from moisture, increasing the shelf life. The following steps include subcoating and smoothing. Subcoating is done by adding calcium carbonate or talc, in combination, with sucrose solution; the latter providing smoothing as the rough layers formed in the previous step are removed. For aesthetic purposes, coloring can then be done with the corresponding pigments. The tablet is thereafter polished with a thin coating of wax to give a characteristic shine, and imprinted with a logo.

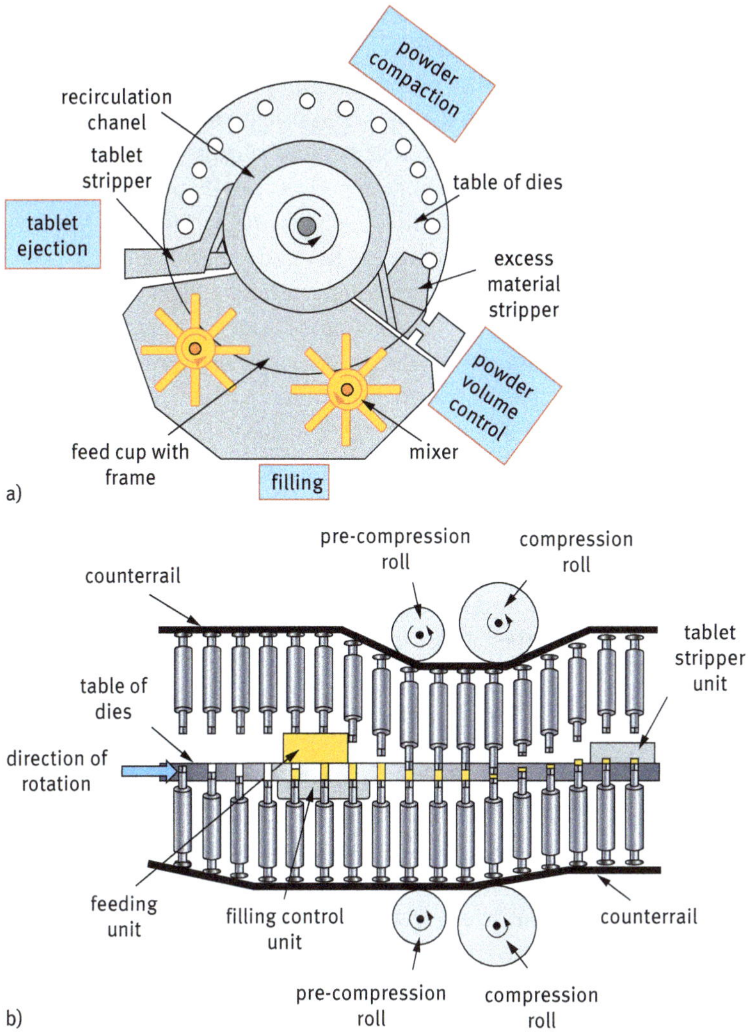

Figure 4.27: Operation of rotary press (**a**) top-view and (**b**) side view [25].

The challenges in sugar coating include long processing time of up to 6 to 7 days and the multistep operation (sealing, subcoating, smoothing, coloring, polishing, etc.), which requires skilled operators as some processes are difficult to perform (e.g., coating) without automation. A disadvantage of sugar coating is that a sugar solution is susceptible to bacterial growth.

Film coating is faster than sugar coating – making a thin layer of a polymeric material in a single stage process. It involves spraying a coating solution containing a polymer (e.g., cellulose acetate phthalate for enteric coatings and sodium carboxymethyl cellulose for water-soluble coatings), a solvent, a plasticizer (e.g., poly(ethylene glycols) and phthalates) and a colorant along with opaquant extenders.

Good solubility in aqueous fluids is required from polymers to ensure efficient dissolution of the active ingredient from the finished dosage form. Alternatively, a polymer system with low water solubility or permeability is chosen for a modified release action. A low viscosity of a polymer solution permits better spraying of industrial film coating. Some polymers improve the shelf life, being efficient barriers against water vapor or other gases. Polymers should have adequate strength to withstand impact and abrasion. Finally, the selection of polymers should comply with the relevant pharmacopoeial requirements and regulations. Plasticizers can be used, in the range from 1 to 50 wt% of the film former. Concentration of fine powdered (below 10 μm) colorants depends on the color shade, dye type and amounts of the opaquant extenders, and varies from less than 0.01% for very light shade to above 2.0% for dark shade. Opaquant extenders are inorganic materials (e.g., titania, magnesia and silicates), while for colorants, compounds such as carotenoids, chlorophyll, indigo, flavones, etc., can be used. Organic solvents used in the past were replaced by water because of health and explosion risks.

Spraying of the coating solutions is done on a rotating tablet bed, followed by drying. Film coating does not increase the tablet weight significantly, in contrast to sugar coating.

Efficient film coating requires atomizing the spray liquid, adequate mixing and agitation of the tablet bed, providing sufficient heat input through drying air and setting up appropriate exhaust facilities to remove dust- and solvent-laden air.

A special case of film coatings is functional coatings, which also performs a pharmaceutical function. In enteric coatings, enteric polymers (cellulose acetate phthalate, polyvinyl acetate phthalate and suitable acrylic derivatives) are used. They are capable of forming a direct film of a sufficient weight. Larger weight (two-three times more than a simple film coating) is needed to get an enteric effect, protecting the tablet core from disintegration in the stomach acid environment. As a result, active ingredients will not disintegrate at low pH, or the stomach is protected if a certain drug causes irritation. For controlled-release coatings, polymers with restricted water solubility or permeability are used. For such type of coatings, film coating is followed by filling into hard gelatin shells, or compressing directly into tablets.

Sugar and film coatings are applied in different types of rotating pans. Conventional coating pans include Pelligrini, immersion sword and immersion tube-type pans. For a standard circular metal pan (Figure 4.28a) that is mounted at 40° and rotated on its horizontal axis by a motor, coating solutions are applied to tablets by ladling or spraying. The drawbacks of the standard pans are related to poor mixing efficiency of tablets and low drying efficiency. The latter can be explained by drying taking place only on the surface of the tablet bed. There are also health hazards for operators and risks of explosions when organic solvents are applied in coating solutions. There have been further developments related to introduction of a baffled pan and a diffuser for uniform distribution of the drying air in Pellegrini pans. Limitations in drying capabilities of this modification were overcome by installing immersion sword or immersion tube air handling systems. In the latter case (Figure 4.28b), hot air is delivered into the tablet bed through a long tube, with a spray nozzle at its tip. The drying air flows upward and is exhausted by a conventional duct. This operation mode results in relatively rapid processing times and can be used not only for sugar coating, but also for film coatings.

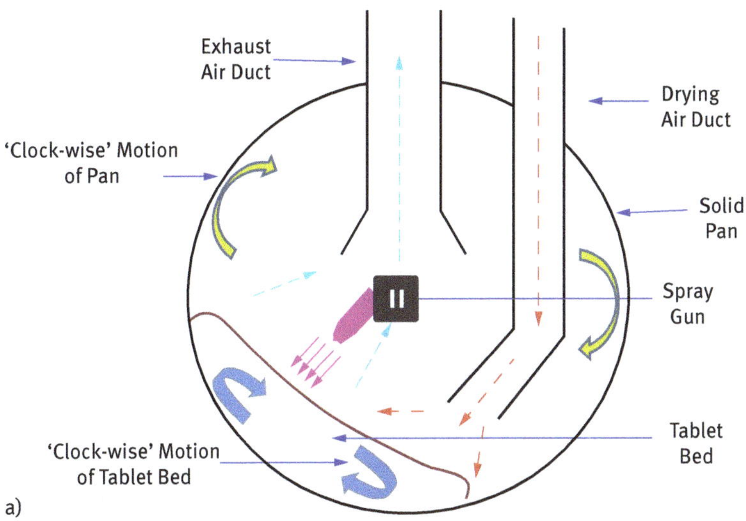

a)

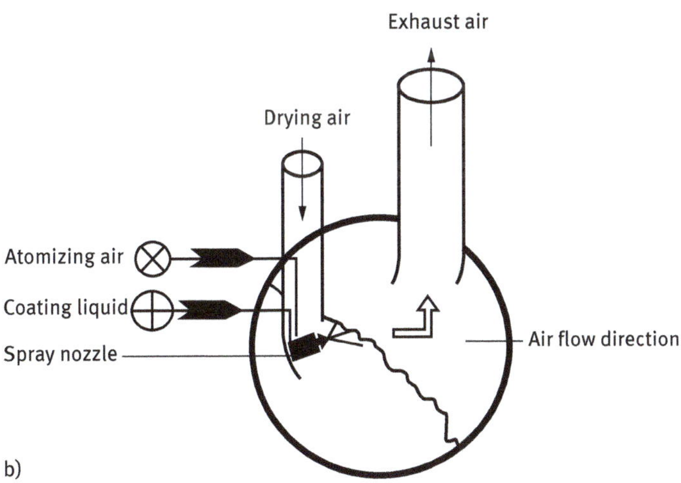

b)

Figure 4.28: Coating pans: (**a**) standard coating pan [26], (**b**) pan with an immersion tube [27], (**c**) Accela cota pan and (**d**) Glatt coater [28].

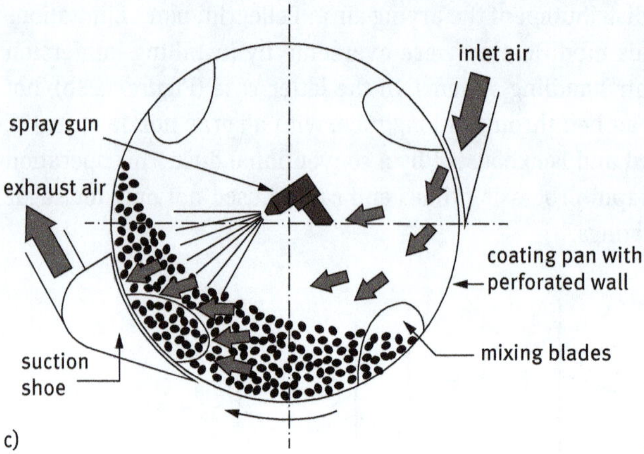

c)

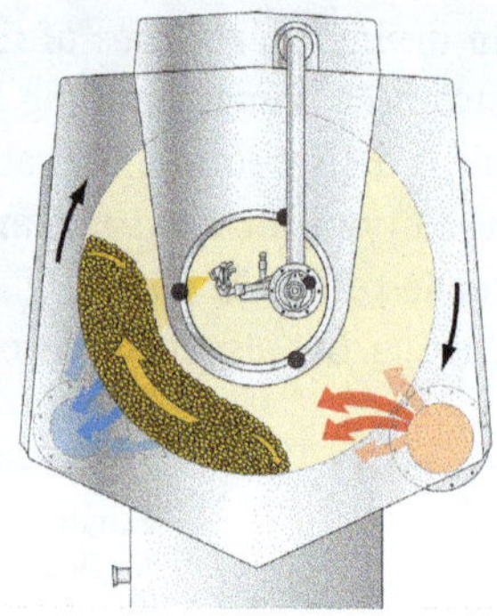

d)

Figure 4.28 (continued)

In perforated coating pans of different types, air is introduced through perforations. Overall, perforated pan coaters allow high coating capacity and efficient drying. These pans are rotated on their horizontal axis. The coating solutions are applied to the surface of the rotating bed of tablets through spraying nozzles. In Accela-cota (Figure 4.28c) and Hi-Coater systems, drying air is directed into the drum, passed through the tablet bed, and exhausted through perforations. In driacoater pans, drying air enters through hollow perforated ribs located on the inside periphery of the drum. These ribs dip into the tablet bed as the coating pan rotates and the drying air passes through. In driacoater pans, the exhaust is from the back of the pan. The Glatt coater (Figure 4.28d), most recently introduced in industrial practice, is similar to that of Accela-cota. Air can be, however, blown beneath the moving tablet bed.

There are several common problems in making coatings (Figure 4.29). Picking and sticking (attachment of tablets to each other) happens when the coating removes a piece of the tablet from the core (Figure 4.29a). It is caused by over-wetting the tablets, underdrying or poor tablet quality. Reducing the liquid and controlling the drying rate as well as changing the formulations can be a remedy. In particular, reducing the amount of the binder or using dry binders, adding high melting point materials and lubricants to prevent softening of low melting point substances from the compression heat, or increasing lubrication by applying fine particles of colloidal silica could be the operation policy to combat picking. If a low-point medication is required in high concentrations, refrigerating the granules and the entire table press can be an option.

Capping and lamination or the partial or complete separation of the top or bottom crowns of a tablet and separation of tablets into two or more distinct layers (Figure 4.29b) happens because of a number of reasons, specifically too large amounts of fines and too dry/very low moisture content. In the former case, fines should be removed by screening, while moisturizing the granules would be the remedy for the latter case. The same behav-

Figure 4.29: Problems in making coated tablets: (**a**) picking and sticking, (**b**) capping and lamination, (**c**) erosion, (**d**) orange peel, (**e**) twinning, (**f**) peeling, (**g**) chipping, (**h**) mottling and (**i**) peeling.

ior is observed when there is insufficient binder or lubricant amounts or even due to the type of these excipients. An obvious solution would be to change the binder/lubricant type and/or their concentration.

Erosion (Figure 4.29c) can be a result of over-wetting the surface by inadequate drying or too soft tablets lacking surface strength. A special coating texture (Figure 4.29d), resembling the surface of an orange (orange peel), is a result of high atomization pressure and too high spray rates. Such an unwanted texture can be prevented by thinning the solution (decreasing the viscosity by adding a solvent) and decreasing the drying rates. Twinning (Figure 4.29e) or sticking of two tablets or capsules together can be avoided by reducing the spray rate or increasing the pan speed. Peeling (Figure 4.29f) occurs when the coating peels away from the tablet surface in a sheet, and can be due to defects in the coating solution, over-wetting or high moisture content in the tablet core. Chipping (Figure 4.29g) or breaking of tablet edges is a result of high pan speed, a friable tablet core or a poor plasticizer in the coating solution. Inadequate color distribution or mottling (Figure 4.29h) happens when improperly prepared coating solutions are used (e.g., a colored drug with colorless excipients or improper mixing with the dye or the colored binder). As remedy, appropriate colorants should be apparently applied, and mixing should be organized in a proper way, diminishing the size of a dye and introducing dry color additives during powder blending. Mottling, originating from a dye migrating to the surface while drying, can be mitigated by changing the solvent system or the binder, decreasing the drying temperature and also using a smaller particle size, as already mentioned above.

Other deficiencies in coatings are peeling (Figure 4.29i) and bridging. The latter occurs as a result of an improper application of the solution when the coating fills in the lettering or the logo on the tablet. Poor logo visibility is related to a high internal stress, which is relieved within the film by detachment of the coating within the region of the logo. The remedy for logo bridging is in improving film adhesion and the tablet punches design, reducing stress within the film by changes in the tablet core or the film coating solution and adjustment of the process conditions.

Coatings can be accomplished in equipment other than rotating pans, such as in pressure coating, involving compaction of the coating material around a preformed core. This more complex method allows the formulation of incompatible materials, as one or more compounds can be placed in the core and others in the coating material. The coating can be a protective or functional, with the initial release part in the coating per se and a sustained release unit in the core. The coating may thus cover a bitter substance, conceal an unpleasant or mottled appearance, serve as a barrier for substances that irritate the stomach and allow an accurate dosing.

Other advantages include better stability of moisture-sensitive drugs, improved patient compliance because of a decreased dosing frequency and a possibility to include flavoring agents in compression-coated tablets. Moreover, multiphasic release is possible when a patient suffers from diseases such as diurnal rhythms that requires variations of drug concentrations during the day. The drug or combination of drugs can be nonuniformly distributed in the coat and the tablet core. By manipulation of the drug loading and the polymer layer in each level of a multilayer system, different drug release patterns can be obtained. For instance, delayed release is obtained when the entire surface of the tablet core is compression coated. The lag time can be controlled by applying different polymeric coats, preventing premature drug release until complete erosion, swelling or rupture of the polymer coating. Degradation of the polymeric compression coating can be done by specific enzymes produced by enterobacteria in the colon, thus allowing microbial controlled release. Microbially degradable polysaccharides, containing glycosidic bonds (e.g., arabinogalactan, arabinoxylan, cellulose and chitosan), can thus be used in coatings.

Alternatively, some enteric polymers as coatings provide site-specific drug delivery, especially for colon and intestine, operating according to the pH-controlled drug release mechanism.

In pressure coating, two tablet presses are used. The first is needed to make the core tablet while the second receives the core and compresses the coating around it.

In air-suspension coating, a cylindrical chamber is used in which the cores are suspended in a stream of air. The bottom-spray fluid-bed method (Figure 4.26) is capable of making a superior film compared with other coating techniques. Spraying of the coating solutions is done either through a spray nozzle at the bottom of the chamber or by nozzles located in the upper region of the chamber. In the latter case, the coating solution is sprayed onto the top of the tablet bed. This coating method can also be used for powders or beads, which, after coating, can be encapsulated or compressed into tablets, providing prolonged-release dosage forms. In tangential coating, centrifugal motion is used for making spherical pallets. The particles are bonded by the binder liquid.

4.2.4 Manufacture of pharmaceutical forms: 3D printing

Recent trends in administration of medication consider the so-called personalized medicine to account for variations in patients' age, weight and overall body conditions, thus adjusting the dose [29–31]. 3D printing has emerged as a procedure that can regulate the dose, mitigating the potential side effects in the standard dosage form. Improper dosing is considered to be responsible for a large share of the adverse conditions resulting from medication.

Moreover, the production technology of pharmaceuticals, involving powder milling, wet granulation, dry granulation, tablet compression, coating, and packaging, requires various excipients for these steps, which can cause serious side effects.

3D printing of pharmaceuticals, being a disruptive approach, has the advantage of avoiding extra operations and the presence of different excipients needed for these operations. Moreover, the drug release rates can be controlled by using complex external and internal structures as well as allowing nonuniform distribution of active pharmaceutical ingredients.

Complex geometry can be printed, for example, cylinder, cube, torus, pyramid and sphere forms, or multifunction tablets with separate compartments, when such compartments regulate the release and absorption of drugs. In addition to such complex geometry and flexibility of drug dosages, 3D printing can offer also on-demand production.

Moreover, 3D-printed patches of chemotherapeutic agents can be implanted directly into pancreatic cancer sites, which are difficult to reach due to low aqueous solubility during intravenous or oral drug administration.

Several methods available for 3D printing, briefly reviewed in Table 4.4, are visualized in Figure 4.30, while examples of complex geometries available through 3D printing are presented in Figure 4.31.

A popular method of 3D printing – fused deposition modeling (FDM) – requires high processing temperature, as the hot-melt extruded filament, with a loaded drug, is used as primer for printing. The method is not suitable for certain temperature-

Table 4.4: Comparison of representative 3D printing technologies in pharmaceutical applications [31]. Reproduced with permission.

Category		Technique	Concept/rationale	Advantage	Limitations
Extrusion-based systems	Material extrusion	Fused deposition modeling (FDM)	Drug-loaded filaments are heated to a critical state, made to a semifluid state, and then extruded from the printing nozzle, according to the model parameters.	– High equipment diversity (multiple nozzles) – Low equipment price – Good mechanical properties of printed dosage forms	– Difficult to scale up – Low drug loading
		Semi-solid extrusion (SSE)	SSE extrudes the paste evenly via a syringe-based print head under pressure or screw gear rotation and deposits layer by layer on the platform for printing	– High drug loading – Mild printing process/condition – Broad range of excipients	– Postprocessing – Low resolution – Low efficiency
		Melt extrusion deposition	MED® converts powder feedstocks into softened/molten states, followed by precise layer-by-layer deposition to produce objects with the desired structures	– No postprocessing – Easy to scale up – High equipment diversity (multiple nozzles	– Low drug loading
Powder-based systems	Binder jetting (BJ)	Ink-jet 3D printing	BJ assembles 3D objects by first preparing a 2D powder-based layer and then ejects a binder solution to pattern and solidify specific regions in the powder bed.	– Easy to scale up – High volume – High drug loading	– Postprocessing – Inefficient powder usage

	Powder bed fusion (PBF)	Selected laser sintering (SLS)	PBF uses a focused power source (e.g., laser or electron beam) to selectively consolidate powder particles into solid objects	– High resolution – No need for support materials	– Risk of drug degradation – Postprocessing
Liquid-based systems	Vat photo-polymerization (VP)	Stereolithography (SLA), digital light processing (DLP)	VP is based on the selective photopolymerization of liquid photosensitive resins using ultraviolet laser source	– High resolution and accuracy – Ability to print micro-sized structure	– Postprocessing – Potential material toxicity – Limited photosensitive resin
	Material jetting (MJ)	Continuous or drop on demand (DOD)	Material droplets are deposited through a print head building and dried by solvent evaporation or by solidification under ultraviolet light and the object is built layer by layer	– High resolution – High-quality surface of printed tablets	– Postprocessing – Limited excipients
Sheet lamination-based systems	Sheet lamination (SL)	Screen Printing Innovational Drug Technology (SPID®)	3D screen printing is based on the transfer of the printing paste through distinct openings of the printing screen onto a given substrate.	– Easy to scale up – High printing speed	– Postprocessing

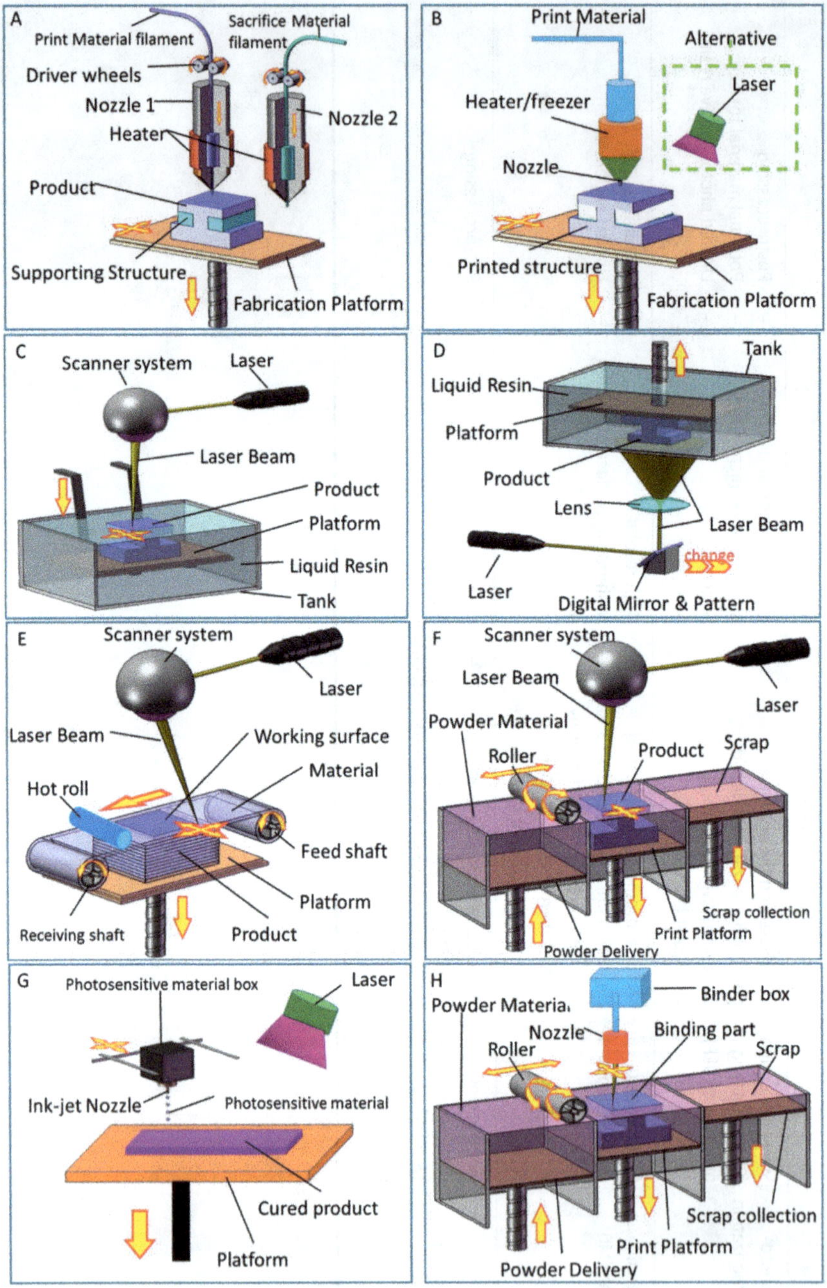

Figure 4.30: 3D printing technologies: (A) fused deposition modeling (FDM); (B) directly ink writing (DIW); (C) stereolithography (SLA); (D) digital light procession (DLP); (E) lamination (LOM); (F) selective laser sintering (SLS) and selective laser melting (SLM); (G) photopolymer jetting (Ployjet) and (H) binder jetting (3DP) (reproduced from [32]).

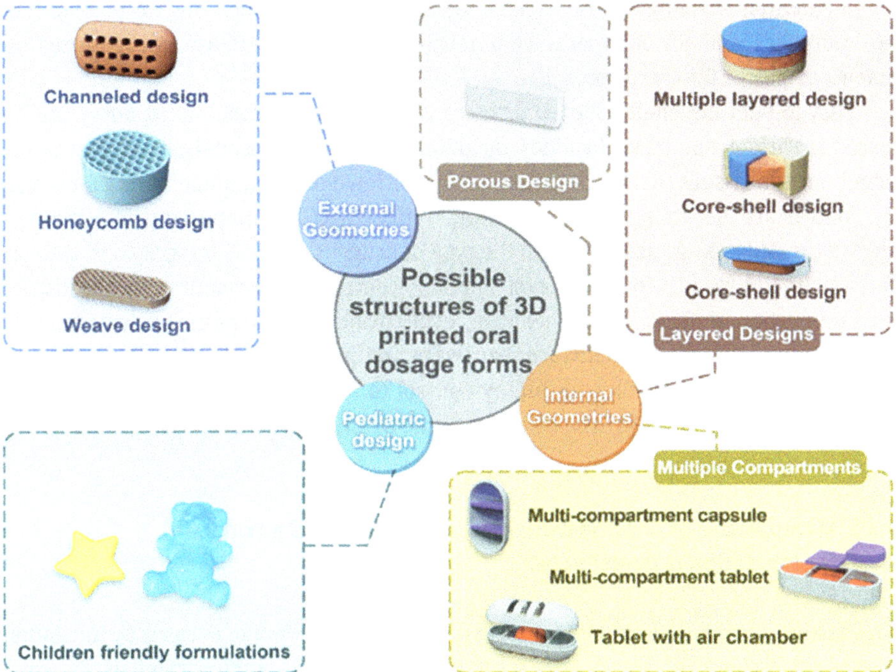

Figure 4.31: Possible structures of 3D-printed oral dosage forms [33]. Reproduced with permission.

sensitive drugs and biologics. For a drug-polymer mixture, thermoplastic polymers like polyvinyl alcohol or polylactic acid can be used as a polymer additive. Application of other materials, such as acrylonitrile butadiene styrene resins, is questionable if they are nonbiodegradable. The advantages of FDM are use in complex geometries, allows high loading, and offers a better resolution as well as good mechanical strength when compared to other methods.

Extrusion deposition methods have limited daily production and are thus mainly viewed as a tool for clinical trials or as personalized medicine. Application of a nozzle array design can overcome such limitations in the future.

Other popular methods in 3D printing include inkjet printing and selective laser sintering (Figure 4.30).

In stereolithography (SLA) or vat photopolymerization, the resin (predominantly, toxic and nonbiodegradable) is hardened by the exposure to ultraviolet light. The limitations of the method are related to availability of just a few biocompatible and UV-curable materials (e.g., polytheyleneglycol diacrylate). In SLA, the drug is dispersed or dissolved inside the resin before printing, making the layer one over the other, which is done by the moving the laser.

A potential challenge in the widespread application of 3D printing for pharmaceutical purposes is a lack of experience in formulations of, for example, drug-loaded filaments or pastes for extrusion.

Easy-to-operate small-scale 3D printers with fine-tuned customized doses can be placed at various non-cGMP facilities, including pharmacies, hospitals, etc. and even patient homes, as issues such as production costs for low-volume manufacturing and quality control are becoming very important. While in cGMP manufacturing, there is always a quality assurance related to the type of drug, its physical integrity and the correct amount, obviously, the same level of quality control will be required for the distributed production of personalized medicines by 3D printing at the patient private homes.

Remote monitoring of software, hardware and printing capacity will be required not only from the viewpoint of quality control to avoid discrepancy in the product quality, but also to avoid misuse or abuse of the printing process and the products.

4.2.5 Manufacture of pharmaceutical forms: amorphous powders for inhalation drug delivery

An effective noninvasive method for local and systemic delivery of drugs is by inhalation of aerosols. A couple of dozens drug substances are currently administrated in this way.

First, it is instructive to discuss the human respiratory tract, which is divided into the upper and lower parts. The upper tract consists of the nose, nasal cavity and pharynx (Figure 4.32a) while the lower part (Figure 4.32b) is constituted of the larynx, tra-

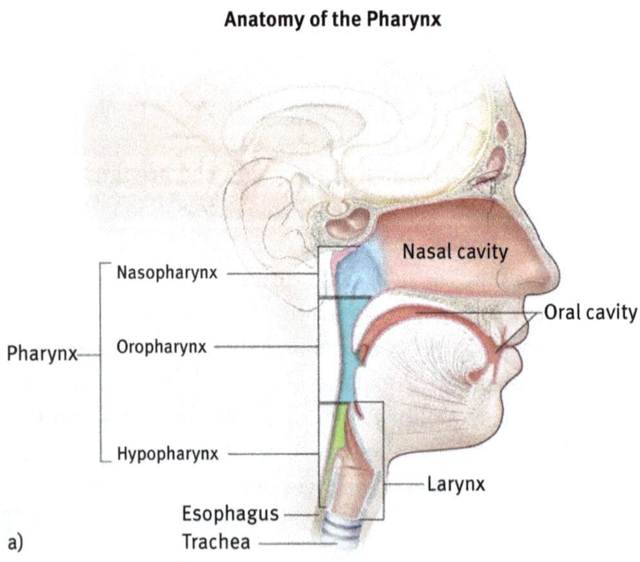

Figure 4.32: Human respiratory tract: (**a**) nasal [34] and (**b**) pulmonary systems [35].

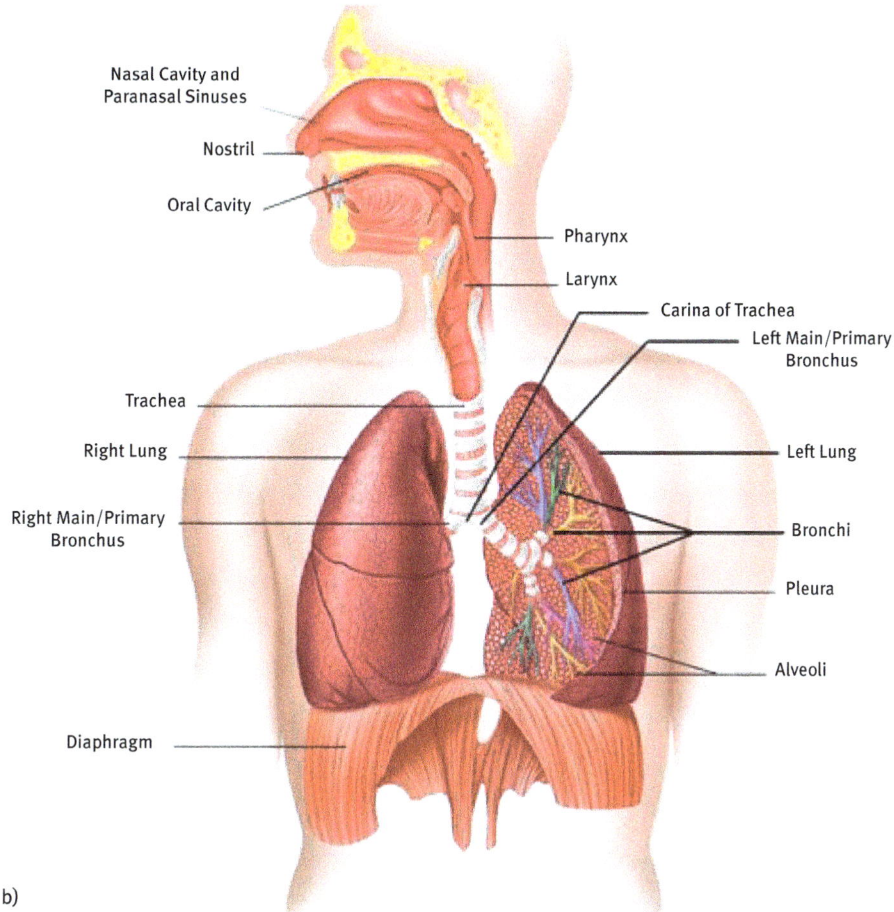

b)

Figure 4.32 (continued)

chea, bronchi and the lungs. The lungs are responsible for gas exchange (e.g., addition of oxygen and removal of CO_2). Alveoli are the terminal parts of the bronchi and are the functional units of the lungs. One of the main reasons for using inhalation as a means of drug delivery is that some drugs can be readily absorbed by alveoli directly; they can enter the systemic circulation because of a very large absorptive surface area (140 m^2) and a highly permeable membrane (0.2–0.7 μm thickness) in the alveolar region.

Other advantages of pulmonary delivery, which can be done for a wide range of substances of different sizes, are related to a possibility of delivering high drug concentrations directly to the disease site, minimizing the risk of systemic side effect and facilitating a rapid clinical response. This method of administration avoids first-pass metabolism in the liver and thus allows a smaller dose than by conventional oral delivery. Moreover, this noninvasive, needle-free delivery and easy-to-administrate method reduces side effects, avoiding gastrointestinal tract problems (e.g., poor solubility, low bioavailability, irritation of the gut and undesired metabolism).

There could be also some disadvantages of pulmonary drug delivery – improper dosing, reproducibility of a dose, lower dose than required for some drugs to be efficient, problems with stability and difficulties to obtain the optimum particle size.

The major challenges in pulmonary drug delivery systems are related to targeting the drug of an optimum particle size to specific receptors in lungs and then retaining them. Too small a size will result in exhaling, while if the particles are too large, they can affect the oropharynx and larynx. The optimum particle size required for pulmonary delivery is 1–5 µm. Besides the size and shape of the particles, other pharmaceutical factors affecting aerosol deposition are aerosol velocity, density and physical stability.

Deposition of particles in lungs (Figure 4.33) is by impaction (happens for large particles due to changes in the air flow), sedimentation (settling of large particles due to low air flow), interception (in small airways) and electrostatic deposition or diffusion (for particles less than 0.5 µm).

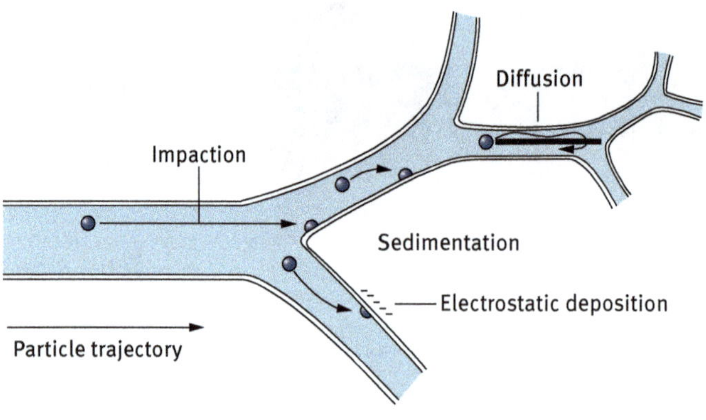

Figure 4.33: Mechanisms of particle deposition [36].

Spray drying, spray-freeze-drying and supercritical antisolvent precipitation are the methods used to achieve the desired particle size. High-energy milling, by introducing mechanical stress, makes crystalline materials partially amorphous and does not allow fabrication of complex engineered structures, e.g., particles that are porous and hollow, coated or encapsulated.

The fundamental aspects of spray drying are discussed in Chapter 2. It is a method that gives a uniform particle size, having a disadvantage of not being applicable for very temperature-sensitive drugs. For the latter case, spray freeze-drying can be used, which is a combination of spray and freeze-drying process.

Pulmonary bioavailability and stability of drugs depend on their physical properties, and could be improved by including various additives, such as buffers, to maintain pH, stabilizers and surfactants to prevent denaturation during storage and diminish chances of aggregation.

For pulmonary route, not only the formulation aspects but also drug delivery devices are very important. Such devices include metered dose inhalers, dry powder inhalers and nebulizers (of jet and ultrasonic types).

In metered dose inhalers (Figure 4.34) only a minor part of the dose (10–20%) is deposited in the lung. Another apparent disadvantage of these devices is a need to educate patients for their proper operation. Fine droplets, with a particle size of less than 5 µm, are generated in metered dose inhalers from a formulation that fills the device and is released by an actuator. The formulations can be either suspensions or solutions. Therefore, such devices contain besides the drug per se, a propellant (e.g., hydrofluoralkanes), a surfactant and a co-solvent or an excipient in the formulation. Several key formulation issues should be considered for this type of inhalers, such as stabilization of the suspension by formation of loose aggregates and their readily redispersion by shaking; absence of particle growth; prevention of drug deposition of the can walls; and ensuring long term stability by protection from moisture.

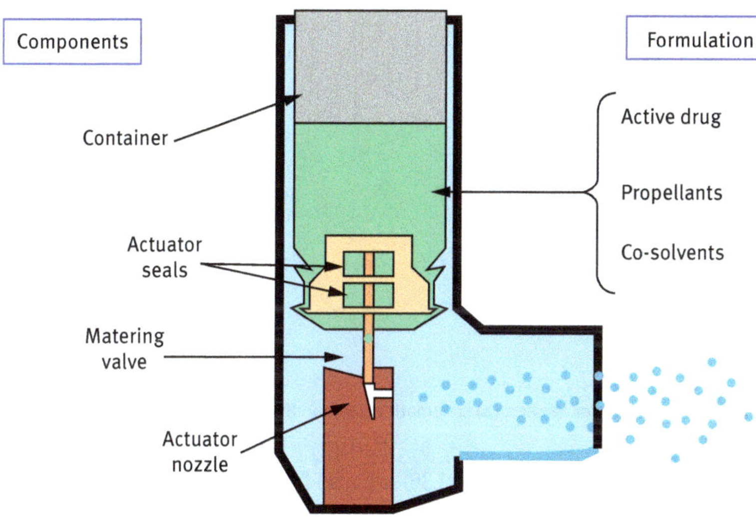

Figure 4.34: Metered dose inhalers [37].

Dry powder inhalers do not have any environmentally harmful propellants, in contrast to metered dose inhalers. They contain either the drug alone or alternatively, drugs mixed with suitable carriers (e.g., lactose or glucose) to increase the flow properties of the drug. The drug and lactose carrier sizes are ca. 1–2 µm and 20–65 µm, respectively. Such a type of inhalers has a higher storage stability for crystalline drugs. One of the challenges in the application of dry powder inhalers is related to balancing the stability and aerosol performance of amorphous powders. They have higher water solubility and higher dissolution rate than their crystalline counterparts, while being less stable at the same time.

Aerosol deposition in the lungs and the efficiency of the drug depends not only on the design on the inhalation device but also on the creation of the turbulent flow in the inhaler. The aerodynamic performance thus relies on the ability of a patient to inhale the powder at a sufficiently high inspiratory flow rate.

Devices for dry inhalation in general can be designed for single or multi doses. In single-dose devices, a capsule containing the powder is placed in a holder and is opened within the device, followed by inhalation of the powder. In the case of multi-dose devices of different designs, the drug located within a storage reservoir is dispensed into the dosing chamber by simple back and forth twisting actions. One example of dry powder inhalers is shown in Figure 4.35a, while the operation schematics is presented in Figure 4.35b.

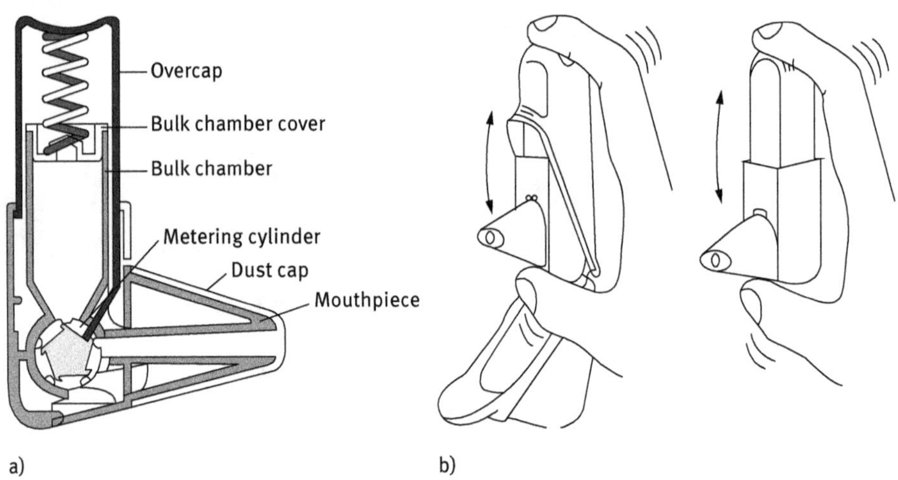

Figure 4.35: One type of a dry powder inhaler: (**a**) schematics [38] and (**b**) mode of operation [39].

Finally, jet and ultrasonic nebulizers (Figure 4.36) are also frequently used for inhalation.

Nebulizers allow inhalation of medical substance during the so-called quiet breathing of a patient; it does not require special effort from the most severe patients who are not able to use other types of inhalers, as well as from children and elderly patients. Nebulizers can be used for delivery into the respiratory tract, a variety of preparations and, if necessary, high doses of drugs.

Depending on the type of energy that turns a liquid into an aerosol, there are two main types of nebulizers: jet – using a gas jet (air or oxygen) and ultrasonic – using vibration energy of the piezocrystal. The latter is considered to be inefficient for delivery of drug suspensions. Another disadvantage of nebulizers is a temperature in-

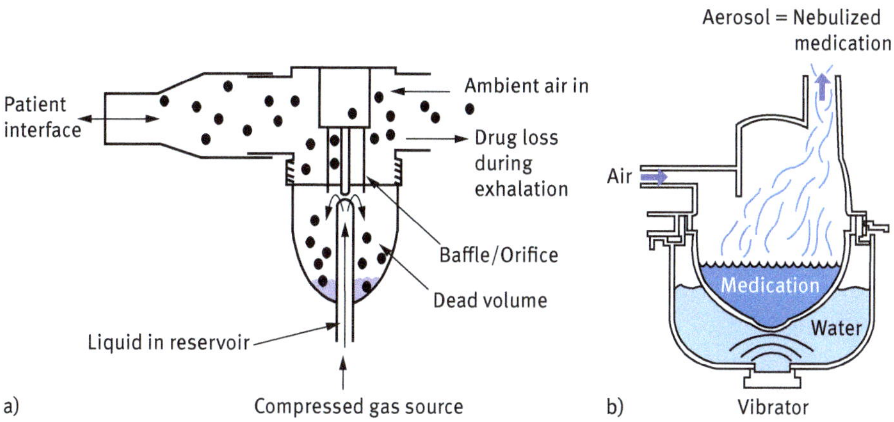

Figure 4.36: Nebulizers: (**a**) jet [40] and (**b**) ultrasonic [41].

crease during inhalation, which can damage the active pharmaceutical ingredient. Moreover, a substantial amount of drug is wasted.

In jet-type nebulizers, compressed gas with a flow of > 6 L/min is used to spray the liquid as fine droplets. Large droplets are retained because of the baffles in the chamber. The main operating principle in jet nebulizers is the Venturi effect, namely that the velocity of an incompressible liquid increases as the diameter in which the fluid passes through, increases proportionately. When the gas passes a narrow Venturi tube, a low-pressure zone is formed, which pulls droplets up through a feed tube from a solution or suspension of a drug into the nebulizer body. A primary aerosol is formed, with the size of 10–500 μm, which is then reduced to the size of 0.5–10 μm, which is capable of passing the orifice and entering the respiratory tract of a patient. The major part of the aerosol is returned back to the internal chamber of the nebulizer.

The droplets obtained in this way are very small in size; therefore, they cannot penetrate deep into the respiratory system, but they are quite capable of providing an effective therapeutic effect on the lower and upper respiratory tracts. Most models of the jet-type nebulizers have an adjustable spray system, which allows choosing the exact direction of the drug flow. Depending on the set particle size, the jet-type nebulizers can affect different parts of the respiratory system. Treatment of diseases of the upper and middle respiratory tract, such as tracheitis or bronchitis, can be efficiently done using particles with a size of 6 μm.

A simple design of jet nebulizers that ensures its reliability and a long service life, has disadvantages such as high levels of noise, disturbing children and older patients as well as requiring a particular posture (sitting) that is not always suitable for children or patients with physical disabilities. Moreover, only 10% of the dose is absorbed in the lungs using the jet-type nebulizers.

An ultrasonic type (Figure 4.36b), more expensive and relying on assistance of healthcare personnel, is based on the application of high-frequency vibrations of a piezoelectric crystal to generate aerosol droplets. The main disadvantage of the ultrasonic inhaler is the inability to adjust the size of droplets and the limited therapeutic formulations with which it can work. Most substances with a complex chemical structure (e.g., essential oils, hormones or antibiotics) decompose or lose their medicinal properties under the action of ultrasound.

4.3 Personal care products

4.3.1 General

For personal care, a broad range of products is used, including skin care products, toothpastes, shaving and bath preparations, various shampoos, hair conditioners and hair-styling aids in the form of foams, suspensions, emulsions or their combinations.

Cosmetic products used externally contain an active substrate (e.g., vitamins or plant extracts) and the base of a cosmetic preparation. In fact, the base should be carefully selected, as the components of the base (e.g., polyols, lipids, emulsifiers, thickeners and preservatives) can also have an impact on the skin. The ingredients of complex formulation products for skin care, with ca. 25–40 ingredients, are thus chosen by first considering their medical safety.

Cosmeceuticals, i.e., cosmetic products with bioactive ingredients possessing some medical benefits, should not contain perfumes or dyes, to minimize potential allergic reactions.

Elements of product design and product engineering for cosmetic products at a holistic level include a range of features as illustrated in Figure 4.37.

As can be seen from this figure, development of skin cosmetic agents used in the care, cleansing and decoration of human skin requires not only appropriate performance and application convenience, but also certain aesthetics including color, form or smell. These, in addition to other requirements (availability, costs, compatibility with environmental regulations and with other ingredients in the formulation, patent status, skin feel and compatibility, etc.), determine the selection of the raw materials.

An often-used marketing tool is to introduce the label “natural” or “herbal,” by sometimes adding herbal extracts to address the consumer perceptions of better safety and even efficacy. Sometimes, the label “natural” is applied to, for example, surfactants, which implies that the surfactant taken from a natural source is modified through some chemical transformations.

Skin care products can be divided into three functional groups: drugs to prevent diseases, cosmetics to improve the feeling aspects of normal skin and finally cosmeceuticals, i.e., cosmetic products that can enhance the function of the skin.

Rheology of cosmetic creams and lotions is important for the performance and consumer acceptance. Cosmetic products for skin moisturizing, cleansing and protecting require rapid deposition of continuous, protective oleic films of a certain thickness.

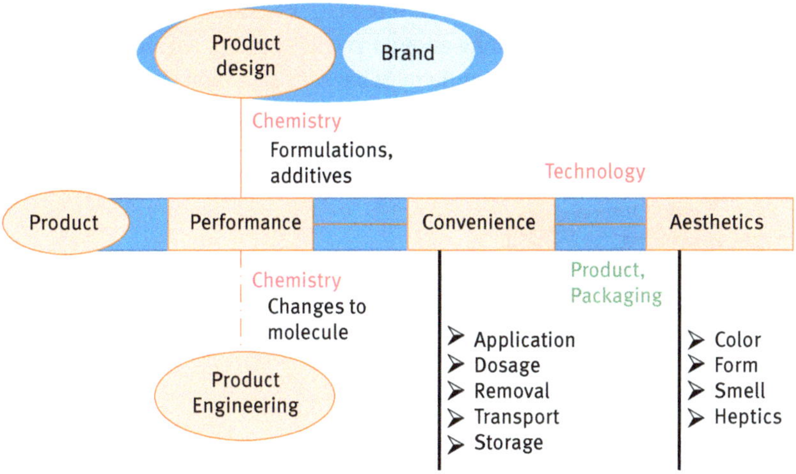

Figure 4.37: Parameters in product design and engineering of skin care products (reproduced with permission from [42]).

Cosmetic creams and lotions are typically non-Newtonian fluids, possessing shear thinning behavior. Creams should be thixotropic with low viscosity during application under shear and high viscosity after application at rest. Lotions are less viscous than creams; they still require some viscosity to enhance their suspension stability. Lotion bases typically include suspensions of cetostearyl alcohol or stearic acid particles. Skin care products are formulated to have a yield stress; therefore, stability against aggregation can be rather easily achieved, while it might be less straightforward to ensure stability against coalescence.

Convenience of a product includes its stability from the time of manufacturing to the time of opening the package as well as stability after opening. Compounds prone to oxidation, such as vitamins and natural oils, require either the presence of antioxidants or chemical modifications of such compounds by introducing protecting groups. To ensure microbiological stability, water should be of the required quality, which is achieved by sterilization or introduction of preservatives. Heavy metals can be removed by ion exchange.

Products for skin care are typically needed for moisturizing and blocking UV radiation, providing smoothness, pleasant feel and certain appearance. They should be stable chemically and microbiologically. Moreover, the skin, whose structure is shown in Figure 4.38, should not be irritated. Hair, also illustrated in the same figure, is a protein filament, growing from the follicles in the skin.

Lubrication of the hair is done by the secretions of the sebaceous glands situated beside the hair follicle. If a protective hydrolipid film on the skin is removed (degreased) by a solvent such as an alcohol, water evaporates from the swollen cornified layer, leading to the skin drying out. Such hydrolipid cover is also removed along with dirt when washing with a detergent, even if the latter is adsorbed on the surface. Application of soaps is beneficial for long-enough swelling of the cornified layer, allowing regeneration of the hydrolipid mantle by sebaceous secretion.

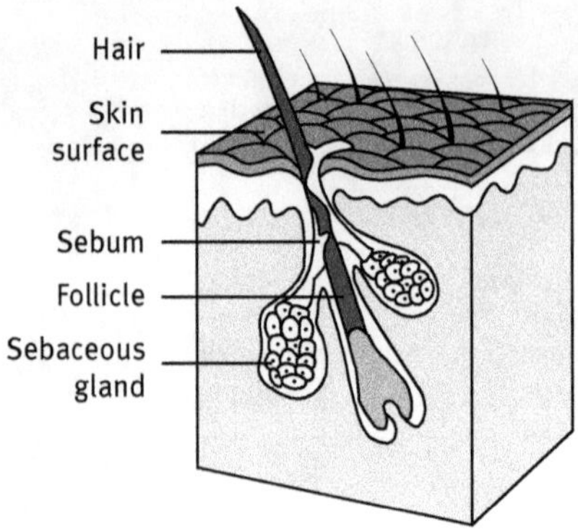

Figure 4.38: General structure of the skin [43].

In cleansing agents, the emulsifier is present in a large excess; nevertheless, cleaning is not effective, compared to lipid emulsions, when a lipid layer is formed, preventing water evaporation. Even better are those W/O emulsions that distribute not only lipids but also water and emulsifier onto the skin.

With time, the human body changes, and legs, elbows, and forearms become longer. Aging also results in wrinkles and age spots because of UV radiation. This, along with overuse of soaps, perfumes, hot bathes and deodorants that cause low humidity, require the application of moisturizers to reduce such water loss. Moisturizers or emollients make the external layers of the skin (skin surface or epidermis) softer, increasing the water content in the skin by diminishing evaporation. Other types of antiaging products include ointments (mixtures of W/O), creams (mixtures of O/W) and less viscous lotions (O/W emulsions). Age spots mentioned above can look like cancerous growths, though they are harmless. At the same time, skin cancer is one of the most common cancer, accounting globally for 40% of cases and being caused mainly by UV rays.

Skin types are categorized into normal, dry and oily skin. The normal skin has an equal balance between water and oil. With such a balance and medium-sized pores, it is well hydrated, thus being smooth and firm to touch. Dry skin lacks hydration, has small pores and is flaky and rough to touch. Overproduction of sebum (an oily or waxy matter) secreted by sebaceous glands (Figure 4.38) to lubricate the skin and hair results in oily skin. Such skin has the appearance of an orange peel and possesses large hair follicle.

A number of factors influence the production of sebum; they comprise triglycerides, free fatty acids, wax esters and cholesterol. In addition, genetic and hormonal factors such as diet, stress, age, weight and pregnancy can be mentioned. Maximum secretion of sebum is at ca. 20 years, decreasing by 28% after each decade.

Skin is the body organ that reflects the physical wellbeing, showing the first signs of stress. As a result, a substantial amount of money is spent on skin care products, which are needed to protect the skin from wind and cold, insects and pollution, and UV and IR radiation. In addition, skin care products help to repair dry skin, as well as fight against sunburns, acne, puffy eyes, wrinkles, age sports, pigmentation and even against cellulite and skin cancer. In fact, skin care products is the largest sector of cosmetics market (ca. 27%), outperforming hair care and makeup (20% each).

4.3.2 Skin care products

4.3.2.1 Assortment

A brief overview is presented here, while more details about particular products are described in subsequent sections.

Skin care products include, among others, various moisturizers, hand and body lotions, after-sun lotions, sunscreen emulsions, body oils, makeup products (e.g., eyeliners and eyeshadow, and lipsticks), antiperspirants and deodorants, bath oils and foams, and hair care and dental care products. A special type is skincare products for babies, which is related to more delicate skin properties.

As apparently clear from the above, not only creams, lotions and emulsions, but also other formulations, such as aerosols and sticks, are an essential part of the skin care repertoire.

Stick formulations comprise, for example, lipsticks, eyebrow pencils, scent and deodorant sticks. They have as base, a wax or grease mass for makeup or a shape-stable gel for deodorants, with a melting/softening temperature nonmarginally exceeding the body temperature (by ca. 15 °C). Sodium stearate can be used as a base for deodorant sticks, forming a solid gel when mixed with an alcohol. Addition of essential oils and menthol leads to a cooling and refreshing feeling of the stick formulation.

Mixtures of liquid oils or suspensions of solid oils in a liquid one, with a variety of waxes (beeswax, carnauba wax, etc.), are used for lipsticks and lip balms. A shiny film, when the lipstick is dried after application, is related to utilization of solid oils (e.g., lanolin and palm oil). The composition of various lipsticks, depending on the duty and preferences of the customers, will be discussed below. Here, it is worth mentioning that apart from liquid oil required for the easy application, other ingredients are added to give the color, shade, thermal stability during storage, moisturizing properties and protection from UV as well a rheological behavior of a viscoelastic solid. The latter implies that when the stress is low, there should be small deformations. With its recovery, the stress is removed. To ensure that lipsticks are not easily removed, hydrophilic solvents (glycols or tetrahydrofurfuryl alcohol) are added to the lipstick formulation.

Hair care products, in addition to conventional and special (antidandruff) shampoos and conditioners, also comprise hair-styling products (creams, pomades, hairsprays, dyes, decolorizers, etc.).

For a shampoo, besides its primary goal of mild detergency, other requirements are important, addressing either the habits of customers (foaming during application and conditioning) or their perceptions and concerns (aesthetics, hair preservation). The detergency is provided by the surfactants, some of them are also responsible for foaming properties. Anionic surfactants are applied in combination with amphoteric or nonionic surfactants to reduce the eye and skin irritation. Such an approach, giving a non-ideal mixing, moreover decreases the critical micelle concentration. In addition to surfactants, thickeners and emulsifiers are also applied along with a range of many other additives, as will be discussed in a section on hair care products.

Nail care products include nail polish, which in fact is not used for polishing, despite the name, varnishes and their removers, hardeners, sprays and creams. The nail polish, to be discussed below, in essence, represents mainly pigment suspensions in a volatile nonaqueous solvent, even if there are also recent attempts to use aqueous formulations. From a rheological viewpoint, nail polish should be thixotropic, exhibiting a proper flow for an even coating but at the same time, being not too thin, as otherwise the nail polish will start to drip, falling in drops. Gelling should proceed with favorable kinetics. Too fast gelation results in an uneven coating (so-called brush marks), while dripping occurs when even gelation is too slow. A control of the relaxation time requires application of surfactants.

Shaving aids, per se soaps, creams and foams as well as products for preshave and aftershave treatment have specific properties. A high potassium soap content in shaving soaps aids in foaming, which also requires the utilization of a foam stabilizer. Humectants, drawing water into skin, are also added to shaving aids. Dry skin requires nonfoaming shaving creams to be applied without a shaving brush.

Degreasing the skin for shaving is achieved by using preshaving aids with a high alcohol content. Spreading of the oil layer is facilitated by adding lubricants to such formulations. The role of the aftershave products is to neutralize the alkaline environment created on the epidermis by shaving. Small injuries made by less experienced users and sometimes even by experienced ones require their (injuries, not users) healing and thus the presence of the corresponding active substances in the formulations of aftershave lotions.

The role of antiperspirants, common in the USA, is to inhibit not only odor but also diminish sweating. More common to Europe, deodorants have the main function matching their name. An unpleasant odor develops when the skin is decomposed by the bacteria present on the skin. The strategy to remove odor ranges from simple masking with perfume oils to more complex approaches. The latter include biological (inhibiting bacterial flora) or chemical ones, for example, adsorption of the compounds that exhibit a bad smell, by finely dispersed ion exchange resins or oxidation of such compounds with strong oxidants (peroxides). Precipitation of proteins by the corresponding compounds is the basis of the action of antiperspirants. The detailed composition of antiperspirants/deodorants will be discussed separately. Here, it can be just mentioned that they contain 60–80% water as well as 5–15% lipid (e.g., stearic acid or beeswax), in addition to many other ingredients to provide the antiperspirant effect, viscosity, odor or good skin feel, to name a few. From the rheological viewpoint, there should be particle sedimentation and good spreadability.

Formulations of skin care products take into account the skin thickness (feet, hands and face), conditions (dry or normal) and the intended use (e.g., sun protection and antiaging).

Some skin care products are formulated to address changes of the skin's performance with age. Younger skin allows light to partially pass through it. Such translucent behavior results in a shine or glow. With age, the layer between the epidermis and dermis becomes smaller, resulting, along with some other reasons, in a dull and uneven appearance.

The way light is reflected back from skin can be altered by formulating the skin care products with interference pigments (0.1–2.0 wt%), such as, for example, mica (sheet silicates, containing Al, Si and Fe, in addition to alkali and alkali earth metal) coated with thin films of iron oxide or titanium dioxide. Visual appearance of age spots can be diminished by using an appropriate particle size, which, along with some color combinations, gives the skin a radiant glow.

In the example above, introduction of pigments does not bring healing to an aging skin. Generally speaking, depending on the recipe, cosmetics can also cure some slight dermatological diseases. These types of products are coined cosmeceuticals, as mentioned in the previous section. They are mainly W/O emulsions and have a healing potential because physiologically active substances (natural oils, vitamins/provitamins, proteins and peptides) are available in their formulations in sufficient amounts to provide a physiological effect. Obviously, the auxiliaries should be reduced to a minimum level and utilization of perfumes or dyes should be avoided to minimize the risks of allergic reactions and other undesired effects, as already mentioned.

Biologically active substances are selected from those chemicals that do not require any legally binding regulations, because cosmeceuticals are, in fact, cosmetic products and not medicines. In addition to various substances extracted from natural plants/trees (e.g., polyphenolic antioxidants) and possessing a range of desired properties (healing, antiseptic, anti-inflammatory, antioxidant and UV-protective), different types of vitamins are used, including vitamins A (antiaging), B3 (circulation stimulator), B5 (healing ingredient), B6 (regulator of sebum production) and C and E (antioxidants). Proteins and peptides can provide, for example, improvements in skin hydration and elasticity. Vitamins, fragrances, proteins or botanical extracts can be encapsulated into such delivery vehicles as neat (chitosan, agar, alginate and lecithin) or modified biopolymers (e.g., carboxymethylcellulose) and used in many different products, such as, for example, shower and peeling gels, antiaging and anti-wrinkle creams or anticellulite formulations. Facial masks (concentrated suspension of, e.g., clays) can also contain besides other substances (essential oils, vitamins, scents, etc.), cosmeceutical additives such as salicylic acid to treat mild acne.

4.3.2.2 Emulsions

Different types of cosmetic emulsions are applied, namely, W/O, O/W, multiple (e.g., W/O/W), nano- and microemulsions.

For a greasy feel, the emulsions have oil as an external phase. The same type of emulsions is used in emollient and cleansing creams, allowing a longer lasting residue. On the contrary, water as an external phase is utilized in hand cream, when a greasy feeling is not required.

While W/O emulsions correspond to the skin's physiological conditions, O/W emulsions exhibit initial cooling because of water evaporation, thus being more popular.

Such O/W emulsions allow 1–10 μm of large oil drops to quickly penetrate the skin without leaving an oily film on the skin's surface. After partial evaporation of the continuous phase (water), the substances left on the skin smooth it and provide the required protection. O/W emulsions are applied in common moisturizers, hand and body lotions or, in other words, in cases when the product should not be too greasy and the oil volume fraction does not need to be too high.

Another important consideration is the release rate discussed above in the context of pharmaceutical products. In the case when physiologically active therapeutic ingredients are present in a cream, the release can be prolonged using W/O emulsions.

Formulating creams such as emulsions allows easy spreading and adsorption and ensure that both water and oil-soluble components are present in a single product.

The aqueous phase of O/W emulsions (ca. 60–80 vol.%) contain ingredients such as humectants (hydroscopic substances) to retain moisture and prevent water loss (e.g., glycerin, sorbitol and propylene glycol), co-solvents for solubilizing fragrances or preservatives; chelating agents (e.g., EDTA), water soluble surfactants for stabilization and germicidal activity, viscosity builders (e.g., gums, cellulose and clay minerals), proteins, water-soluble vitamins and minerals. In the oleic phase (ca. 20–40 vol%), the following components are present: oils and waxes (e.g., silicone or mineral oil, paraffin or lipids), dyes and perfumes, and oil-soluble surfactants.

In general, nonionic uncharged surfactants of low skin sensitization potential are preferred. Phosphoric acid esters, being similar to the phospholipids that constitute the natural building blocks of the stratum corneum, are also applied in some cosmetic formulations. Other biomimetic surfactants are triglycerides, as they are important ingredients of the sebum, which is the skin's natural lubricant. Macromolecular surfactants (e.g., ABA block copolymers and polyglycosides) and carbohydrate-based surfactants (sorbitan and sucrose esters) are also applied. Even emulsifier-free emulsions with cross-linked acrylic polymer derivatives can be formulated, minimizing the overall irritation potential.

The formulations should be non-irritant, easy to spread over the skin and rub in without "soaping up". Moreover, there should be a soft skin feeling and a pleasant smell. The basic components of moisturizing products (e.g., day and night preparations, hand and body lotions, and barrier creams) are moisturizers, emulsifiers, water, preservatives and antioxidants.

Day moisturizing products are light, O/W emulsions, based on stearic acid. The oil phase is partially neutralized (16–20%) with bases such as triethanolamine or potassium hydroxide. The alkali-stearate soaps can result in skin dryness, which is prevented by the introduction of secondary emulsifiers (e.g., nonionic surfactants: glyceryl monostearate, polyethylene glycol esters and fatty alcohols or fatty alcohol ethers).

A typical composition of the O/W night cream is presented in Table 4.5. Such compositions have the advantage of less pores blocking and not that shiny appearance, compared to W/O emulsions. Besides emulsifiers, emollients are also introduced in the composition for a smooth lubricating effect. Liposomes, small artificial spherical vesicles with at least one lipid bilayer, composed most often of phospholipids (e.g., phosphatidylcholine) and cholesterol, can also be included for the purpose of preventing skin irritation. This is achieved by the formation of the liquid crystalline structures on the skin's surface.

Table 4.5: Composition of O/W night cream [44]. Reproduced with permission.

Ingredients	Concentration (%)	Function
Orange roughy oil	8	Conditioner
Dimethicone silicon oil (200–300 cS)	1	Emollient
Cetylacetate/acetylated lanolin alcohol	1	Emollient
Myristyl myristate	3	Emollient
PEG-24 stearate	3	Emulsifier
Cetearyl alcohol (a mixture of predominantly of cetyl and stearyl alcohols)	2	Emulsifier
Glyceryl stearate	7	Emulsifier
Propylene glycol	3	Moisturizer
Soluble collagen in water (0.3%)	11	Moisturizer
Water	To 100	

W/O emulsions, with oil as a continuous phase, are applied when large amounts of oil are required, providing a long-lasting residue. In this case, oil remains on the skin as a sticky film, which is acceptable for cold creams, cleansing or evening (overnight) creams.

From the viewpoint of product formulation, development of a W/O emulsion, which is stable, safe for the skin, smooth and spread easily, is challenging, as apparently a compromise should be reached. Stable emulsions and those that are more pleasant to use as they have good skin compatibility might be uncomfortable for the skin and can be unstable at higher temperatures.

Night creams can also be formulated as W/O emulsions, being less cosmetically attractive, with a feeling of stickiness on the skin. They provide, however, the vital nutrients to the skin that are lost during the day. The continuous oil phase is in direct contact with the skin, right after application, forming a protective film without any surfactants. Beeswax, borax and mineral oil systems serve as a basis for traditional night creams.

As discussed in Chapter 2, stabilization of emulsions is required; otherwise, two immiscible liquids will separate. Stabilization is achieved by incorporating surfactants and additional polymers, solids, and liquid-crystalline phases. Limitations with applications of microemulsions are connected with higher concentrations of surfactants, which are needed for the stabilization of microemulsions, thereby increasing the cost and the potential skin irritation.

The advantages of microemulsions, with the small size of 100–500 nm, is their skin penetration ability, transparency and thermodynamic stability, being reasons for their utilization in, for example, various gels (sunscreen, perfume and skin care). From the manufacturing viewpoint, mixing of the water and oil phases should be done above 90 °C to achieve transparency. Viscosity can be regulated by adding long-chain fatty alcohols; however, the chain length of the emulsifier should not exceed the corresponding value for the oil per se.

In addition to W/O or O/W emulsions, multiemulsions, such as W/O/W, which also afford a less oily feeling along with moisturizing properties of W/O emulsions, are used in formulation of various cosmetic products, including hand and shaving creams, sunscreens or makeup cleansers. In a W/O/W emulsion, water droplets are dispersed in oil drops, which in turn are dispersed in water. Two emulsifiers, along with thickeners, are required in their composition. The latter are needed for sufficient stability. Different compartments in W/O/W emulsions allow the utilization of incompatible or unstable ingredients for the prolonged hydration of the skin and even formation of gelled structures. Preparation of W/O/W multiple emulsion starts from making first a W/O emulsion with a surfactant that exhibits a low HLB number (5–6), followed by subsequent emulsifying in an aqueous solution, using a high HLB number (9–12) surfactant and much less intensive stirring, as in the first step (Figure 4.39).

The role of the first emulsifier is to prevent coalescence of the water droplets, while the second one is needed to avoid flocculation and coalescence of the multiple emulsion drops. Too high a content of the first emulsifier might lead to formation of a W/O emulsion, while an O/W emulsion can be formed if the amount of the second emulsifier in the aqueous phase is too high.

Particular attention is given to baby care products, as the skin is much more sensitive, being more prone to absorption of toxic compounds and to bacteria (e.g., streptococci or staphylococci). Moreover, a thick film of grease, sticking strongly to the surface and thus resulting in skin drying, is required to protect against wetness. For this purpose, fine particles of zinc oxide, which will be discussed later in connection with sunscreens, are added. Table 4.6 provides an approximate composition of a W/O baby cream.

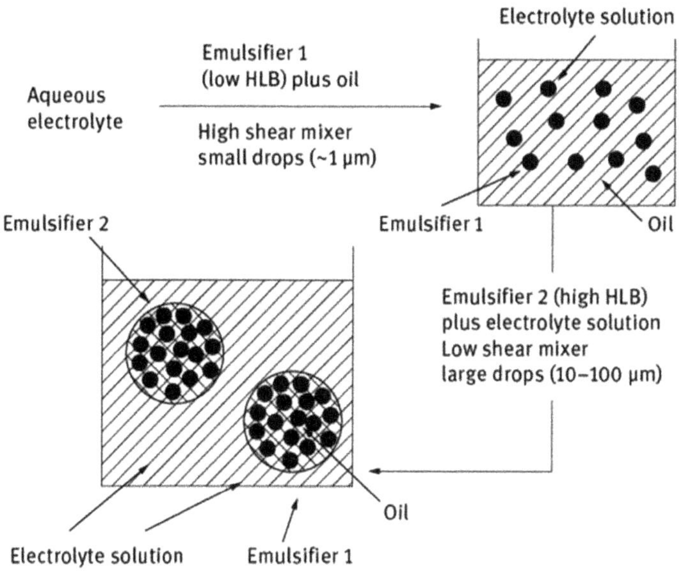

Figure 4.39: Preparation of a stable W/O/W emulsion [44]. Reproduced with permission.

Table 4.6: Composition of a W/O baby cream [44]. Reproduced with permission.

Ingredients	Concentration (%)	Function
Lanolin alcohols	2	Emulsifier
Lanolin	4.5	Emollient/moisturizer
Mineral oil (70 cS)	17	Emollient
White petroleum jelly	13.3	Emollient
Butyrated hydroxytoluene	0.01	Antioxidant
Glycerine	5	Emollient
Zinc oxide	7	sunscreen
Water	To 100	

Different types of moisturizing ingredients are presented in Tables 4.5 and 4.6, as well as in Table 4.7. Some examples of emollients (softening agents) include hexyl laurate, sucrose oleate and isopropyl myristate. Occlusives are introduced in the formulations to create a thin film over the skin, serving as a barrier against water loss. Among occlusives, propylene glycol, squalene, lecithin and silicone derivatives (e.g., dimethicone and cyclomethicone presented in Figure 4.40) can be mentioned.

Hands and body products are similar to day moisturizers in composition, being formulated as lotions and, sometimes, aerosol mousses. To increase the water phase viscosity, agents such as carbomers (polymers of primarily acrylic acid) and cellulose derivatives are added. Healing or soothing agents (urea and lanolin) that have some barrier properties can also be incorporated in these products.

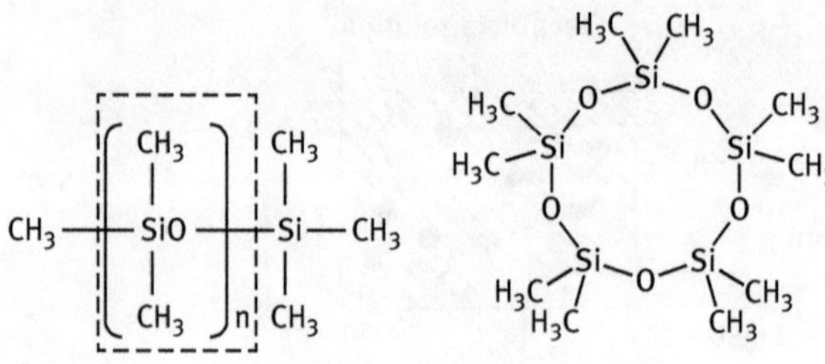

a) b)

Figure 4.40: Structure of silicon derivatives used as occlusives: (**a**) dimethicone and (**b**) cyclomethicone.

Table 4.7: Formulations of moisturizing cream lotions.

Ingredients	F1	F2	F3	F4	F5
	% *w/w*	% *w/w*	% *w/w*	% *w/w*	% *w/w*
Aqueous Phase Material					
Disodium EDTA	0.025	0.025	0.025	0.025	0.025
Glycerin	8.000	3.000	3.000	3.000	3.000
Acrylic Acid Copolymer	-	2.000	-	2.000	2.000
Propylene Glycol	0.500	0.500	-	1.000	1.000
Aloe vera extract	2.500	2.500	2.500	2.000	2.000
Carbopol ETD 2020	0.500	0.400	-	0.500	0.200
Oil Phase Material					
Mineral Oil	6.000	12.000	12.000	10.000	5.000
Glyceryl Stearate	3.000	2.500	1.500	4.986	2.493
PEG-100 Stearate	5.000	4.500	3.500	5.014	2.507
Almond Oil	2.000	2.000	3.000	2.000	1.000
Cetyl Alcohol	6.000	4.000	3.000	4.500	2.250
Post Emulsification Material					
Dimethicone	1.500	1.500	1.500	0.010	0.010
Dimethiconol	1.500	1.500	1.500	0.190	0.190
Alpha Tocopherol	0.100	0.100	0.100	0.500	0.500
Phenoxyethanol and Ethylhexylglycerin	2.000	1.500	1.000	1.000	1.000
Sodium Hydroxide	Q.S to adjust pH 5.0–6.0	Q.S to adjust pH 5.0–6.0	Q.S to adjust pH 5.0–6.0	Q.S to adjust pH 5.0–6.0	Q.S to adjust pH 5.0–6.0
Purified Water	Q.S to 100	Q.S to 100	Q.S to 100	Q.S to 100	Q.S to 100

Barrier creams per se are also used for protective purposes, preventing water, solvents, soil and dirt from skin permeation. In particular, water repellence is provided by, for example, application of lanolin (wool wax), consisting mainly of sterol esters; while for oil repellence, non-oil-soluble film formers (e.g., gum acacia or sodium alginate) can be used. Glyceryl monostearate is introduced in the formulation to facilitate washing off after use. As a plasticizer for gum, acacia glycerin is applied.

The so-called all-purpose creams and lotions used as moisturizers or cleansing creams on the face, hands and body can be formulated with emollients (e.g.. polypropylene glycol stearyl ether) and emulsifiers (e.g., stearyl alcohol).

4.3.2.3 UV protection

Sunscreens are nowadays used not just to provide tanning without sunburn, but also to prevent premature skin aging and skin cancer. Ultraviolet radiation can be of different types, influencing the skin in different ways. UV radiation, being a part of the light, has wavelengths shorter than visible light. UVA, with the longest wavelength at 320–400 nm, constitutes 95% of the total UV reaching the Earth and UVB of 290–320 nm wavelength (ca. 5% of total UV) plays an important role in premature skin aging, causing eye damage (including cataracts) and skin cancers. In particular, DNA damage immunosuppression, sunburn and skin cancer are associated with UVB, while UVA is responsible, for example, in premature skin aging and cancer. Shorter UVC rays of 100–290 nm cannot penetrate the atmosphere, being blocked by the ozone layer, and thereby do not reach the Earth.

For people with white skin, living in the tropics, protection from UVB and UVA in fact is needed throughout the year. Efficient protection requires maximum absorption of UVB and UVA at relatively low dosages, nonvolatility of the active agents with chemical and physical stability, sufficient resistance to removal by respiration as well as solubility/dispersibility in the oil or aqueous phases. Moreover, the active agent should not penetrate the skin and should be devoid of dermato-toxicological effects. Finally, the active ingredient should be compatible with the compounds in the sun protection product.

Sunscreens are formulated to protect the skin by both absorbing and scattering UV light. This can be achieved by using organic and inorganic compounds that absorb UV radiation of a particular wavelength, in addition to reflecting UV radiation. The latter compounds have an advantage of a much broader spectrum of wavelengths being, in addition, non-irritant. As illustrated in Figure 4.41, ZnO, in particular, adsorbs in a broad UV range.

Inorganic materials such as zinc oxide or titania absorb and scatter light, with the efficiency depending on the size of fine particles. For titania, maximum UV attenuation is achieved at 30–50 nm (Figure 4.42). Low levels of titania utilization for cosmetic purposes were associated with whitening, inherent for this well-known pigment, as a result of visible wavelengths. Subsequently, titania was applied in sun protection products only for babies, as these customers do not complain about aesthetics. More recent advances in modification of surfaces of titania and in control of the particle sizes, improving dispersability and compatibility with other ingredients, paved a way in using titania for attenuation of UV in sunscreens, without a negative impact on aesthetics.

Besides the basic requirements mentioned above regarding UV attenuation and other physio-chemical and consumer related features, other issues such as registration, important for global distribution, and patent freedom should be considered.

A typical composition of a sunscreen is shown in Figure 4.43 for a product with the SPF (sun protection factor) of 30. This value corresponds to the fraction of sunburn-producing UVB rays reaching the skin. Therefore, SPF 30 blocks 97% of UVB rays with 3% reaching the skin, while SPF 50 blocks 98% of UVB rays, allowing 2% of UV rays to influence the skin.

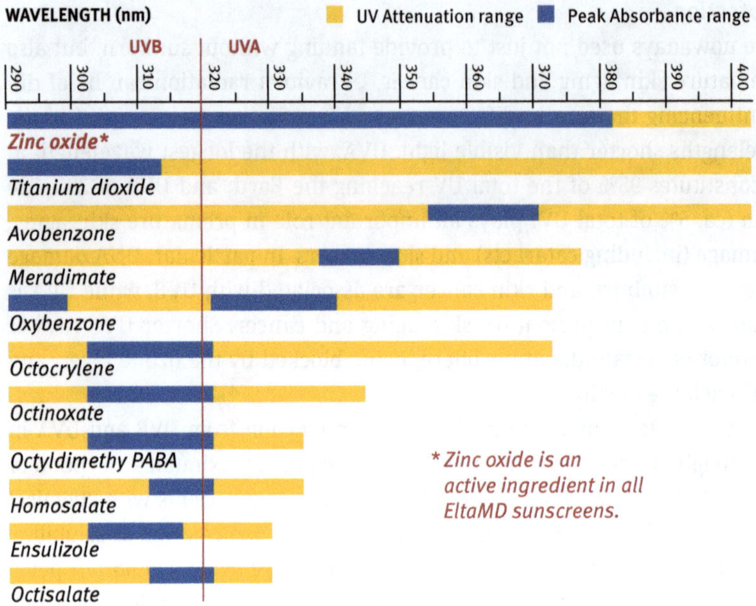

Figure 4.41: Performance of sunscreen ingredients [46].

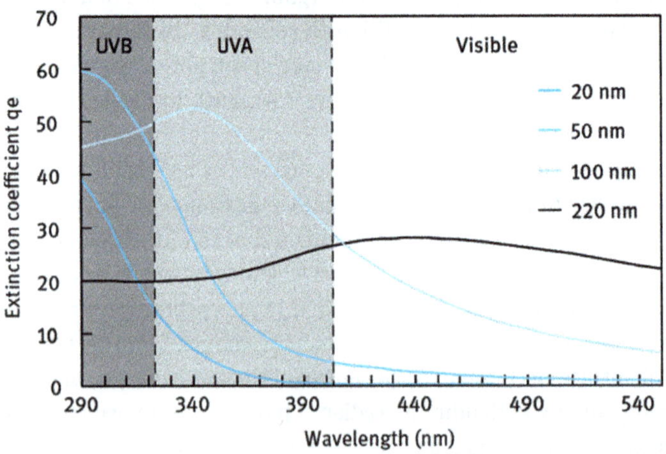

Figure 4.42: Effect of particle size on UVA and UVB absorption [44]. Reproduced with permission.

A broad range of organic compounds can be used as UV filters. Figure 4.44 shows some of the sunscreen ingredients. Some disadvantages of organic UV filters are related to too good solubility in oil and thus a need to introduce large amounts of solubilizers and potential coloration of the product due to the chromophore nature of these organic compounds. Moreover, coloration of the fabric can also occur because of sweating.

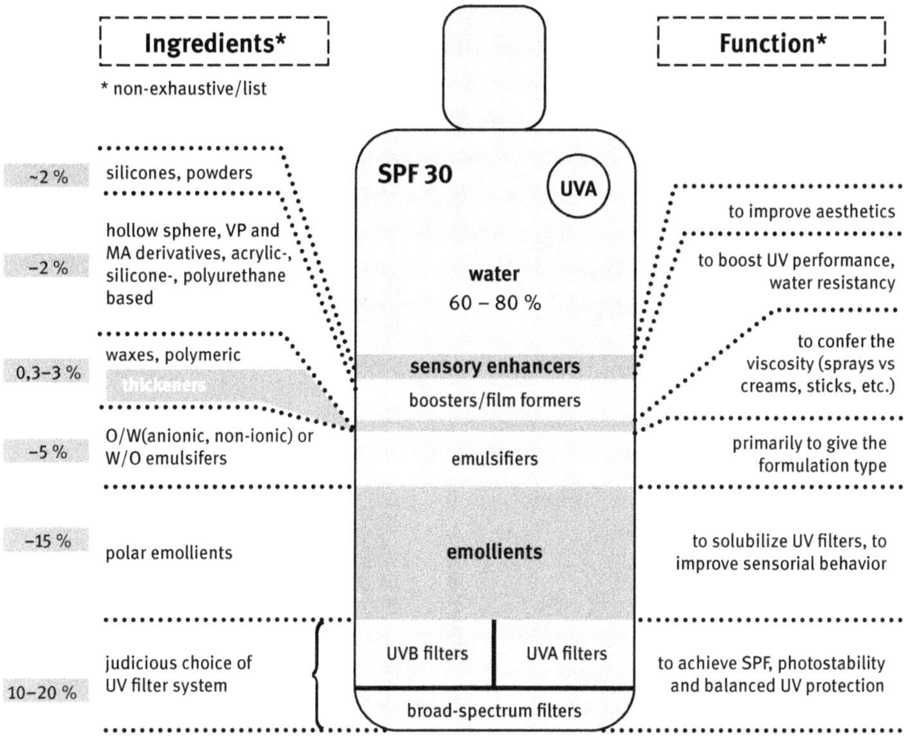

Figure 4.43: Ingredients of a typical sunscreen and its functions (1). SPF, sun protection factor; O/W, oil-in -water; W/O, water-in-oil; UV, ultraviolet (from [47]).

Organic UV filters presented in Figure 4.44 require low concentrations for good efficacy, providing a good skin feel with no powdery appearance. As can be, however, seen in Figure 4.41, most of them have a narrow spectrum and some are not photostable. As a result, mixing of different UV filters is required to get a broad spectrum and high SPF. For example, a combination of butyl methoxy-dibenzoylmethane (avobenzone) with octyl methoxycinnamate (or more precisely ethylhexyl methoxycinnamate) and octocrylene

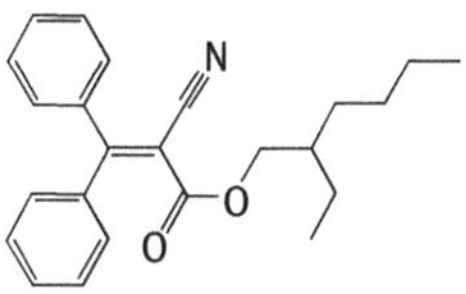

Octocrylene

provides protection across a broad spectrum. Moreover, some of the organic UV filters are questionable from the viewpoint of irritancy and environmental impact.

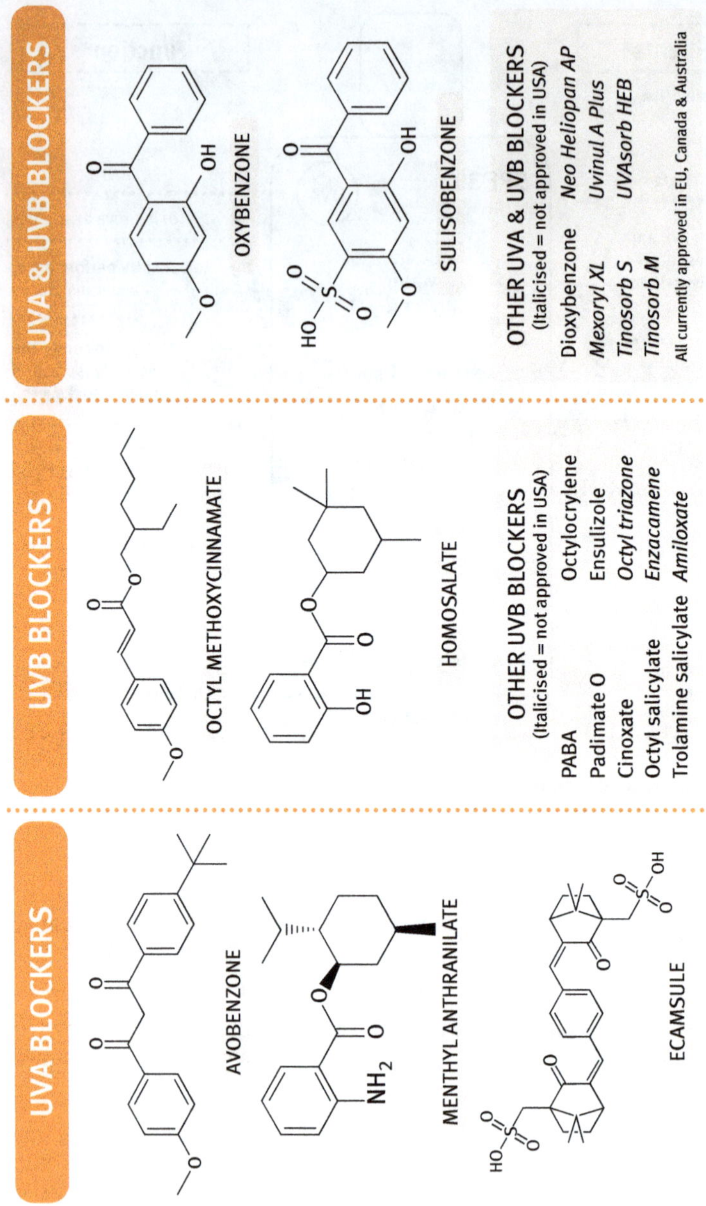

Figure 4.44: Some organic UV filters (adopted from [48]).

Development of sunscreens is commonly done by increasing the sun protection factor of an already existing sunscreen, preferably without changing other characteristics. Just adding more UVB filter also requires addition of more UVA filter to keep a balance between those filters. To prevent the subsequent crystallization of the UVA filter, some adaptation of the oil phase might be needed, which translates into adjustment of the emulsifier concentration and, possibly, some other alterations in the formulation.

An apparent nuisance of powder sunscreen is a whitish appearance, which results in the application of lower amounts of sun care products than needed to achieve a required SPF. Subsequently, a range of sun care products have been developed, in addition to those containing O/W emulsions, namely, W/O lotions as well as spray or stick-type products. The former should be waterproof and durable during physical activities on, for example, a beach. The latter have an advantage of a particular point in skin protection.

4.3.3 Decorative products/makeup

The main purpose of the decorative makeup cosmetics is to make the face and skin attractive by hiding imperfections (stains or wrinkles), adjusting skin tone, decorating the body or countering for a more attractive and healthy-looking appearance. These products are also important from the phycological viewpoint, improving the inner sense of well-being and self-satisfaction, in addition to attracting attention of the others. Obviously, the reasons for wearing makeup, the style and type depend on the age of the users.

Creams, lotions or powders (popular in Asian markets) are used as a base for facial makeup. The most basic foundations in Europe, North and South America are O/W foundations, containing thickeners and stabilizers for the powder/oil phase–water phase systems.

An important issue with the decorative makeup is the influence of light, as sunlight and artificial light possess different brightness and spectra. In daylight, some makeup might look whiter than indoors. This effect can be mitigated by introducing photochromic powders, which reversibly change color in the presence and absence of sunlight due to photochemical reactions. For example, doping titania with iron oxide harmonizes the color of titania, outdoors and indoors.

Durability of makeup depends on the pigment's resistance to coloring with time as well as stability against sweat and body oils. Although the application of methyl polysiloxane-coated powders and pigments as well as surface treatment with fluoride are beneficial for mitigating problems with stability originating from by body oils, such materials can have low affinity with the oil bases in foundations. This leads to weakening of coloring or making the makeup uneven. Subsequently, high oil-absorbing powders are required in the formulations, exhibiting large surface areas or very small size of particles. As makeup is mainly applied during daytime, UV blockers of both types (organic and inorganic), discussed in the previous section are added to makeup.

4.3.3.1 Eye makeup

Mascaras, eyeshadows and eyeliners constitute eye makeup. An intense look of the eyelashes is achieved in mascaras by their thickening, separation and lengthening. A precise line at the base of eyelashes is drawn with the help of eyeliners, while the color of the iris is highlighted by the application of eyeshadows.

From a customer perspective, mascara should be easy to apply, devoid of harmful compounds, contain conditioning substances for nourishing and moisturizing, not too thick or watery, waterproof and easy to remove. The choice of mascara and its style depends on the length, curvature, color, density of eyelashes and, apparently, on the desired effect (changes in the curvature, length or thickness). Some examples of different looks with the corresponding brushes responsible for the volume, precision and curve are shown in Figure 4.45. Obviously, not only the composition of mascara,

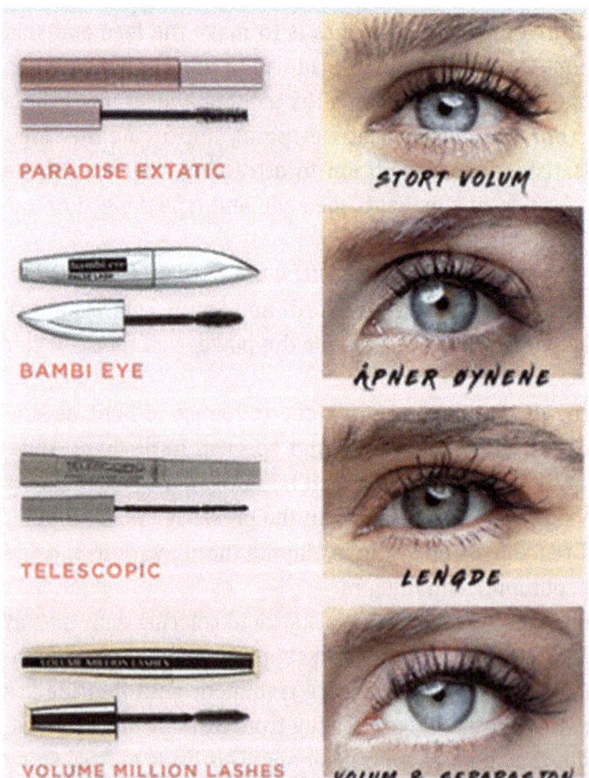

Figure 4.45: Comparison of mascaras and the brushes [49].

but also the type of brush and the diameter of the aperture of the mascara tube influence the performance of the product.

A typical composition of mascara, with the corresponding concentration ranges, is given in Table 4.8.

Table 4.8: Typical mascara ingredients and ranges of concentrations [50].

Ingredient	Concentration (wt%)	Examples
Color	1–12%	Iron oxide, mica pigments
Emulsifier	3–10%	Myristoyl, penta/hexapeptides
Emollients	25–50%	Oils (jojoba, wheat, argan)
Texture modifiers	5–20%	Hydroxycelluose, waxes
Preservative	0–1.5%	Phenoxyethanol, propylparaben
Film formers	1–8%	Silicones, dimethicone
Solvent	Q.S	Water or petroleum distillates

Liquid mascaras can be based on water, other solvents or a hybrid of those. Waxes (e.g., beeswax, carnauba wax, and synthetic wax) help to thicken the eyelashes after water evaporation from water-based mascara. The content of waxes for so-called waterproof mascara can be even higher than mentioned in Table 4.8, reaching 30%. Pigments are required to color eyelashes; the vast majority of mascaras contain carbon black, giving that color to the product. Other ingredients present in mascara are needed to avoid the smell of fatty compounds, minimize the risk of infection or improve the overall performance.

Eyeshadows (cosmetic applied to the eyelids for the purpose of attracting attention) are formulated similar to those of compact powders for the face; however, they have a wider color range. In addition to a base filler, which is typically cosmetic-grade talc or mica, binders (kaolin clay or derivatives of zinc or magnesium) are required for adherence to the skin. So-called "slip" components (silica, boron nitride, bismuth oxychloride or dimethicone) are added to assist the powder in sliding more easily over the lid. The growth of bacteria is mitigated by adding preservatives (e.g., glycol and tocopherol).

4.3.3.2 Lips protection

Expose of the thin skin of the lips to dry air, cold temperature or winter results in a tendency to dry and develop small cracks. Lipsticks are thus needed for protection as well as for purely decorative purposes. From the consumer's perspective, the lipstick should satisfy the expectations with respect to ease of application; neutral taste; color and its uniformity; stability and adherence; overall appearance, without visible separation of small oil droplets; mechanical stability and shelf life. From the application viewpoint, a well-adhering film of fat to the lips should be formed without being too greasy, sticky, or smeared.

Lipstick and lip balm are usually concentrated suspensions (60 mass%) of solid oils and/or wax in a liquid oil or in a mixture of liquid oils (40%). A lipstick must also be temperature-resistant as the formulation is done at relatively high temperature – above the melting point. Subsequent cooling allows the development of a significant yield stress. An approximate composition of lipstick is shown in Figure 4.46.

Lip balm and lipstick often contain wax-like substances such as beeswax, carnauba wax, petroleum jelly (mixture of hydrocarbons with 25 + carbon atoms), etc., giving a lipstick its shape, and providing an occlusive layer on the lip surface. This layer is needed to relieve chapped or dry lips, sealing moisture in lips and protecting them from external exposure. The base of a lipstick, besides solid waxes, also contains softening agents such as lanolin, palm oil, mango butter or cocoa butter, giving the lipstick its tough and shiny film when it dries. In the continuous phase, liquid oils (castor, olive, sunflower or almond oil) are used, ensuring ease of application.

Flavorants and fragrances (e.g., linalool or menthol), dyes (oil soluble or insoluble) and pigments (e.g., titania), sunscreens (e.g., oxybenzone), preservatives, antioxidants and various medicaments (e.g., salicylic acid) provide additional benefits.

Different product types require alterations in the relative content of wax, oil and pigment. Higher proportions of wax and pigment are introduced in matte lipsticks, while more oil and less wax are needed for gloss lipsticks. UV-absorbing features are provided by adding such dispersed solids as titania and zinc oxide, discussed above in the section on sunscreens. Other solid materials such as silica or bismuth oxychloride (synthetic pearl), also applied for eye makeup, are used, respectively, for shimmering and frosted lipsticks. For hot countries, there is a need to hydrate and protect lips for longer time. Lipstick for such countries is thus enriched with botanical extracts, and contains some color pigments. Transparent lipstick uses soluble or solubilized dyes and suitable solvents like di(propylene) glycol methyl ether. A composition of such a lipstick could be: polyamide resin (average MW 800) – 20%, polyamide resin (average MW 600–700) – 5%; propylene glycol monolaurate – 28%, castor oil – 12.6%, eosin – 0.3%, di(propylene) glycol methyl ether – 8%, ethoxylated lanolin ether – 10%, ethoxylated lanolin alcohols – 10.1%, isopropanol – 5% and perfume – 1%. Comparison between lipsticks of different types is provided in Table 4.9.

4.3.3.3 Nail protection

Nail care products are applied probably by billions of people to maintain smooth and shiny nails, motivated not only by self-esteem aspects, but also rising awareness on fungal infection and thus, hygiene requirements. The world market in nail care products is estimated to exceed US $22 billion, with a steady growth. The driving forces in the growing market are related to an increase in literacy in developing regions regarding nail cleaning and hygiene as well as demands for nail polishes of bold and dark colors.

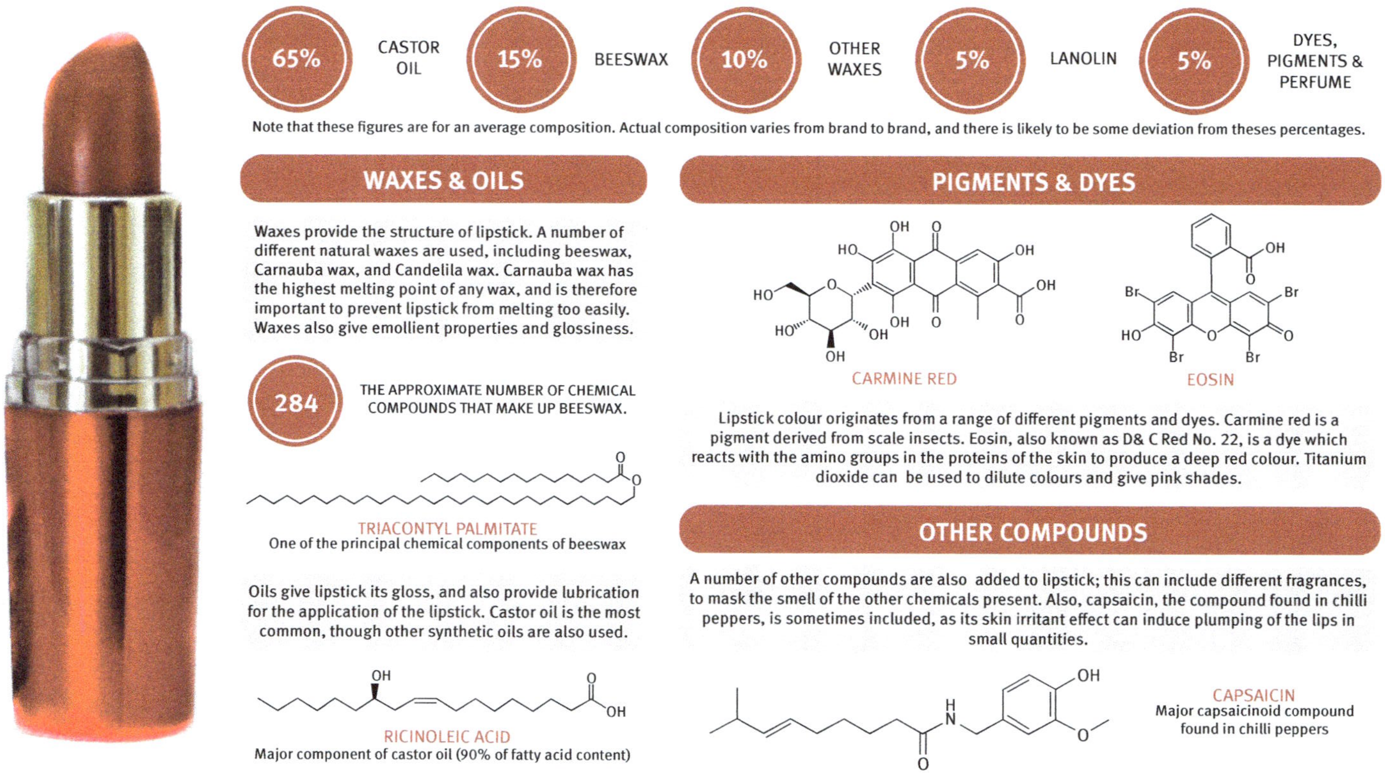

Figure 4.46: Composition of lipstick [51].

Table 4.9: Composition of lipsticks for different purposes.

Type	Lipstick with optical effects	Lipstick for hot countries	High-gloss lipstick
Brand	Revlon	Lakme 9 TO 5	L'Oreal
Composition	Trioctyldodecyl citrate, polybutene, ethylhexyl palmitate, octyldodecyl neo-pentanoate, penta-erythrityl tetraiso-stearate, bis di-glyceryl polyacyl-adipate, vitamin C, silk powder, silica	Trioctyldodecylcitrate, ethylhexyl palmitate, octyl-dodecylpentanotetraiso-stearate, caprylic/capric triglyceride, C10-30 cholesterol/lanosterol esters, B15-diglycerylc polyadipate-2, calcium sodium bromosilicate, paraffin	Sesame seed oil, oleyl erucate, disteardimonium hectorite, tocopheryl acetate rosa canina fruit oil, calcium aluminium borosilicate, silica, argan oil, geraniol, hexyl cinnamal, hydroxy-citronellal, calcium sodium borosilicate, synthetic fluor-phlogopite, methyl-2 octynoate, polyethylene terephthalate, polymethyl methacrylate
Staying time	6–7 h	8–9 h	7–8 h

Nail polishes, per se, are essentially organic coatings, containing, besides the polymer and the solvent, some other additives. In fact, no actual polishing is involved in the application and the product (nail lacquer rather than nail polish) is dried to a hard glossy coat. Nail polish, nail varnish, nail paint or nail lacquer thus refer to the same product, which is applied by painting the nail with a brush.

In best manicure practices, several layers are introduced. The base coat or the first layer, comprising a resin, a solvent and a plasticizer, is needed to increase the adhesion of the colored nail lacquer to the nail. The composition of the base coat is somewhat different from the subsequent ones, as more resin is added in the first coat to improve adherence of the varnish to the nail, while no colorants are required. Another difference is the amounts of thinners and solvents in those coats.

On the top of a base coat, two coats of colored nail lacquer are used, followed by a topcoat layer, increasing the chip resistance. Formulation of nail polish thus comprises, besides the lacquer base, colorants, thixotropic agents and color stabilizers (UV-absorbing materials such as benzophenone). The topcoats contain more nitrocellulose and plasticizers for better nail varnish resistance.

The main ingredients in the nail polish are illustrated in Figure 4.47.

Beside the most common polymer nitrocellulose (12%), others polymers, e.g., methacrylates and vinyl polymers are used, as will be commented below. When nitrocellulose is applied alone, the formed films tend to shrink and become brittle, exhibiting insufficient adhesion to the nail surface. Subsequently, secondary resins, including, e.g., formaldehyde or polyester, are needed to enhance adhesion and flexibility. The content of organic solvents (e.g., ethylacetate and butylacetate) is ca. 70–80%. When applied, the

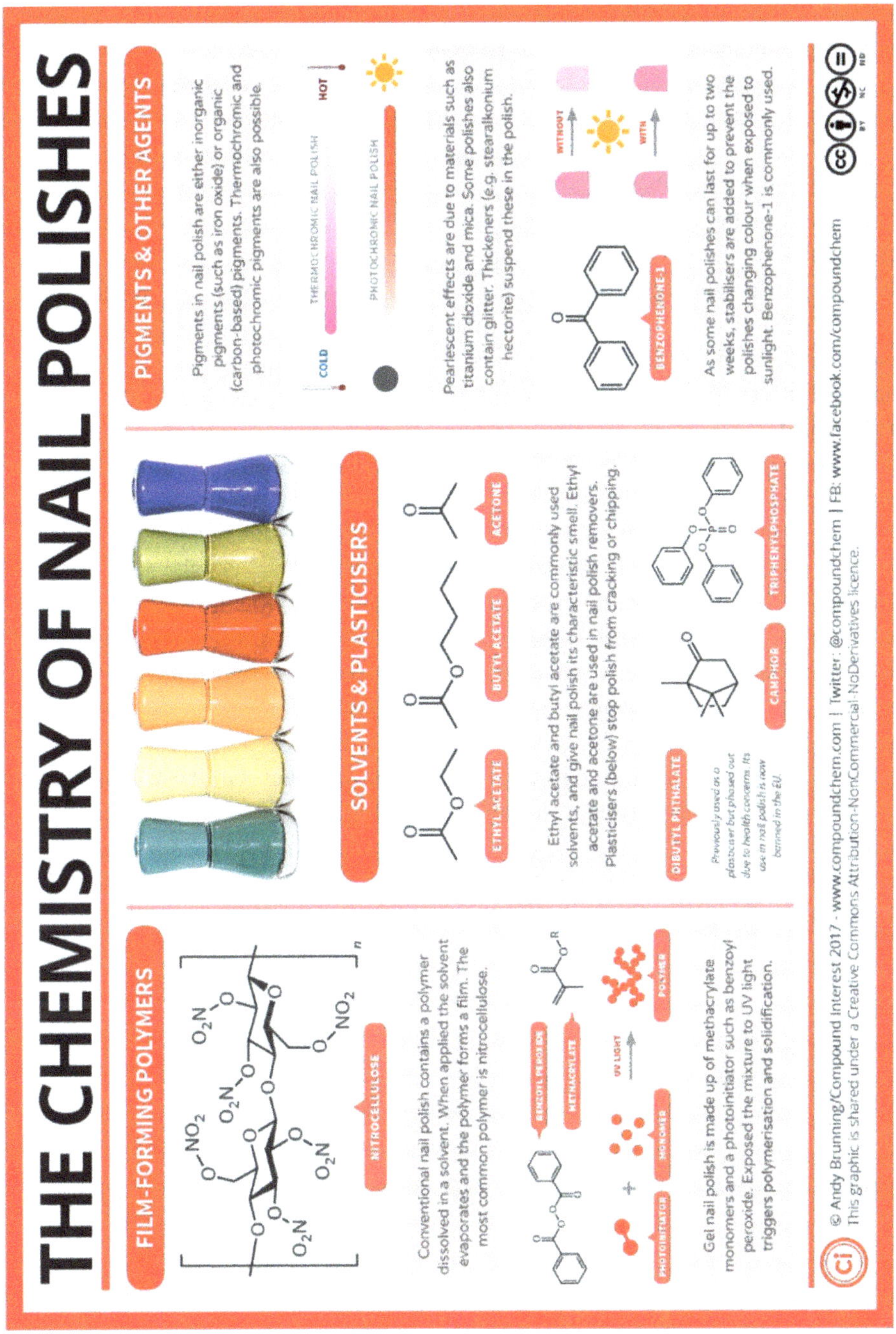

Figure 4.47: The main compounds in nail polishes and their chemical function [52].

solvent evaporates and the coating is formed. The ratios of solvents can be modified, influencing the film characteristics, its flexibility and the drying time.

The lacquer base contains, besides resins and the solvent, also of a plasticizer (up to 1%). In the past, dibutylphthalate was applied as a plasticizer, which is needed to stop the polish from cracking or chipping. After the phase-out of phthalates in EU, camphor and triphenylphosphate are applied for this purpose. Either inorganic (e.g. iron oxide) or organic pigments (ca. 5%) are used. The latter are adsorbed or co-precipitated on inorganic materials (silicates or oxides). Titania or mica can be added to allow pearlescent effects. Nail polish should be viscous enough for good adherence to the applicating brush. At the same time, shear-thinning is required to provide an even coating on the nails. Flow control is thus provided by thixotropic agents, such as clay derivatives. The combined concentration of suspending agents and fillers can be up to 9%. Stabilizers such as benzophenone-1 or etocrylene are introduced to prevent changes in the color when nail polishes are exposed to sunlight. Antifungal agents or perfume can be also added to the nail polishes formulations.

As follows from Figure 4.47, besides nitrocellulose-based nail polishes, UV-curable formulations can be used containing monomers derived from acrylic acid. They undergo UV polymerization and require incorporation of an initiator in the formulation, which generates free radicals upon exposure to UV light, forming acrylic polymers. Because of the complicated methodology and the specific requirements of the film in terms of softness and brittleness, application of these products in mainly done by nail technicians/manicurists.

A market demand led to development of nail varnishes with color-changing effects, based on thermo- or photochromy, i.e., activation by, respectively, temperature or light. In the latter case, the organic molecule changes the configurations present in a formulation (for example, non-conjugated to conjugated), upon exposure to light (Figure 4.48). For thermochromy, liquid crystals or leuco dyes are used.

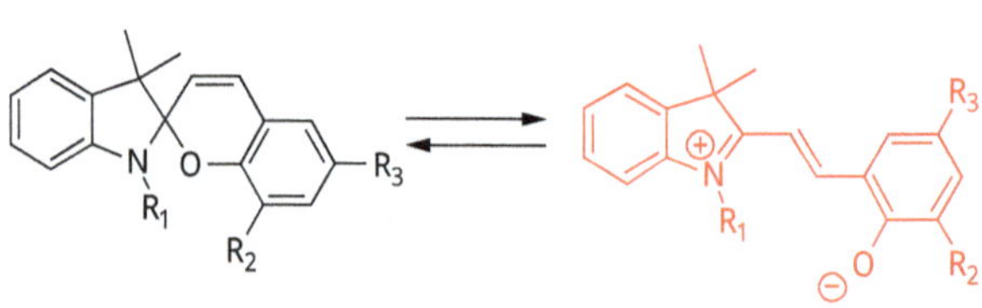

Figure 4.48: Reversible interconversion between ring-closed spiropyran and ring-opened merocyanine forms [53].

Besides color-change varnishes, textured nail polishes are used with a varying texture (e.g., from satin to matte), which is achieved by adding silica for satin and matt nail polishes or polyethylene terephthalate fibres for a 3D woven fiber effect to nails. Utilization of larger silica particles gives a velvet effect, while “caviar” finishes on nails are obtained by applying glass or plastic microbeads of different size and color to the nails. The latter can be done, for example, by inserting the nails into a small container of microbeads or pouring the beads onto the nail. After the varnish is dried, the beads are fixed to the nail with a topcoat.

Widespread utilization of nail polish has raised a number health and environmental concerns related to the nondegradable nature of acrylics, the toxicity and flammable character of the nail polish per se (containing, e.g., formaldehyde) and nail polish removers, and environmental pollution associated with the extraction of mica and pollution of water because of glitter-containing microplastics. More sustainable nail polish should be made of the so-called 10-free formulations – without formaldehyde, toluene, dubutylphtalate, formaldehyde resin, camphor, ethyl tosylamide, xylene, parabens, acetone, triphenylphosphate, tert-butyl hydroperoxide and gluten.

In addition to solvent-based nail polishes, water-based ones have also appeared, being popular with young age groups and in some geographical areas (e.g., Japan). The presence of water makes them prone to microbial contamination. Moreover, water-based nail polishes are rather brittle and do not stay for a long time. On the other hand, a wider variety of water-soluble additives can be added, especially those that can be used for treating nail pathologies. The main advantage of water-based nail polishes is considered to be the complete absence of smell.

4.3.4 Hair care, styling and dyeing

Shampoo are used for hair cleansing/cleaning (i.e., removal of dirt, oil, dandruff, various contaminants, etc.) without stripping out too much hair, while also keeping the hair healthy, soft and shiny. Therefore, besides surfactants per se (10–50% of the formulation), other ingredients (opacifiers, clarifying agents, antifreezes, conditioning and finishing agents, sequestrants, thickening agents, proteins, foam builders and antidandruff agents) are applied, with their levels ranging from 1 to 10%.

From the customer's perspective, any shampoo product, in addition to cleaning, should afford lathering in hard or soft water. In fact, foaming being, probably, a must for kids to have fun while taking a bath, might not be needed for the cleaning purposes, and is used more as a marketing tool. Other expectations include an easy and complete lather removal, without a residue, safety and non-irritating or eye-damaging performance, chemical and physical stability with a long shelf life, a certain viscosity for pouring of more precise amounts, solubility in water for clear shampoo, pleasant fragrance, low toxicity, good biodegradability, slight acidity (pH less than 7), no damage to hair and repair of the damage already done to hair, to name a few.

In the design of shampoo products, these wishes are incorporated by selecting a proper detergent and myriad of additives.

Selection of surfactants is limited by a need to keep a certain amount of sebum (waxy matter) secreted by the skin for softening purposes. Too little sebum results in dry and straw-like hair, while when the level of sebum is too high, the hair looks greasy and matted. Mild surfactants (e.g., an anionic surfactant sodium lauryl sulfate) are able to remove the greasiness while still keeping the shine.

The isoelectric point of hair is ca. 3.7, implying that its surface is negatively charged at neutral pH, repelling anionic surfactants, which otherwise will deposit on the hair, bringing it to an unmanageable condition.

During application, anionic surfactants are adsorbed on hydrophobic surfaces of the undamaged hair by their hydrophobic tails, with the negatively charged heads orientating outwards. Nonpolar materials are then displaced from wetted surfaces. Removal of semipolar compounds, by rinsing, is possible because they are solubilized into micelle structures. For particulates, adherence to hair through ionic and van de Waals forces is diminished in the presence of surfactant systems. Thus, the hair can also be cleaned by rinsing. Mixing an anionic sulfonate with, e.g., a nonionic ethoxylate or betaine

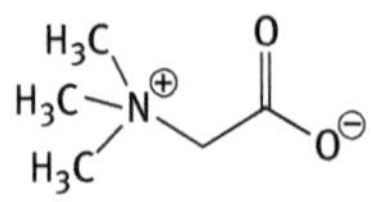

lowers the critical micelle concentration, thereby decreasing the surfactant monomer concentration and subsequently diminishing eye and skin irritation. As a result, surfactants are not used alone but with some other surfactants. The amphoteric surfactants are not very effective in conditioning the hair. Subsequently there was a quest for the development of more effective additives, resulting in the introduction of cationically modified water-soluble polymers.

Another direction in the development of shampoo is related to sulfate-free and silicon-free products, reflecting perceptions of the consumers of a more environmentally friendly nature of such shampoos, along with lower chances of skin and eye irritation or contact allergy.

Several classes of derivatives of amino acids such as glutamic and aspartic, as well as glycine, methylalanine or methylaurine have been tried as low-irritation surfactants to replace sulfate-based surfactants. While these surfactants are milder from the irritation viewpoint, they are more expensive and less efficient, exhibiting relatively low viscosity, compared with the traditional shampoos. Thickening with cationic cellulose or alternatively, amphoteric or nonionic, surfactants as cosurfactants is done to achieve the required shampoo rheology.

In general, thickeners or viscosity modifiers (e.g., sodium chloride, lauramide diethylamine, carbomers, cellulose polymers, guar and polyethylene glycol) are added in 1–3% to provide a richer texture to otherwise runny liquids. These thickeners, besides cationic polymers mentioned above, also include electrolytes, as viscosity of many surfactant systems increases upon addition of electrolytes at an optimum concentration, e.g., the most commonly used sodium or ammonium chlorides as well as sodium sulfates, etc.

Extension of the shelf life is ensured by incorporating preservatives of ca. 0.2% (e.g., methyl and propyl parabens), preventing microbial infestation of the shampoo medium, which can otherwise happen in warm and humid conditions of a bathroom. Other ingredients include antimicrobial agents (e.g., salicylic acid, cadmium or selenium sulfides) to treat dandruff, fragrances or color additives, employed in amounts less than 1%. Figure 4.49, as an example, provides a list of components used by one of the manufacturers.

Water is deionized to minimize degradation by bacteria. A mixture of surfactants is applied to achieve the desired properties. Fatty oils, alcohols, waxes and silicones are added to shampoos as conditioners (5–12%) for the purpose of moisturizing dry hair, increasing its strength and volume, diminishing surfactant irritation, and positively influencing the foam behavior. The link between foaming and cleaning properties is, however, not straightforward and very effective shampoos can be formulated without strong lathering properties. This can be even more beneficial from the viewpoint of conditioning or less irritation. However, from a perspective of (uneducated) consumers, foaming is needed; thus, foaming agents are added, increasing the price of the shampoo, rather than for the sake of performance.

A shampoo also contains pH-adjusting agents, since hair needs a pH between 4 and 6 (ca. 5) to exhibit the maximum wet strength, while most surfactant-water mixtures display pH values above 10, thereby being strongly alkaline. The most common additives to shampoo to lower pH are rather mild citric and phosphoric acids. The products marketed as nonalkaline, pH-controlled or pH-balanced are in fact acidic, which is typically not mentioned.

Hair fibers (Figure 4.50) consist of a central cortex, constituting the major part of the fiber, and is responsible for the hair's tensile properties.

The cortex is surrounded by a cuticle (i.e., 8–10 layers of overlapping cells), which determines such consumer-related properties as hair feel, shine, combability, etc. Medulla, the inner most layer, composed of large loosely connected keratinized cells, is influenced by coloring hair. Hair surface can also be damaged because of stress when hair grooming (e.g., brushing) and by heat treatment with blow dryers and/or curling irons. Chemical treatment (bleaching, and permanent dyeing) can also cause significant damage to the hair fiber by oxidation of cystine to cysteic acid and transforming the fiber surface from hydrophobic to hydrophilic.

The virgin hair surface is hydrophobic because of a layer of fatty acids covalently bound to the outermost surface of the cuticle (epicuticle). The hair surface has, however, an isoelectric point near 3.67, as already mentioned, which means that it is negatively charged at the pH levels of hair care products. The type of conditioning agents, which bind to the surface, is then determined by this combination of hydrophobicity and hydrophilicity. The negative charge density on virgin hair has a gradient that increases from the root to the tip, as the tip part is exposed to UV radiation for a longer time, being more hydrophilic. The tip part is also exposed to more combing.

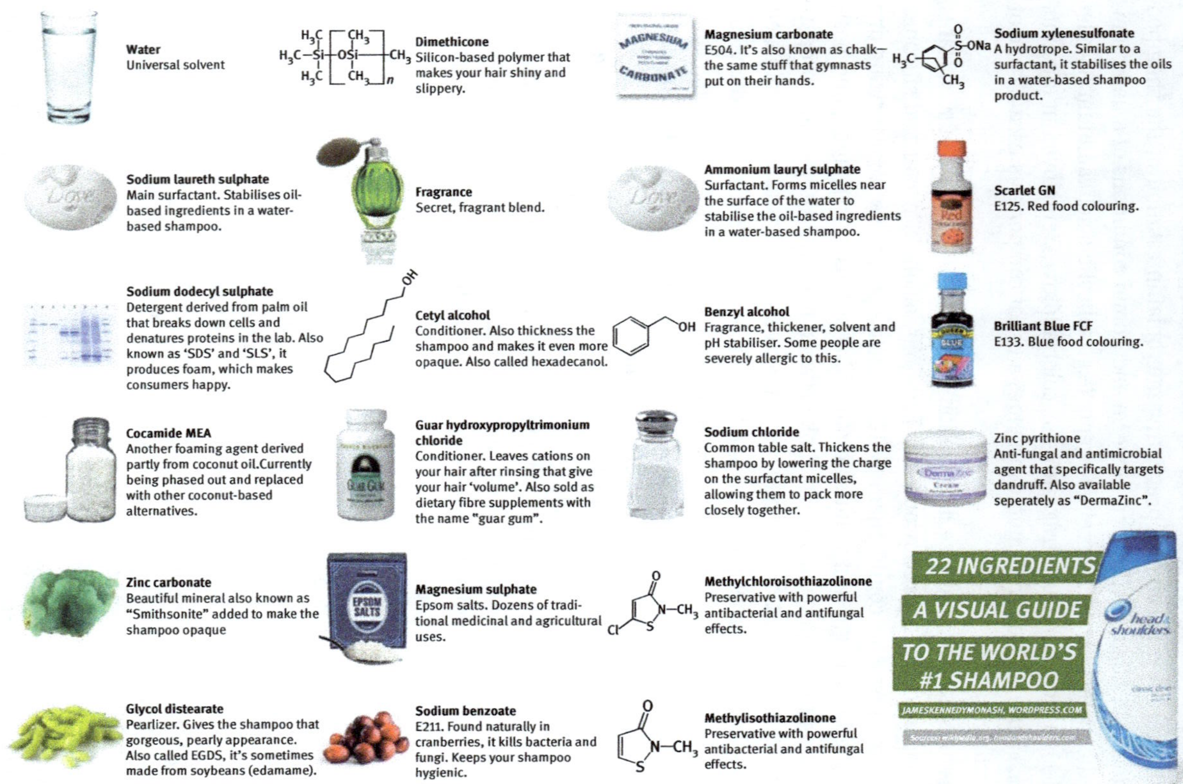

Figure 4.49: Ingredients used in shampoo [54].

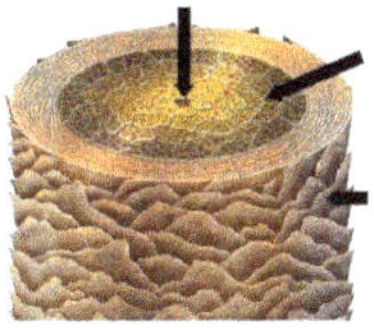

Figure 4.50: Structure of hair fibers [55].

During bleaching, the bleaching agent penetrates the hair cortex, oxidizing the natural pigment and making it colorless. Overall, bleaching leads to dry, brittle, inelastic, more porous hair, exhibiting split ends. Such hair is prone to other chemical and non-chemical processes. Customers are nevertheless interested in bleaching despite all the apparent damages to hair because of it.

Due to various reasons discussed above, there is an apparent need for products that would help in prolonging the health of the hair fibers, especially of chemically processed ones. When a hair conditioner is used as part of the grooming process, it helps to diminish the magnitude of the forces associated with combing or brushing of, especially, wet hair. This is achieved through deposition of conditioning agents that lubricate the hair fiber, decrease surface friction and make the hair feel softer and more moisturized. Reduction or prevention of flyaway hair makes the hair fibers to be aligned in a more parallel configuration, increasing the hair shine without an increase in the shine of the individual fibers. Certain ingredients in hair conditioners also provide more color retention for color-treated hair.

The often-claimed advantages of hair conditioners (e.g., reparation and strengthening of hair) should be actually taken with care as they might not be supported by data from actual usage.

Commercial hair conditioners that are produced in different forms (creams, gels, mousses, lotions, spray mists) can be categorized as regular rinse-off conditioners, intensive treatment conditioners, and leave-in products. Instant conditioners of the first category are the most common and popular ones because of easier styling being applied, immediately after shampooing. Thereafter, they are left for a short period of time (ca. 5 min), prior to rinsing. Such conditioners can be applied on a daily basis with minimally damaged hair. Deep conditioners or intensive hair conditioners can be left on the hair for 20–30 min. These more concentrated conditioners, sold typically as thicker creams, can be used for extremely dry hair as well as before chemical treatment, providing a higher degree of conditioning. Leave-in products are typically lighter and can stay on the hair until the next shampoo.

Conditioners consist, besides water, of cationic surfactants (cetyltrimethylammonium chloride, behentrimonium or propyltrimonium, stearamidopropyl dimethylamine), polymers (mono and polypeptides), bodying agents and thickeners (fatty alcohols, e.g., stearyl alcohol, waxes and salt), emollients (natural or synthetic oils) and auxiliary emulsifiers (ethoxylated fatty alcohols). Cationic surfactants and fatty alcohols are considered bifunctional, being applicable to both wet hair and dry hair. Key ingredients in hair conditioners are presented in Table 4.10.

In quaternary ammonium compounds (quats) used as cationic surfactants, different counter anions, such as chloride, bromide and methosulfate, can be used. Cetrimonium chloride

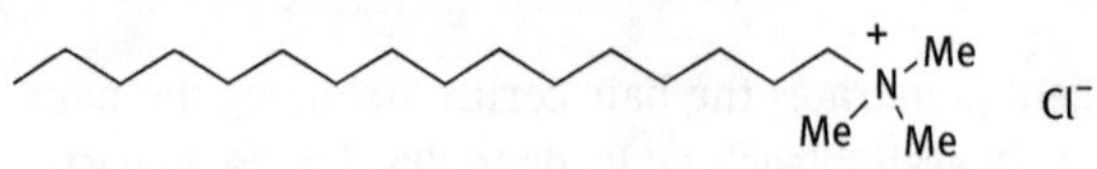

provides light to medium conditioning, while stearalkonium chloride and dicetyldimonium chloride provide heavier conditioning. These compounds, typically not used in shampoos, being incompatible with common anionic detergents, are applied together with long-chain fatty conditioners, allowing for enhanced conditioning.

Table 4.10: Ingredients in hair conditioners [Modified from 56]. Reproduced with permission.

Ingredient	Function	Example	Amount
Water	Solvent	Deionized water	Up to 90%
Cationic surfactant	Surfactant, conditioning agent	Behetrimonium chloride, Steartrimonium chloride, stearylamidopropyldimethylamine	0.5–3%
Fatty alcohol	cosurfactant, conditioning agent	Cetyl alcohol, stearyl alcohol, behenyl alcohol	0.5–10%
Silicone	conditioning agent	Dimethicone, dimethiconol, amodimethicone	0–10%
Oil	conditioning agent	Cetyl esters, mineral oil, oleyl alcohol, ceryl ethylhexanoate, squalene	0–10%
Polyol	Solvent, humectant	Glycerin, propylene glycol, dipropylene glycol, butylene alcohol	0–5%
Polymer	Rheology modifier	Hydroxyethyl cellulose	0–2%
Organic acid	pH adjuster	Citric acid, lactic acid, glutamic acid	0–1%
Scalp care active	Scalp care	Zinc pyrithione, octopirox	0–1%
Preservative	Preservation	Methylparaben, phenoxyethanol Methylisothiazolinone	<1%
Perfume	Fragrance	–	0.1–1%
Color	Appearance modifier	–	0–0.5%

Important examples of cationic polymers are polyquaternium-10, or cellulose 2- hydroxyethyl 2-[2-hydroxy-3-(trimethylammonio)propoxy]ethyl 2-hydroxy-3-(trimethylammonio)propyl ether chloride (Figure 4.51a) and polyquaternium-7, a copolymer of diallyldimethylammonium chloride and acrylamide (Figure 4.51b)

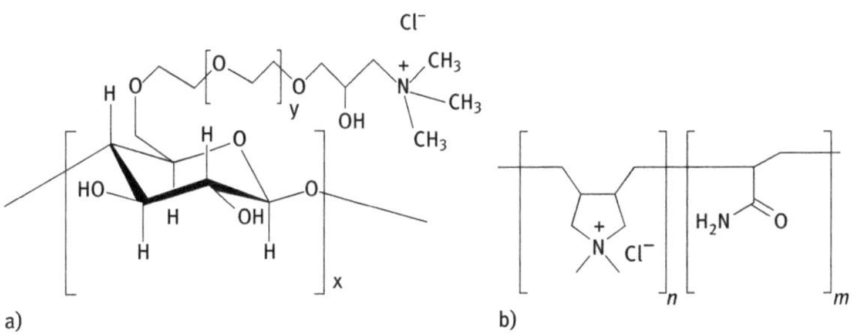

Figure 4.51: Polyquaternium polymers: (**a**) PQ-10 and (**b**) PQ-7 [57].

Jojoba oil or olive oil is used as a natural oil, while synthetic oils are mainly based on various silicones, such as dimethicone (Figure 4.40a) or cyclomethicone (Figure 4.40b). Other silicones include amodimethicone (amino-substituted silicone) and silicone quats, containing permanently quaternized ammonium groups. They allow even better performance than the natural counterparts. These compounds have, however, increased substantivity; therefore, special attention should be put to prevent their build up over time. Concentration of silicones in the formulations depends on the geographical location. Larger diameters in Asian hair, compared to the Caucasian hair, means that higher concentrations of silicone are used (3–5% vs 0.5%). Insolubility of many silicones requires their emulsification; thus, in addition to neat silicones, emulsions performed of different type (anionic, cationic, or nonionic), size (microemulsion or macroemulsion) and manufacturing method (mechanical or emulsion polymerization) are available. Smaller size of the silicone particles typically gives a more stable product emulsion.

As mentioned above, a range of additives exemplified in Figure 4.52 for different shampoo is included in the formulation of commercial conditioners for functional as well as marketing and aesthetic reasons (preservatives, thickeners, humectants, emulsifiers, fragrances, dyes, pearlizers, herbal extracts or vitamins).

Ethanol, applied in concentrations of 20% or more, does not require any additional preservatives to ensure microbiological stability. Different types of thickeners can be added to conditioning products, such as, for example, cationic polymers, polyamides or nonionic hydroxyethylcellulose. Glycols and glycerin can attract moisture and are thus added as humectants. Among emulsifiers, nonionic ones are mainly selected, for example, ethoxylated fatty alcohols, ethoxylated fatty esters and ethoxylated sorbitan fatty esters.

Conditioners are formulated for a target population. For instance, ethoquats provide light conditioning, together with very good static control. They are introduced in formulations that primarily control flyaway, thereby being attractive for persons with short, straight hair in good condition. Trialkyl quats exhibit improved hair feel and allow easy combing; thus preferred for long, heavily bleached hair. Some people might prefer a 2-in-1 shampoo because of application convenience. Conditioning of such shampoos is overall lower than of stand-alone conditioners, which is primarily seen during rinsing. At this stage, the shampoo emulsion is broken, releasing one of the silicones used as conditioning agents (e.g., dimethicone, amodimethicones, dimethiconols and copolyols silicone) for deposition on hair.

INGREDIENTS: WATER(AQUA), CETEARYL ALCOHOL, DIMETHICONE, STEARAMIDOPROPYLDIMETHYLAMINE, FRAGRANCE (PARFUM), BEHENTRIMONIUM CHLORIDE, LACTIC ACID, DIPROPYLENE GLYCOL, AMODIMETHICONE, POTASSIUM CHLORIDE, GLUCONOLACTONE, DISODIUM EDTA, TREHALOSE, DMDM HYDANTOIN, PEG-7 PROPYLHEPTYL ETHER, CETRIMONIUM CHLORIDE, SODIUM SULFATE, METHYLCHLOROISOTHIAZOLINONE, METHYLISOTHIAZOLINONE.

HEALTHY SHINE CONDITIONER: AQUA/WATER/EAU, CETEARYL ALCOHOL BEHENTRIMONIUM CHLORIDE, PEG-180, AMODIMETHICONE, CETYLESTERS, ISOPROPYL ALCOHOL, PARFUM/FRAGRANCE, GLYCERIN, HYDROXYETHYLCELLULOSE, TRIDECETH-6, CITRIC ACID, CHLORHEXIDINE DIGLUCONATE, CETRIMONIUM CHLORIDE, PUNICA GRANATUM SEED OIL, ROYAL JELLY F.I.L. #D49752/3

Water, cetearyl Alcohol, Isopropyl Myristate, Cetrimonium Chloride, Hydrolyzed Keratin, Hydrolyzed Silk, Sodium PCA, Octyldodecyl PCA, Dimethicone, Panthenol, Glycerin, Hydroxypropyl Guar Hydroxypropyltrimonium Chloride, Lactic Acid, Alcohol Denatured, Glycoprotein, Methylparben, Propylparaben, Fragrance, Orange 4 (Cl 15510), Yellow 5 Lake (Cl 19140), Blue 1 Lake (Cl 42090), Red 33 (Cl 17200) (Cl 17200)

1119595 - INGREDIENTS: AQUA / WATER ● CETYL ALCOHOL ● GLYCERIN ● AMODIMETHICONE ● CETYL ESTERS ● CETEARYL ALCOHOL ● CETRIMONIUM CHLORIDE ● MYRISTYL ALCHOHOL ● BEHENTRIMONIUM CHLORIDE ● DIPALMITOYLETHYL HYDROXYETHYLMONIUM METHOSLUFATE ● HYDROXYETHYLCELLULOSE ● TRIDECETH-6 ● BEHENTRIMONIUM METHOSULFATE ● ISOPROPYL ALCOHOL ● ROSA CANINA FRUIT OIL ● QUATERNIUM-33 ● CHLORHEXIDINE DIGLUCONATE ● CITRIC ACID ● ALCOHOL ● LYCIUM BARBARUM FRUIT EXTRACT ● SAFFLOWER GLUCOSIDE ● GENTIANA LUTEA EXTRACT / GENTIANA LUTEA ROOT EXTRACT ● PARFUM / FRAGRANCE. (F.I.L. C165819/2).

Figure 4.52: Ingredients in different commercial formulation, showing a diversity of components and their chemical nature.

The hair styling market consists of a range of specific products, with aerosol hairsprays applied for setting the style and aerosol mousses, hair gels, creams and pomades for style creation.

Hairspray exhibits typically low to moderate amounts of shine, giving a strong hold and leaving the hair feeling stiff. They are a good option for balding men but can also be used for longer hair that is falling apart. Styling sprays should obviously dry easily after application and are often applied on dry hair. These requirements define the composition of styling spray, which usually comprises ethanol-water solutions of styling polymers. An alcohol needed for fast drying is present in concentrations of 40–80%.

The polymers are attached to the hair surface, forming thin films between hair fibers, and after the evaporation of ethanol, bind the hair together. Apparently, the polymer type and the size of polymer solution droplets define the performance, with larger droplets favoring a strong sticky styling effect, with a longer drying time. On the contrary, smaller droplets are less efficient on styling, however, without the hard hair feel, as with larger droplets. An important issue is a possibility to remove a styling spray after hair washing.

Mousses, which can be alcohol free or contain ethanol in amounts of less than 20%, appear as a foam after leaving a container, and have a very weak hold, compared to other products. They allow both shiny and matte finish but most on the matte, low shine side. Most of mousses, due to low surface tension, can be evenly distributed on wet or dry hair, leaving a soft and smooth texture, with the optimum efficiency for thinning hair.

Creams providing a weak to moderate hold, typically leave very little shine and provide a flexible, smooth, naturally looking texture.

Most pomades have a moderate hold, while some might be slightly stronger or weaker. Pomades work best with shorter styles; allow a shiny, wet finish to the hair and leave a smooth, greasy feel, when applied in a correct way to moist hair.

Waxes usually have very strong holds, leaving a dry, matte finish, and should be properly warmed up, prior to application. The most effective waxes are applied for medium and shorter length hairstyles, and can also be used on thicker hair.

Gels, based on water and without alcohol, have a rather strong hold, leaving a wet and shiny finish to the hair. Gels provide controlled, sleek looking hairstyles for thick and medium length hair or shorter. Gels with a strong hold should not be used on thin or fine hair. Hair gels are most commonly used to add texture and structure to hair without greasiness or waxiness.

Styling gels typically utilize a combination of two or more polymers with different functions, namely one contributing to the styling properties (e.g. polyvinylpyrrolidone or vinylpyrrolidone/vinyl acetate resins), while the other one acing as a rheology modifier. The amounts of the styling polymers in a styling gel and a mousse are in the range of 0.5–4%, while their content in styling spray is 2–12%.

One of the most efficient and widely used thickeners in hair styling is carbomer,

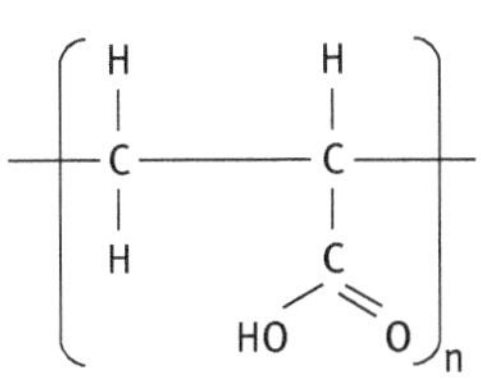

used in combination with PVP. Low setting power, humidity resistance, running of the gel, on application, and visible flaking on hair call for improvement in the performance of this combination. Preferably, the right product can exhibit both functionalities, allowing substantial benefits in formulation (lower flaking and better humidity resistance) and simpler manufacturing of hair styling gels.

Improvement of the thickening behavior of a polymer, having a high molecular weight, for better hair fixation can be achieved by adding chemical cross-linking via trifunctional monomers and/or long alkyl chains, creating physical cross-linking through hydrophobic/hydrophobic interactions. Swelling via osmotic pressure provides, as a result, thickening in water-based formulations. One example of such compound is the acrylates/C10-30 alkyl methacrylate copolymer, which functions in 0.25–1% amounts as a thickener, improving the overall setting performance of the styling formulation, in comparison to formulations with carbomer. Clear and effective styling gels can be made using this co-polymer, in combination with other nonionic, anionic and cationic styling polymers, such as polyvinylpyrrolidone (PVP), vinylpyrrolidone (VP)/vinyl acetate, vinyl-pyrrolidone/ methacrylamide/vinyl imidazole copolymer. Rheology profile in these formulations is, however, different to systems with carbomers, exhibiting more shear thinning. Anionic styling polymers and thickeners require neutralization prior to use, which is done with inorganic neutralizing agents (sodium or potassium hydroxides) or organic amines (triethanolamine or 2-amino-2-methyl-1-propanol).

For emulsifying ingredients, which are otherwise not soluble in water, nonionic surfactants are added to products with a high water content (e.g., mousse and gel). Other ingredients include plasticizers (e.g., 0.1–0.5% glycols), improving the flexibility of the styling polymer film, propellants (dimethylether, isopentane, etc.) in ca. 6–12% for mousse and 15–35% for styling spray for better spraying, preservatives (0–1%), perfume (0.1–0.5%), antirust agent (0–1%) to prevent rusting in a can, etc.

Finally, in this section, hair dyes will be discussed. Being similar to textile dyes, they should not obviously damage the scalp. Nitro- and anthraquinone dyes as well as azo and quinonimine compounds, with a quaternary acyclic ammonium group, can be used as direct dyes without any further treatment. A large size of the dyes prevents them from penetrating the hair; thus, they have a semipermanent rather than permanent character and can be removed, as opposed to permanent dyes. Such dyes, similarly to vat dyes for textiles, are converted to their final forms with the aid of oxidation with hydrogen peroxide and a coupling reaction (Figure 4.53). This figure illustrates that a coupler and a developer are applied to the hair, together with the oxidation agent, using a suitable formulation (e.g., a cream or a gel). Obviously, diffusion of both organic compounds should be of a similar magnitude to achieve the desired color. The composition of the dying formulation includes, e.g., stabilizers, alkaline pH adjusters and fragrance to mask the amine smell.

Toning products are needed to refresh hair color, also contain other compounds such as rheology modifiers or surfactants.

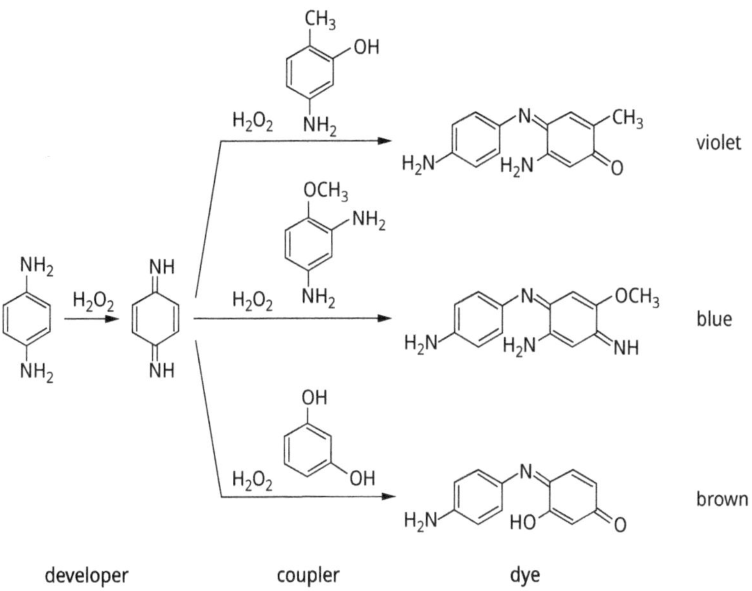

Figure 4.53: Oxidative coupling in hair dyes [58]. Reproduced with permission.

4.3.5 Hand sanitizers

Hand sanitizers became very popular after an outbreak of the COVID-19 pandemic. While hand disinfectants are easier to use in numerous public places (shopping malls, governmental buildings, banks, schools, universities, etc.), they are less effective at killing certain kinds of germs, and simple hand washing using water and soap is the preferred option. On the other hand, alcohol-based hand sanitizers are considered to be more effective at reducing bacteria and are often used in various healthcare facilities.

Hand sanitizers, manufactured as gels, sprays and foaming products, contain minimum 60 vol. % of either ethanol, more effective against viruses, or isopropanol, which is mildly more effective against bacteria. Application of either of the alcohols requires their drying on hands without wiping off. The concentration of the alcohol in water can be as high as 95% and thus flammability should be carefully addressed.

In addition to alcohol, other compounds (ca. 1%) can be added, including glycerol or other humectants (butylene glycol, sodium lactate or aloe vera), as moisturizers, to prevent drying of the skin; surfactants, for preparation of agglomerate-free dispersions or fragrances. Fragrances such as sugared lime and citron, matcha tea or aloe are often used in hand sanitizers. Obviously, those fragrances that might cause allergic reactions should not be used.

Utilization of surfactants comes along with application of thickeners such as low-density carbopol homopolymers, representing acrylic acid, cross-linked with allyl sucrose or allyl pentaerythritol.

Other components of hand sanitizers include denaturants, with bitter taste, to discourage drinking of sanitizers; hydrogen peroxide that does not cause skin allergy in small amounts (< 0.5%) for the purpose of deactivating germs; emollients (e.g., isopropyl myristate, aloe vera or tocopheryl acetate) to maintain the skin healthy; vitamin E as an antioxidant and behentrimonium chloride as a conditioner for skin softening.

One of the key steps in the manufacturing process is preparation of a clear bright gel, essential for manufacturing an agglomerate-free dispersion, after neutralization. Introduction of a thickener into liquids requires efficient stirring, which is also needed for minimizing aeration, as difficult-to-remove entrained air results in clouding in clear gels. Moreover, poor mixing leads to longer process times for obtaining a uniform dispersion as well as stratification, especially when there is a substantial difference in the viscosity of the phases. Of late, high shear blenders are used for complete emulsification and homogenization, to ensure an agglomerate-free and uniform gel/liquid solution.

After the introduction of a thickener, as a powder, to water and the formation of a viscous solution, alcohol is added to give a uniform product. This is followed by the addition of solubilizing agents, humectants, fragrances and other additives, After an efficient mixing, the mixture is filled in bottles, capped and labeled.

4.3.6 Antiperspirants and deodorants

Antiperspirants are applied to influence the functioning of the body, limiting eccrine sweat production and thus underarm wetness. Antiperspirants, containing aluminum and/or aluminum zirconium salts, were introduced in the beginning of the twentieth century, with aluminum chloride as the basis. It forms an occlusive plug of metal hydroxide in the eccrine duct.

Currently, more complicated formulations are used. For example, aluminum zirconium chlorohydrate – glycine complexes are obtained by reacting buffered aluminum salts with zirconyl chloride in the presence of glycine. The latter acts as a buffering agent, giving very complex polymeric structures in water. The efficacy of the active ingredient (so-called antiperspirant active) is lower in anhydrous formulations, compared to water.

Antiperspirants are usually applied as sticks, roll-ons and aerosols. A slightly sticky feeling of the active ingredients solutions in water requires reduction of tackiness by silicone oils (cyclomethicone) or ester oils like di-(2-ethylhexyl) adipate). To minimize the visible white residues on skin or clothing because of aluminum powders in aerosols and suspension sticks, liquid emollients such as (PPG)-14 butylether are added. The downside of hydrophobic ingredients, like emollients, is that the effectiveness of an antiperspirant active is lower when a cosmetic oil phase covers the pores of the eccrine duct. Suspending aids (quaternium-18 hectorite or quaternium-18 bentonite) help to prevent settling of the antiperspirant active ingredient while also thickening the cyclomethicone oil phase.

In a suspension stick, the powder form of the antiperspirant active (ca. 20%) is suspended in a silicone oil phase (54%), with stearyl alcohol (20%) acting as the hardening agent. Other ingredients include PPG-14 butylether as an emollient (2%), quaternium-18 hectorite as a settler, talc (2%) and a fragrance (1%).

Clear anhydrous gel sticks, popular in the United States, usually contain a high (ca. 45 wt.%) level of alcohol and/or polyols (e.g., propyleneglycol and dipropyleneglycol), in addition to the antiperspirant powder per se (50%). A typical gelling agent is dibenzylidene sorbitol (2%). The composition can also include dimethicone copolyol and di-isopropyl sebacate (2% each) and glycine (1%).

O/W emulsion roll-ons, popular in Europe, use ethoxylated surfactants (e.g., PEG-40 stearate) for solubilization of the oil phase. The antiperspirant active is dissolved in the outer phase. In contrast to Europe, anhydrous suspension roll-ons or transparent water-in-silicone emulsions are more popular in the US.

Antiperspirant aerosols, also applied in Europe as well as in Asia, contain a substantial amount of a propellant (75 wt%), namely, propane, butane and isobutane. Other ingredients, in addition to the antiperspirant powder (ca. 10%), are volatile silicone (ca. 13%), quaternium-18 hectorite and ethanol. Hectorite clay is needed to minimize agglomeration of solid particles and for settling of actives.

Manufacturing of an aerosol includes: preparing a mixture of hectorite clay with the emollient, in the presence of a polar activator, followed by shearing such mixture; introduction of the antiperspirant active, giving a uniform agglomeration-free suspension; filling the concentrate into an aerosol can and adding the propellant.

Contrary to antiperspirants, which are treated by the US Food and Drug Administration as drugs, deodorants applied to diminish underarm odor, are considered as cosmetics.

Deodorants might contain antiperspirant active compounds such as aluminum chlorohydrate in lower amounts to combat secretion of eccrine sweat or can be odor-masking, odor-neutralizing or odor-quenching deodorants. The strategy of masking the odor is based on introducing 0.2% to 1.5% fragrance, which would mask the underarm odor. The latter can be neutralized by adding sodium bicarbonate for the purpose of reacting with odorous fatty acids. Binding odorous fatty acids, amines and mercaptanes with zinc salts of ricinoleic acid is the strategy behind utilization of odor-quenching deodorants. Other options include utilization of inhibitors for exoenzymes from the underarm bacteria by applying zinc glycinate or application of antimicrobial active-containing deodorants. The latter apply ethanol and other antimicrobial actives (famesol, phenoxyethanol, etc.).

Typical applications for deodorants are similar to antiperspirants, including sticks, roll-ons, creams, aerosols and gels. In the case of deodorant sticks that are based on ethanol or propylene glycol, solidification is done with sodium stearate.

An example of a deodorant composition is presented in Table 4.11.

Table 4.11: Revlon Signo Active deodorant composition [59].

Compound	Function	Content (wt%)
Aluminium zirconium tetrachlorohydrex GLY	Active	16–19
Dipropylene glycol	Volatile carrier	7–10
Dimethicone volatile	Carrier	0.5–2.5
Water	Carrier	5–10
Cyclopentasiloxane	Refractive index adjuster	45–55
PEG-12 dimethicone copolymer	Emollient, antiwhitening	0.2–0.5
PEG/PPG-18/18 dimethicone	Volatile carrier	5–7.5
Glycerin	W/o emulsifier	1–5
Fragrance	Scent	qs
PEG/PPG-19/19 dimethicone	Hydrogen bonding enhancer	0.25–1.0

In the case of a spray deodorant, a liquefied propellant (propane, butane or isobutane) is added to the solution of an antimicrobial active in an ethanol and/or propylene glycol, rather than a suspension, as in the case of antiperspirant aerosols.

The following parameters influence the performance of a deodorant spray: spray rate and shape, pressure in the can, particle size, amount of propellant and the concentration of fragrances and other compounds.

4.3.7 Production technology of skin and hair care products

The different types of emulsions, suspensions, gels, etc. clearly imply a large variation in the preparation processes. A few selected examples will be discussed below.

Technology development in the case of skin and hair care products is done in combination with the product design, as illustrated in Figure 4.54. The rules of good manufacturing practice (GMP) are followed in designing modern production facilities, ensuring quality and hygiene and defining specific requirements for raw materials, equipment, manufacturing facilities, water quality, packaging, etc., as explained in more detail below.

Several important issues should be mentioned concerning the manufacture of cosmetic products in general. Careful control of the temperature profiles, amounts and type of emulsifiers, proper sequence of adding formulation ingredients, sufficient energy input, protection from UV radiation, oxygen, pH changes and microbial contaminations are essential in preventing destabilization of emulsions. Special care

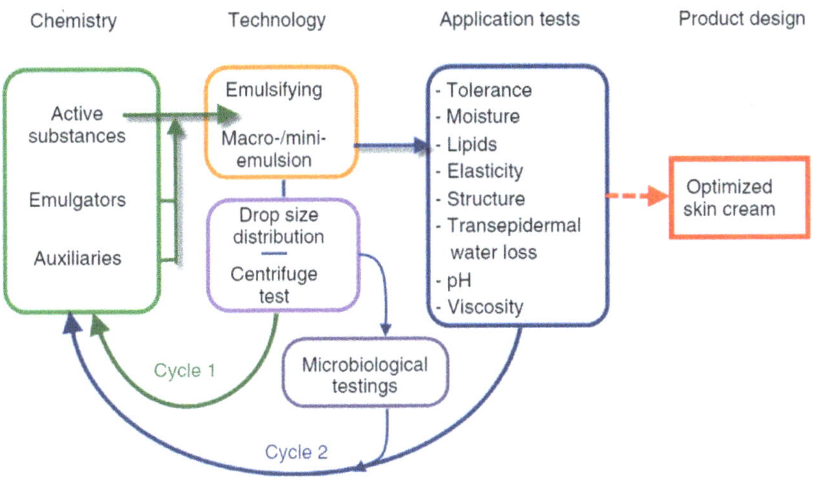

Figure 4.54: Cycles for a skin cream design optimization [42].

should be taken regarding water quality, especially microbiological purity, because water is a very important raw material for the cosmetic industry requiring careful purification. Production, filling and packaging require clean, disinfected equipment, following cosmetic good manufacturing practice requirements on cleanliness. Moreover, in the cosmetic industry, there are also requirements of protective clothing. For disinfection of the inside surfaces, several oxidation chemicals can be used, such as, for example, peracetic acid, hydrogen peroxide, sodium hypochlorite or glyoxal. For batch processes, typical in the manufacturing of cream and lotions for cosmetic purposes, each cycle can start first with cleaning the equipment by 2-propanol.

The technology for the manufacture of emulsion creams consists of the following operations: water phase preparation; heating and melting of the oil phase; emulsification; structuring with cooling; addition of biologically active substances, vitamins, etc.; packaging and packing.

Emulsification is carried out using homogenizers (e.g., rotor-stator type) and various agitators, as discussed in Chapter 2. It can also be done using ultrasound. Homogenizers, with external recirculation, are often used (Figure 4.55).

Typically, the oil phase, after heating to 75–80 °C and subsequent melting, is emulsified with water heated to the same temperature. An alternative low-temperature method used to reduce energy consumption and shorten the process time relies on stepwise addition of room temperature water to the hot oil phase, under continuous homogenization.

The batch time can be ca. 5–15 h and includes loading, heating, homogenization per se (from 5 to 25% of the overall time) and cooling, depending on the batch size and the formulation recipe. The operation schedule includes three runs, with the first and the third run starting from the emulsifying vessel (Figure 4.56), while in the second run,

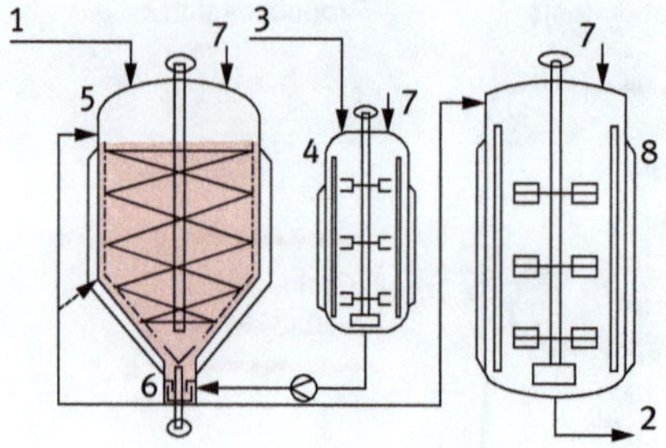

Figure 4.55: Flanged emulsifier, with recirculation of the product: (1) supply of water and hydrophilic substances; (2) to filling; (3) supply of lipophilic substances; (4) stirred tank for temperature-controlled dissolution of lipophilic substances; (5) emulsifier with stirring; (6) rotor–stator machine; (7) nitrogen supply; and (8) temperature-controlled stirred reservoir for filling (reproduced with permission from [42]).

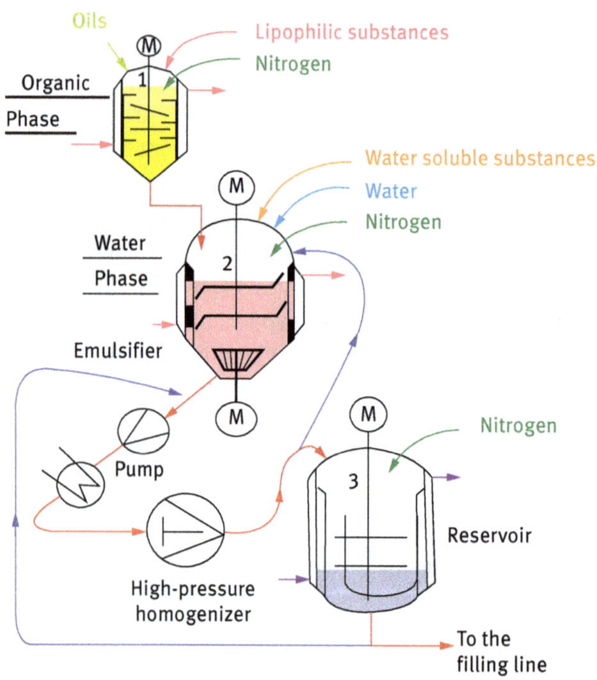

Figure 4.56: Batch plant with a high-pressure homogenizer for the production of oil/water emulsions: 1, stirred vessel for the preparation of the organic phase; 2, vessel for dissolution of the water-soluble compounds acting as an emulsifying tank, as well as the reservoir for the first and third runs through the homogenizer; 3, reservoir for the second run and for the filling line (reproduced with permission from [42]).

the emulsion goes back from the reservoir 3 to the emulsifying tank 2 through the high pressure homogenizer. After the third run, the cream is taken to the filling line for bottling. Such an operation procedure is needed to ensure a stable distribution, with very few or no micrometer droplets. The finished cream is pumped into a container, which is followed by quality assurance and thereafter by packaging.

Somewhat more complicated is the process flow diagram for the manufacture of water-oil-water emulsion cream (Figure 4.57), which comprises feedstock and petroleum jelly (in some countries, the Vaseline brand is used as a generic for petroleum jelly), emulsification per se and then postprocessing, including cooling, adding fragrances and other sensitive ingredients, and finally packaging of the cream.

Through a pipe, lanolin, beeswax and other fat and fat-like components are introduced to vessel 9, having a heating jacket and a propeller mixer. After the mixture is brought to 70–75 °C and properly mixed, hot water and water-soluble compounds (i.e., methyl ester of paraoxybenzoic acid) are added from the measuring tank 1 to the molten mixture. Vegetable oil is fed through the measuring tank 2 while other ingredients, present in small quantities, are separately added to the mixing vessel 9 (line D).

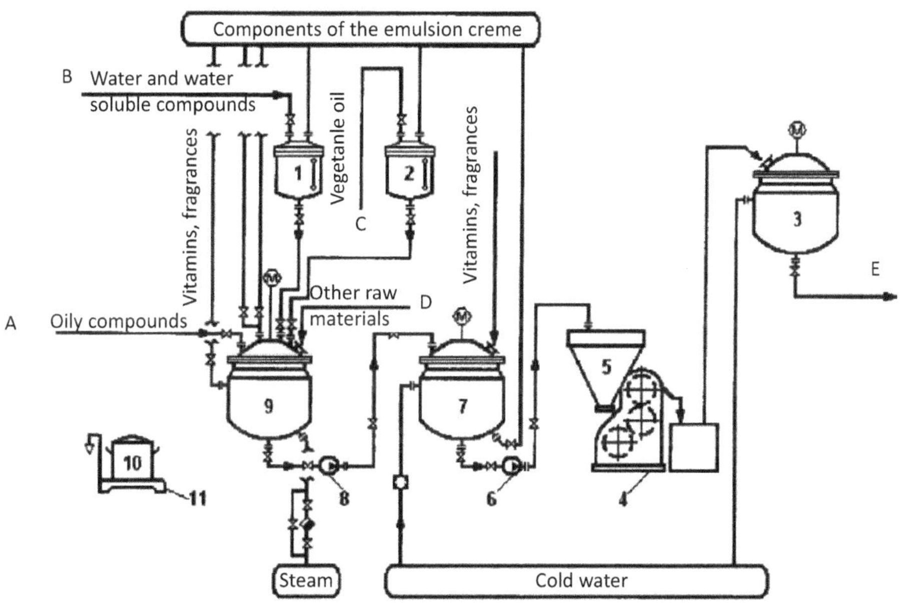

Figure 4.57: Technological scheme for the manufacturing of an emulsion of the water-oil-water emulsion cream [60]: 1, 2, tanks; 3, vessel operating under vacuum; 4, roller; 5, bunker; 6, 8, pumps; 7, 9, vessels; 9, boiler; 10, tank; 11, scales.

Emulsification is done for 10–15 min at 70–75 °C, followed by cooling the emulsion in vessel 7, with a cooling jacket and a frame mixer, which also contains a scraper. After cooling for 20–25 min, without adding cold water to the cooling jacket, cold water is supplied to aid cooling. When the temperature is sufficiently low (ca. 40–45 °C), vitamins and fragrance are added and cooling of the emulsion continues to 30–32 °C.

Finer droplets can be achieved in, for example, roller 4, where double rolling of the cream mass at 32–34 °C improves emulsification, increasing its viscosity. An alternative option to improve the product quality is to use colloidal mills or mechanical emulsifiers. Finally, the finished cream goes to vessel 3 and is then fed (E) for packaging.

Besides batch process, continuous production of cosmetic emulsions can also be arranged, as illustrated in Figure 4.58, featuring mechanical homogenization. In general, other types of homogenization can be used. Lipophilic compounds enter one of the emulsifiers through the line 1 for preliminary emulsification. By using the switch 5, there is a possibility to alternate two emulsifiers. The same approach is used for lipophobic water-soluble compounds, which are mixed in mixer 3 and regulated by a switch (pos. 10). Exact amounts of components are regulated by the control unit 16 through the dosing pumps 6 and 9. The dosed quantities of aqueous and oil phases, after passing through filters 7 and 8, are emulsified in the mechanical emulsifier 2. The obtained emulsion is pumped into one of the reactors 4 that is equipped with three stirrers to facilitate further emulsification and temperature stabilization, in addition to mixing. Alternative operation of these reactors is regulated by the switch 12. The care-

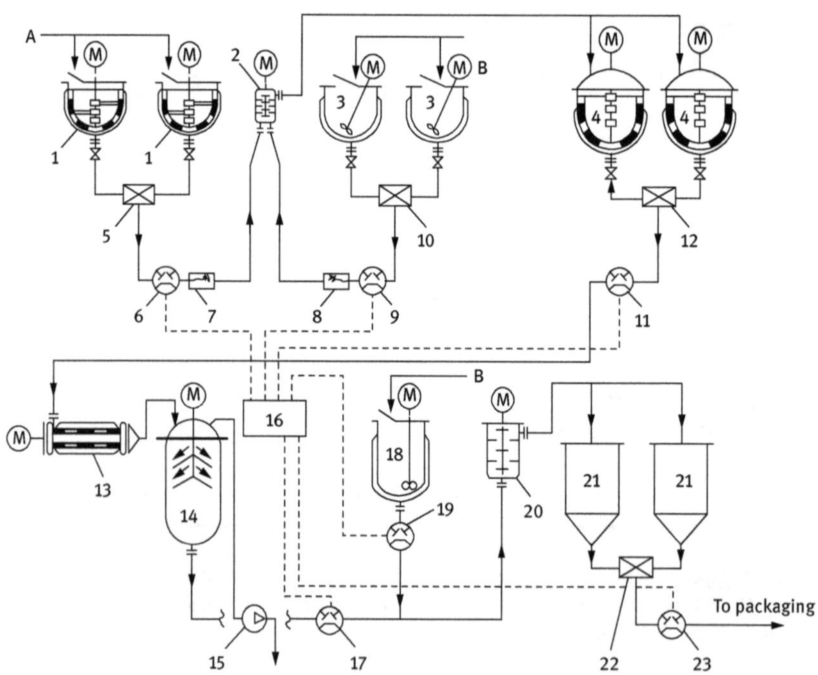

Figure 4.58: Continuous production of emulsion creams [60]: 1,emulsifiers; 2, mechanical emulsifiers; 3, mixers; 4, reactors; 5, 10, 12, 22, switches; 6, 9, 11, 17, 19 and 23, dosing pumps; 7, 8, filters; 13, cooler; 14, deaerator; 15, vacuum pump; 16, control; 18, reservoir; 20, emulsifiers; 21, vessels.

fully mixed and homogenous emulsion, in terms of temperature, is pumped (pos. 11) through a cooler (pos.13), decreasing the temperature to 35–40 °C and then pumped to deaerator 15. The latter, operating under vacuum generated by the vacuum pump 15, is needed to remove air, which is detrimental for the quality and stability of the cream. After deaeration, the emulsion is pumped (pos. 17) to the last emulsifier 20. Biologically active additives, extracts, etc. are introduced to the emulsion by the pump 19 from the line B through the reservoir 18. The final emulsifier 20 allows formation of a highly dispersed phase with size of 1–5 μm, which is routed to vessels 21 for maturation and storage. Thereafter, the finished product goes to packaging.

Shower and bath preparations are mass products requiring rather rapid preparation of homogeneous mixtures. Continuous production can be employed when large amounts of the product are needed, and all the raw materials are pumpable at room temperature. In the case of more time-consuming mixing or melting, as well as solids incorporation, batch processes should be used. Simple, propeller-type stirring equipment is therefore unsuitable. Guided jet mixing turbines are preferred to ensure efficient mixing.

The formulation strategies, including composition of shampoo, have been addressed above. In some formulations, the amount of water is 70–75%, while the fraction of surfactants is 20–25%, the rest being various additives (preservatives, salts, perfume, extracts, etc.). A block diagram for the manufacture of shampoo is given in Figure 4.59. As follows from Figure 4.59, shampoos are produced by mixing all ingredients at ca. 80–90 °C. Components that easily vaporize, such as perfume, are added at lower temperature after cooling. Premixing is followed by mixing in the main mixer at ca. 32 °C. Thereafter, the liquid is transferred to storage tanks before filling into bottles.

Preparation of mascara is done using two methods: anhydrous or emulsion (Figure 4.60).

In the former, the ingredients are weighed and mixed, followed by melting the mixture and stirring it until it is semisolid. In the latter, water and thickeners are mixed to create a lotion or a cream base, while mixing and heating of the waxes and emulsifiers is done separately. Pigments are added prior to combining both mixtures in a high-speed homogenizer, giving, as a result, a semisolid substance that is then ready for packaging.

Lipstick manufacturing consists of several operations, including grinding, melting and mixing, molding and packaging. A standard lipstick manufacture consists of four stages: mixing waxes, oils and extenders at high temperature; preparation of a concentrated dispersion of the pigment using an oily base (e.g., castor oil); adding the pigment mass to hot wax until uniform color and consistency is obtained and then pouring the melted mass into a mold, upside down. After cooling, the lipstick is separated from the mold.

Application of a film-forming polymer, in combination with cyclomethicone, helps to mitigate an otherwise rather fast deterioration of the fresh appearance. However, incorporation of the volatile oil in the formula leads to a loss in the application gloss with time. Glossy lipstick is formulated with a high level of the waxy phase

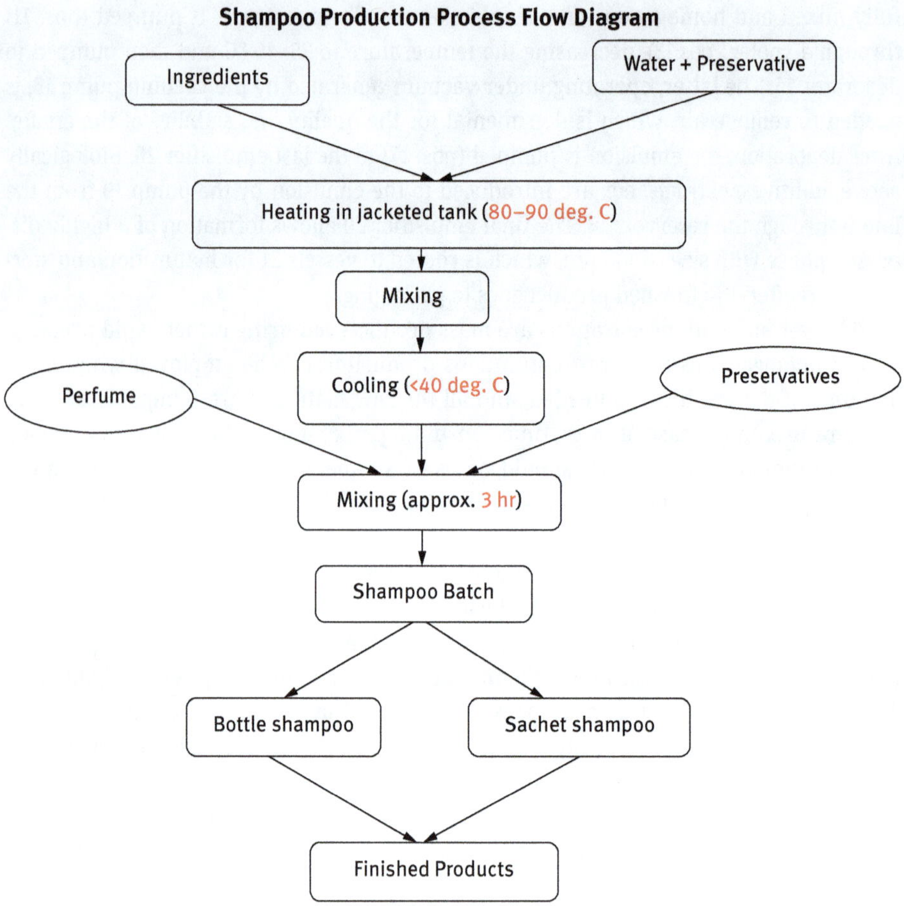

Figure 4.59: Block diagram for shampoo manufacturing [61].

(ca.80%), with less pigments (0–5%), while talc, silica or polymethylmethacrylate are used for matte lipstick, giving a soft touch.

One of the potential defects during molding is coring at the surface of a mold. Such deformations are noticeable in softer formulations. There could be several other molding-related problems, including laddering, when the lipstick does not look smooth or homogeneous, or if it has cratering and a mushy failure because of the granularity of carnauba wax. Other defects in lipstick can be related to problems with the ingredients. Thus, sweating (Figure 4.61) is caused by a high oil content or inferior oil binding, while bleeding and streaking are respectively related to separation of colored liquids from the waxy base and titanium dioxide.

Nail polish technology, typically kept as a secret by manufacturers, consists first of mixing pigments with nitrocellulose and a plasticizer using a two-roll differential speed mill with increasing speed as the pigment is ground down (Figure 4.62). When a

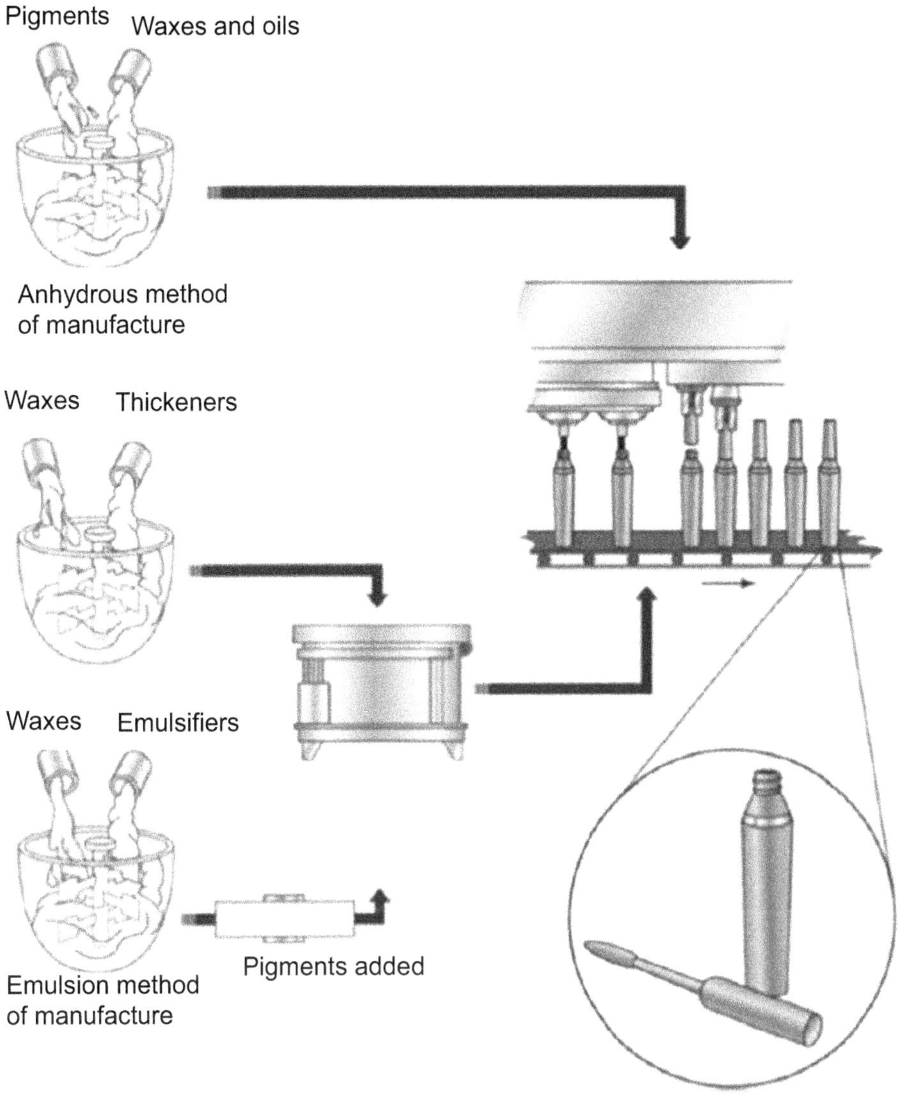

Figure 4.60: A schematic view on the production of mascara [62]. In a closed-lid homogenizer used in the emulsion method, the lotion base created by water and thickeners is combined with the wax mixture. The closed lid keeps out air and eliminates evaporation.

dispersion is produced, this is followed by mixing with the solvent in stainless steel jacketed kettles under careful temperature control to avoid fire and explosion, as nitrocellulose is extremely reactive in the presence of iron.

At the end of the process, the mixture is slightly cooled, prior to adding other ingredients (perfumes and moisturizers). Thereafter, the mixture is pumped into drums

Figure 4.61: Sweating defects in lipstick manufacturing [63].

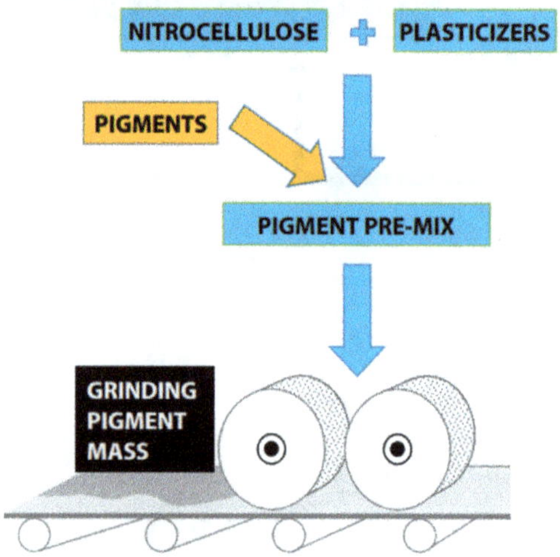

Figure 4.62: Processing of nitrocellulose with plasticizers and pigments for nail polish.

and then into smaller bottles for the retail market. Other manufacturing technologies include the use of single tanks with a high shear mixer.

References

[1] J. M. F. Mendoza, F. D'Aponte, D. Gualtieri, A. Azapagic, Disposable baby diapers: Life cycle costs, eco-efficiency and circular economy, J. Clean. Prod., 2019, 211, 455.

[2] https://vectortoons.com/wp-content/uploads/2014/12/baby-diapers-collection-0021.jpg, accessed on 23.12.2017.

[3] https://www.adhesivesmag.com/articles/93540-disposable-hygiene-one-core-adhesive-does-not-fit-all, accessed on 23.12.2017.

[4] http://nptel.ac.in/courses/116102006/17, accessed on 23.12.2017.

[5] https://www.slideshare.net/vishalgawande/superabsorbent-polymers, accessed on 23.12.2017.

[6] J. M. F. Mendoza, S. A. Popa, F. D'Aponte, D. Gualtieri, A. Azapagic, Improving resource efficiency and environmental impacts through novel design and manufacture of disposable baby diapers, J. Clean. Prod., 2019, 210, 916.

[7] http://www.singhvaid.com/wp-content/uploads/2016/06/slide_21.jpg, accessed on 23.12.2017.

[8] https://archive.kyivpost.com/wp-content/uploads/2020/08/000_1WO165-1-800x520.jpg

[9] http://drug-dev.com/Main/Back-Issues/SOLID-DOSAGE-FORMS-Better-Treatments-Through-Innov-633.aspx, accessed on 23.12.2017.

[10] https://www.dreamstime.com/stock-illustration-asthmatic-airway-cells-include-lung-fibroblastsbronchial-epithelial-bronchial-smooth-muscle-cells-vector-graphic-image89558839, accessed on 5.10.2023.

[11] https://basicmedicalkey.com/wp-content/uploads/2017/02/F500193f10-06-9781455703845.jpg, accessed on 5.10.2023.

[12] https://oscestop.education/wp-content/uploads/2022/02/Screenshot-2022-02-23-at-09.22.39.png, accessed on 5.10.2023.

[13] S. S. S. Mak, P. E. Hui, W. M. R. Wan, C. L. P. Yih, At-home chemotherapy infusion for patients with advanced cancer in Hong Kong, Hong Kong J. Radiol., 2020, 23, 122–129.

[14] Neulasta® (pegfilgrastim) Delivery Kit: Healthcare provider instructions for use. http://pi.amgen.com/united_states/neulasta/neulasta_ifu_hcp_pt_english.pdf. Accessed on 3.02.2015

[15] https://www.maximillianburton.com/max-frog/amgen-enbrel-auto-injector

[16] C. T. Huynh, D. S. Lee, Controlled Release, in Encyclopedia of Polymeric Nanomaterials, Springer, Berlin, Heidelberg, 2015, 439–449.

[17] https://o.quizlet.com/oWuwpqY0ZKt5SahQmUusgw.png

[18] www.idc-online.com/technical_references/pdfs/chemicahal_engineering/Introduction_to_smart_drug_delivery_systems.pdf

[19] https://www.intechopen.com/source/html/45635/media/image1.jpeg., accessed on 23.12.2017.

[20] M. K. Iqubal, A. Iqubal, M. Shuaib, M. Shahryar, An exquisite technology of pharmaceutical science: Nanotechnology. Indo American J. Pharm. Res. 2015, 5, 3528–3540.

[21] M. Janssen, G. Mihov, T. Welting, J. Thies, P. Emans, Polymers, Drugs and polymers for delivery systems in OA joints: Clinical needs and opportunities, 2014, 6, 799–819.

[22] https://www.slideshare.net/ibtihalosman/pharmaceutical-industry-and-unit-process, accessed on 23.12.2017.

[23] Patent WO2014118112A1 Pharmaceutical composition comprising Erlotinib hydrochloride, Synthon b.v., https://patents.google.com/patent/WO2014118112A1/en

[24] L. Fries, S. Antonyuk, S. Heinrich, S. Palzer, DEM-CFD modeling of a fluidized bed spray granulator, Chem. Eng. Sci., 2011, 66, 2340–2355.

[25] http://www.tankonyvtar.hu/hu/tartalom/tamop412A/2011-0016_01_the_theory_and_practise_of_pharmaceutical_technology/ch24.html, accessed on 23.12.2017.

[26] http://kpmpharmamachineries.com/tabletcoatingpan.html, accessed on 23.12.2017.

[27] S. Salar-Behzadi, S. Toegel, H. Viernstein, Innovations in coating technology, Recent Pat. Drug Deliv. Formul., 2008, 2, 209.

[28] https://www.glatt.com/fileadmin/user_upload/content/pdf_downloads/GC_SMART_Prospekt_e.pdf, accessed on 23.12.2017.

[29] L. R. J. Chakka, S. Chede, 3D printing of pharmaceuticals for decease treatment, Front. Med. Technol. Sec. Nano-Based Drug Deliv., 2023, 4, 1040052.

[30] R.Pawar, A. Pawar, 3D printing of pharmaceuticals: Approach from bench scale to commercial development, Fut. J. Pharm. Sci., 2022, 8, 48.

[31] https://news.curifylabs.com/transforming-patient-tailored-pharmacy-preparations, accessed 11.08.2023.

[32] T. Tracy, L. Wu, X. Liu, S. Cheng, X. Li, 3D printing, innovative solutions for patients and pharmaceutical industry, Int. J. Pharmaceut., 2023, 631, 122480.

[33] W. Xu, X. Wu, X. Guo, B. Kong, M. Zhang, X. Qian, S. Mi, W. Sun, The boom in 3D-printed sensor technology, Sensors, 2017, 17, 1166.

[34] https://www.cancer.gov/types/head-and-neck/patient/oropharyngeal-treatment-pdq, accessed on 23. 12.2017.

[35] http://respiratorysystem.organsofthebody.com/, accessed on 23.12.2017.

[36] http://www.ilocis.org/documents/chpt10e.htm, accessed on 23.12.2017.

[37] https://www.hindawi.com/journals/isrn/2013/102418/fig1/, accessed on 23.12.2017.

[38] http://img.medscapestatic.com/fullsize/migrated/446/869/cdi446869.fig1.jpg, accessed on 23.12.2017.

[39] http://www.laakeinfo.fi/PILImages//fbe87943-aa9d-4bb2-b155-716f17361ce6.JPG, accessed on 23.12.2017.

[40] http://itgsuperstarwallpaper.blogspot.fi/2012/10/jet-nebulizer.html, accessed on 23.12.2017.

[41] http://www.healthcare.omron.co.jp/english/overview/neu.html, accessed on 23.12.2017.

[42] W. Rähse, Design of skin care products, in *Product Design and Engineering: Formulation of Gels and Pastes*, Ed. U. Bröckel, W. Meier, G. Wagner. Wiley-VCH Verlag GmbH & Co. KGa, Weinheim, 2013, doi: 10.1002/ 9783527654741.ch10, pp. 273–313.

[43] https://upload.wikimedia.org/wikipedia/commons/thumb/4/4d/Hair_follicle-en.svg/310px-Hair_follicle-en.svg.png, accessed on 23.12.2017.

[44] T. W. Tandros, *Formulation Science and Technology, Volume 3: Pharmaceutical, Cosmetic and Personal Care Formulations*, De Gruyter, Berlin, 2018, (table 8.2).

[45] M. Alam, M. Algahtani, M. Zaki Ahmad, J. Ahmad, Investigation utilizing the HLB Concept for the development of moisturizing cream and lotion: In-vitro characterization and stability evaluation, Cosmetics, 2020, 7, 43.

[46] https://eltamd.com/wp-content/uploads/2013/07/Updated-Think-Zinc-Chart1.jpg, accessed on 23.12.2017.

[47] https://www.carecreations.basf.com/docs/default-source/press-center-files/personal-care-in-the-media/2014/0714_sofw_osterwalder_e.pdf?sfvrsn=4, accessed on 11.08.2023.

[48] http://www.compoundchem.com/2014/06/05/sunscreenchemicals/, accessed on 23.12.2017.

[49] https://i.pinimg.com/564x/d5/cb/2c/d5cb2cdaa54fddb292cd7c4248f27aa5.jpg

[50] https://connect.in-cosmetics.com/formulation/colour-cosmetics-an-introduction-to-formulation-and-approaches-for-mascaras/

[51] http://www.compoundchem.com/wp-content/uploads/2014/08/The-Chemistry-of-Lipstick.png, accessed on 23.12.2017.

[52] https://i0.wp.com/www.compoundchem.com/wp-content/uploads/2017/04/The-Chemistry-of-Nail-Polish.png?ssl=1, accessed 2.8. 2023.

[53] M. Mandal, D. Banik, A. Karal, S. Kumar Mann, A. Kumar Mahapatra, Spiropyran–merocyanine based photochromic fluorescent probes: Design, synthesis, and applications, ACS Omega., 2022, 7, 36988–35007.

[54] https://jameskennedymonash.files.wordpress.com/192013/12/visual-ingredients-head-and-shoulders-shampoo-v2.jpg, accessed on 23.12.2017.

[55] http://boldbarber.com/hair-and-hair-types/, accessed on 23.12.2017.

[56] J. Yang, Chapter 36-Hair Cair Cosmetics, in *Cosmetic Science and Technology: Theoretical Principles and Applications*, 2017, 601. Elsevier, Amsterdam, http://dx.doi.org/10.1016/B978-0-12-802005-0.00036-7.

[57] https://en.wikipedia.org/wiki/Polyquaternium-7#/media/File:Polyquaternium-7.svg, accessed on 23.12.2017.

[58] H. Mollet, A. Grubenmann, *Formulation Technology: Emulsions, Suspensions, Solid Forms*, Wiley-VCH Verlag GmbH, Weinheim, 2001, doi: 10.1002/9783527612925. (Fig. 11.7).

[59] https://www.cosmeticsandtoiletries.com/formulas-products/article/21834359/the-anatomy-of-a-formulaantiperspirant-sticks-soft-solids-and-gels, accessed on 10.08.2023.

[60] N. V. Menshutina, Y. V. Mishina, S. V. Alves, *Innovative Technologies and Equipment for Pharmaceutical Production*, Binom, Moscow, 2012.

[61] https://www.slideshare.net/PratapJungRai/fmcg-ojt-presentation-uniliver-nepal-limited, accessed on 23.12.2017.

[62] http://www.madehow.com/images/hpm_0000_0003_0_img0114.jpg

[63] https://dealstoogoodtopassup.com/2015/08/02/how-to-clean-lipsticks-and-is-this-lipstick-moldy-or-sweating/, accessed on 23.12.2017.

Index

https://doi.org/10.1515/9783110797961-005

www.ingramcontent.com/pod-product-compliance
Lightning Source LLC
Chambersburg PA
CBHW061156240926
51846CB00006B/1033
9783110788440